Systems Biology
Constraint-based Reconstruction and Analysis

Recent technological advances have enabled comprehensive determination of the molecular composition of living cells. The chemical interactions between many of these molecules are known, giving rise to genome-scale reconstructed biochemical reaction networks underlying cellular functions. Mathematical descriptions of the totality of these chemical interactions lead to genome-scale models that allow the computation of physiological functions.

Reflecting these recent developments, this textbook explains how such quantitative and computable genotype–phenotype relationships are built using a genome-wide basis of information about the gene portfolio of a target organism. It describes how biological knowledge is assembled to reconstruct biochemical reaction networks, the formulation of computational models of biological functions, and how these models can be used to address key biological questions and enable predictive biology.

Developed through extensive classroom use, the book is designed to provide students with a solid conceptual framework and an invaluable set of modeling tools and computational approaches.

Detailed lecture slides, along with MATLAB^{TM} and Mathematica^{TM} workbooks, are available for download at www.cambridge.org/sb.

Bernhard O. Palsson is the Galletti Professor of Bioengineering and Professor of Pediatrics at the University of California, San Diego. For almost 30 years, his research has focused on the development of large-scale models of biological functions and their use to solve basic and applied problems in the life sciences. He has authored three previous textbooks.

Systems Biology

Constraint-based Reconstruction and Analysis

BERNHARD O. PALSSON

Department of Bioengineering,
University of California at San Diego, USA

CAMBRIDGE
UNIVERSITY PRESS

CAMBRIDGE
UNIVERSITY PRESS

University Printing House, Cambridge CB2 8BS, United Kingdom

Cambridge University Press is part of the University of Cambridge.

It furthers the University's mission by disseminating knowledge in the pursuit of education, learning and research at the highest international levels of excellence.

www.cambridge.org
Information on this title: www.cambridge.org/9781107038851

First published 2015

Printed in the United States of America by Sheridan Books, Inc.

A catalog record for this publication is available from the British Library

Library of Congress Cataloging in Publication data
Palsson, Bernhard, author.
Systems biology : constraint-based reconstruction and analysis / Bernhard O. Palsson.
  p. ; cm.
Includes bibliographical references.
ISBN 978-1-107-03885-1 (Hardback)
1. Title.
[DNLM: 1. Models, Biological. 2. Systems Biology. 3. Metabolic Networks and Pathways–physiology. QU 26.5]
QH508
571.7–dc23   2014031793

ISBN 978-1-107-03885-1 Hardback

Additional resources for this publication at www.cambridge.org/sb

To SHIREEN and SIRUS

Contents

Preface

The genesis of the bottom-up approach to systems biology was the availability of the first full genome sequences. In principle, these sequences had information about all the genetic elements that underlie the function of the sequenced organism. Enough information was available about the function of subsets of these genes – namely the genes encoding metabolic functions – that an organized assembly of all the biochemical, genetic, and genomic information was achievable. Such an organized assembly is *de facto* a knowledge base, or a k-base, that gives rise to a network reconstruction at the genome-scale. Since such reconstructions are represented with accurate chemical equations, they can be mathematically described. A mathematical description can be used to compute functional states of a network that correspond to observable phenotypes and biological functions. With these elements in place, a new genome-scale science was born that focused on mechanistic genotype–phenotype relationships.

The first genome-scale models of metabolism appeared in 1999 and 2000. In the next half-decade or so, an enthusiastic group of investigators developed many fundamental concepts, *in silico* methods, and algorithms to analyze their properties. At times, and to many, these initial efforts seemed mostly exploratory. Fortunately, in the mid-2000s an abundance of data sets and data types became available to validate and demonstrate the utility of genome-scale models for research and discovery. At the end of the decade several highly curated models were available for model organisms. These models gained predictive power and over the next 5 years or so, a number of prospective uses of genome-scale models appeared. In other words, predictive genotype–phenotype relationships had appeared. These predictions were somewhat limited in scope, but proved useful for a series of applications. Currently the range of possible predictions of biological properties and functions is growing rapidly and it appears that this approach to genome-scale science is in its early stages of development, with a bright future ahead of it.

For most of this fifteen-year history, the focus of genome-scale models has been metabolism. After initial successes with metabolic genome-scale models, it became clear that the same approach that led to their genesis could be applied to any other cellular process reconstructed in biochemically accurate detail. Thus, a vision was laid out in 2003 that the path to whole-cell models was conceptually possible and that such models could be used as a context for mechanistically integrating disparate omic data types. Ten years later, this vision started to be realized and a rapidly growing number of cellular functions are being reconstructed and addressed computationally. Given the fact that the genotype–phenotype relationship is fundamental to biology, this development has a broad transformative potential for the life sciences.

Writing this book was hard. It represents an attempt to summarize the concepts that have been developing over the past 15 years or so, that underlie what has become a true genome-scale science. Looking at the history of the field after the writing process, it is quite remarkable to see its rapid emergence, development, and maturation. Furthermore, looking forward, it appears that numerous areas of microbiology, cell biology, and developmental biology will be influenced by the approach and methods described in this book.

To master this field one needs familiarity with an unusual range of disciplines. One needs to understand the basics of life sciences: biochemistry, molecular biology, genetics, microbiology, and cell biology. High-throughput measurements call for an understanding of basic technological characteristics, such as multiplexing, miniaturization, and automation. The large data sets generated call for proficiency in bioinformatics and a comfort level with big data. Mathematically modeling such data sets from a fundamental standpoint requires familiarity with the mathematical language of linear algebra and logistical relationships. Simulations require the use of constraint-based optimization and an understanding of the evolutionary principles of generation of diversity and selection. Bottom-up systems biology is thus a field with a broad conceptual basis. This book attempts to bring all these concepts from the expert level to the general senior or first-year graduate student level in bioengineering, bioinformatics, and life sciences.

As with all major undertakings, this project could not have been completed without the help of several individuals.

Marc Abrams managed all aspects of the preparation of the manuscript. He tirelessly helped me with preparing the text and the illustrations, assembling the references, correcting LATEX scripts, and interacting with the publisher. Without him this book would not have been completed.

Nathan Lewis and Adam Feist were responsible for the challenging task of managing the original illustrations in the book. Their contribution was immense, making the concepts in the text and the material as a whole more accessible.

The following people were generous with their time and expertise, improving the manuscript with their contributions to the text, figures, or proofreading of the final manuscript. I am very grateful to these individuals:

Ramy Aziz, Aarash Bordbar, Roger Chang, Addiel U. de Alba Solis, Andreas Drger, Juan Nogales Enrique, Gabriela Guzman, Hooman Hefzi, Daniel Hyduke, Neema Jamshidi, Ryan LaCroix, Haythem Latif, Josh Lerman, Douglas McCloskey, Jonathan Monk, Harish Nagarajan, Jeff Orth, Troy Sandberg, Nikolaus Sonnenschein, Alex Thomas, and Daniel Zielinski.

The conceptual framework that this book describes has been under development since the birth of my two children, to whom it is dedicated.

Bernhard Palsson
On the Oracle, August 2014

Abbreviations

ALE	adaptive laboratory evolution
AOS	alternative optimal solutions
BiGG	biochemical genetic and genomic
BOF	biomass objective function
CDS	coding sequence
COBRA	constraint-based reconstruction and analysis
CoSy	community systems
DIET	direct interspecies electron transfer
DIP	di-*myo*-inositol 1,1'-phosphate
DMMM	dynamic multi-species metabolic modeling
EnMe	endo-metabolome
ETS	electron-transport system
ExME	exo-metabolome
FA	fraction of agreement
FBA	flux balance
FCF	flux coupling finder
FIG	Fellowship for Interpretation of Genomes
FVA	flux variability analysis
GAM	growth-associated maintenance
GDLS	genetic design through local search
GEM	genome-scale model
GENRE	genome-scale reconstruction
GOF	gain of function
GPR	gene–to–protein–to–reaction
GUR	glucose uptake rate
HGP	human genome project
HMDB	Human Metabolome Database
HT	high-throughput
I/O	input/output
IDV	isotopomer distribution vector
IEM	inborn error of metabolism
IOFA	input-output feasibility array
k-base	knowledge base
KI	knock-in
KO	knock-out
LIMS	laboratory information management system
LO	line of optimality
LOF	loss of function
LPR	ligand to protein to reaction
LPS	lipid polysaccharide

MDV	mass distribution vector
MILP	mixed-integer linear programming
MOMA	minimization of metabolic adjustment
MS	mass spectrometry
MU	modular unit
NGAM	non-growth-associated maintenance
NMR	nuclear magnetic resonance
NTP	nucleotide triphosphate
ORF	open reading frame
PCA	principal component analysis
PDB	Protein Data Bank
PFL	pyruvate formate lyase
PhPP	phenotypic phase plane
POR	pyruvate oxidoreductase
PPS	pentose phosphate shunt
PVT	pressure volume temperature
QA	quality-assured
QC	quality-controlled
RBR	RNA polymerase binding region
RBS	ribosome binding site
rFBA	regulated flux balance
ROOM	regulation off/on modification
RTS	RNAP-guided transcript segment
SKI	species knowledge index
SNP	single nucleotide polymorphism
SOP	standard operating procedure
SVD	singular value decomposition
TCA	tricarboxylic acid
TF	transcription factor
Tr/Tr	transcription/translation
TRN	transcriptional regulatory network
TSS	transcription start site
TU	transcription unit

1 Introduction

E pluribus unum

Some 60 years ago, the promise of molecular biology held that if we knew and understood the function of the molecules that comprise cells, then we could understand cells and their functions. Although this was true in principle (and in practice in a few cases), the sheer number of molecules made it very difficult to comprehend so many simultaneous functions. The simultaneous measurement of the majority of these molecules became possible over the last 10–15 years through the development of many ingenious technologies. As a result, we now have a growing number of data sets that give us the composition of particular cells and organisms under certain conditions. The chemical interactions between many of these components are now known and this knowledge gives rise to reconstructed biochemical reaction networks on a genome-scale that underlie various cellular functions. Thus, enter (molecular) systems biology.

Systems biology is not necessarily focused on the components themselves, but on the nature of the links that connect them and on the functional states of the biochemical networks that result from the collection of all such links. These functional states of networks correspond to observable physiological or homeostatic states. Completing the relationship between all the chemical components of a cell, with their genetic bases, and its physiological functions is the promise of (molecular) systems biology. This undertaking represents the de facto construction of a mechanistic genotype–phenotype relationship.

1.1 The Genotype–Phenotype Relationship

The concept Through breeding experiments, Gregor Mendel discovered that there are discrete quanta of information passed from one generation to the next that determine the form and function of an organism. These quanta, or packets, of information are now generally referred to as genes. The collection of all the genes and the particular version of them found in a genome of an individual organism is referred to as its genotype. The form and function of an organism is referred to as its phenotype. How the phenotype is related to the genotype represents the fundamental relationship of biology.

For monogenic traits, the genotype–phenotype relationship can be readily understood. One gene confers a phenotype. In the human population, there are now well over 100 of these traits documented and they can be diagnosed at the neonatal stage. However, most phenotypic traits involve coordinated functions of multiple gene products. This makes the genotype–phenotype relationship a challenge to reconstruct and understand. This challenge has two underlying issues. The first comes from the need to know what all the gene and gene products are, and the second comes from understanding the consequences of the complex interactions that can form among a large number of gene products. Today, the former can be addressed using omics data and the latter from the principles of systems analysis applied to biochemistry. The ability to address these challenges has developed over the last 10–15 years and it forms the basis for the bottom-up approach to molecular systems biology that, in turn, ultimately facilitates the realization of the promise of molecular biology.

Towards a mechanistic basis With the publication of the first full genome sequences in the mid 1990s [123], it became possible, in principle, to identify all the gene products that make up an organism. In practice, it has proven difficult to achieve complete or even comprehensive coverage of the genetic elements in simple genomes, but substantial progress has been made. The well-studied biochemistry of metabolic transformations made it possible to reconstruct, on a genome-scale, metabolic networks for a target organism in a biochemically detailed fashion [105,106]. Such metabolic network reconstructions can be converted into a mathematical format yielding mechanistic genotype–phenotype relationships for microbial metabolism [314].

The mathematical format of the underlying biochemical, genetic, and genomic (BiGG) knowledge facilitates the formulation of genome-scale models. Such models are not based on any biophysical theory, but simply represent the reconciliation of the known biochemical properties of the gene products expressed in an organism. Through such large-scale reconciliation, genome-scale models enable the computation of phenotypic traits based on the genetic composition of the target organism [314, 344]. Since the first metabolic genome-scale reconstruction in 1999 and *in silico* models thereof, many more have followed, including that for human metabolism [100]. The scope and content of network reconstructions continues to grow to include the entire transcription/translation apparatus of a cell, for instance [420], and the structural information about the metabolic enzymes [473].

We thus stand at an historical crossroads in the life sciences: the formulation of mechanistic genotype–phenotype relationships has become possible. Given the fundamental nature of this relationship, having mechanistic versions of it is foundational. Today, such relationships are being established for metabolic functions, with an increasing scope in biological content and coverage. Building mechanistic genotype–phenotype relationships and their use represent the scope and content of this book.

1.2   Some Concepts of Genome-scale Science

Paradigm shift   The first full genome sequences emerged in the mid 1990s. At roughly the same time, mRNA expression profiling array and proteomic technologies gave us the capability to determine when a cell uses particular genes. These technologies allow us to achieve a genome-scale view of the contents of target organisms (left side in Figure 1.1). At the beginning of the twenty-first century, this process was unfolding at a rapid rate, driving a fundamental paradigm shift in biology.

The advent of high-throughput experimental technologies forced biologists to begin to view cells as systems, rather than focusing their attention on individual cellular components. Not only did the high-throughput technologies force a systems point of view, but they also enabled the study of cells as systems. What should one do with an available list of cellular components and their properties? As informative as they are, such lists only give basic information about the molecules that comprise cells, their individual chemical properties, and when cells choose to use their components. Such integrative analysis relies on bioinformatics and methods for systems analysis (right side of Figure 1.1).

Thus, at the turn of the century, molecular biology became focused on the systems properties of cellular and tissue functions. These are the properties that arise from the whole, and represent biological properties. In turn, genome-scale science emerged and started to grow.

Genetic circuits and molecular machines   Cellular functions rely on the coordinated action of the multiple gene products. Such coordinated function results from what can

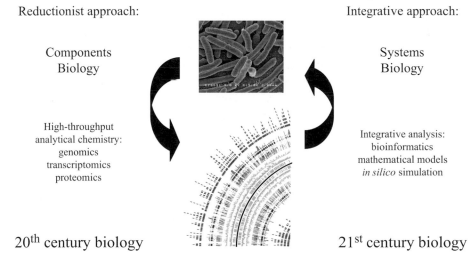

Figure 1.1   Illustration of a paradigm shift at the turn of the century in cell and molecular biology from components to systems analysis. Redrawn from [311]. Top image from Rocky Mountain Laboratories, NIAID, NIH. Bottom image courtesy of Byung-Kwan Cho.

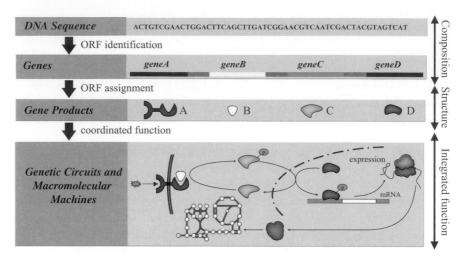

Figure 1.2   Genetic circuits. From sequence, to genes, to gene product function, to multi-component cellular functions. Prepared by Christophe Schilling and Nathan Lewis.

be called a *genetic circuit* (see Figure 1.2). The functions of genetic circuits are diverse, and include DNA replication, translation, the conversion of glucose to pyruvate, laying down the basic body plan of multicellular organisms, and cell motion. Cellular functions are increasingly viewed within this framework, and the physiological function of cells and organisms are viewed as the coordinated or integrated functions of multiple genetic circuits.

The concept of a genetic circuit as a multi-component functional entity – in time or space, or both – is important in systems biology. It is a fundamental factor in the establishment of genotype–phenotype relationships. Individual genetic circuits do not operate in isolation, but in the context of other genetic circuits. The assembly of all such circuits found on a genome produce cellular and organismic functions, and leads to hierarchical decomposition of complex cellular functions and a multi-scale view of the genotype–phenotype relationship.

Genetic circuits function in the context of the entire organism   A commonly used concept in biology is a pathway, an example of a genetic circuit. Figure 1.3 shows an amino acid biosynthetic pathway. One may be interested in various aspects of this pathway; its cellular localization, aggregation of the participating protein, debilitating genetic changes, etc. Such questions represent common pursuits in cellular and molecular biology.

A pathway does not function in isolation, however. It functions in the context of the entire network of interactions in the organism and may interact with many other cell processes. Such interactions can be weak or strong. With the advent of genome-scale network reconstructions, we can now place the function of pathways in the context of all the other known processes in a cell (Figure 1.3).

This genome-scale point of view has proved important in many settings. In the case of metabolic engineering, where new metabolic phenotypes are being built, it is

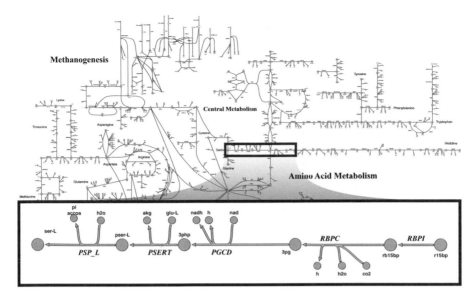

Figure 1.3 Illustration of the functions of a pathway in the context of a whole network. Prepared by Adam Feist.

not enough to identify and express all the genes of the pathway; one must also make sure it functions properly in the network as a whole. An over-expressed pathway may, for instance, drain key biosynthetic precursors, cause an imbalance in redox metabolism, or simply crowd out other cellular functions, leading to a sick or dead host cell if proper balancing of the function of the pathway relative to the whole network is not achieved.

Myriad constraints: the improbability of life The multiplicity of simultaneous molecular and genetic circuit functions occurring in a cell is mind-boggling. However, they all take place in a coherent, organized manner to produce functioning phenotypic states. In a growing bacterial cell, as many as 2.5 million protein molecules are functioning coherently. Ten thousand ribosomes and one hundred thousand tRNAs are busy synthesizing protein molecules. Thousands of RNA polymerases are synthesizing messages to be translated. And all of this happens in a volume of a cubic micron (Figure 1.4). Thus, countless constraints are placed on these functions.

We want to compute these functions with an *in silico* analog of the real cell through a mathematical model. It is popular to state that as the complexity of a mathematical model grows and the number of parameters increases, anything is possible. Inside a cell, nothing could be further from the truth. The numerical range of parameter values that allow these very complex networks to function coherently are severely restricted. Sometimes it is hard to believe that there is even one set of parameter values that allows the living process to take place. Life is indeed improbable.

Evolution and optimization Finding a functioning set of parameter values is the result of a long process of trial and error where the genetic elements change slightly

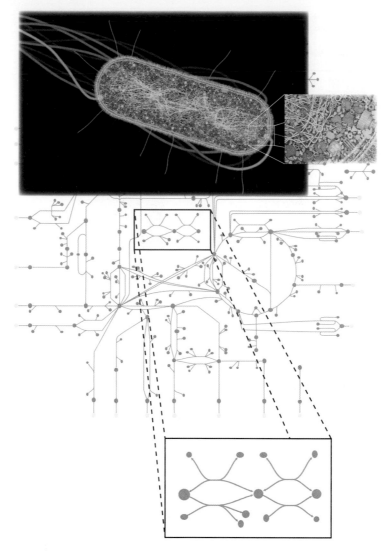

Figure 1.4   A cell operates in a confined space in which many components function simultaneously as a system to produce phenotypic states. Inserted image from [148] used with kind permission from Springer Science+Business Media B.V.

from generation to generation. Evolution is an ongoing optimization process that steadily hones the functions of the existing gene products and adds new ones through mechanisms such as gene duplication and horizontal gene transfer. This process of trial and error is never-ending. Poor choices are punished by extinction while more functional alternatives continue the process.

The change in organism properties with subsequent generations is called *distal causation*. Given the selection process that is at work, key components for describing distal causation are optimization principles. Clearly, one way to generate increased functionality and complexity is through the generation of hierarchy, where new

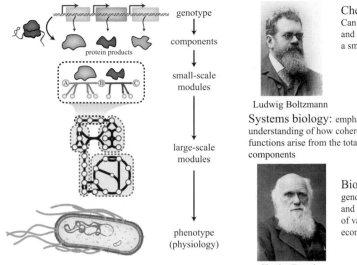

Chemical causation:
Can apply P/C laws
and get causality on
a small scale

Ludwig Boltzmann

Systems biology: emphasis on modules and
understanding of how coherent physiological
functions arise from the totality of molecular
components

Biological causation:
genome-scale changes
and description of 1000s
of variables. Network and
econometric type

Charles Darwin

Figure 1.5 Hierarchy and modularity in biology. Prepared by Nathan Lewis.

functions are built around existing ones. Hierarchy in biology exists in space, time, component abundance, and other properties. Hierarchical organization is key to understanding the genotype–phenotype relationship.

Hierarchical thinking is important in systems biology We are quite familiar with thinking hierarchically about DNA. We think about base pairs as the irreducible unit of DNA sequence. Then we talk about codons, introns, exons, alleles, chromosomes, whole genomes, and other similar measures of DNA size. We can understand them readily, even if we have to scan over nine orders of magnitude in sequence length as in the case of the human genome.

We will need to adapt similar hierarchical thinking techniques to the properties of genome-scale networks. The irreducible elements in a network are the elementary chemical reactions. These can combine into reaction mechanisms, many reactions into modules or motifs, pathways can form and sectors can be defined, as illustrated in Figure 1.5. Currently, coarse-graining of a network relies on objective or subjective definitions [325] of 'modules' that are used to conceptualize a hierarchical network structure.

Our understanding of how to decompose a network hierarchically is likely to improve as we gain a better understanding of the functions of genome-scale networks and our ability to define their properties. Components that always function together in steady or dynamic states normally would fall into modules. Correlated subsets of reactions do appear in the delineation of steady-state properties of networks (Chapter 12). Time-scale separation is often used for temporal decomposition of complex systems, and the stoichiometric matrix does seem to play a role in this formation of dynamic pools [198, 309] that represent the dynamic coarse-graining of a network. Thus, measures of multi-scale thinking are indeed developing in the field.

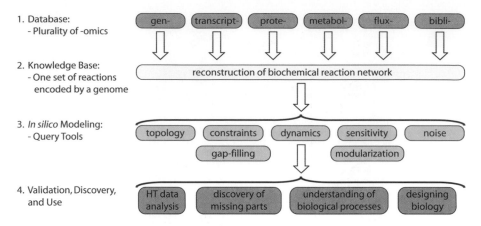

Figure 1.6   The four principal steps in the implementation of systems biology. Note that the second step is unique, while the others are diverse, and it is the interface between high-throughput data and *in silico* analysis. Modified from [118,314].

1.3   The Emergence of Systems Biology

The systems biology paradigm   The ability to generate detailed lists of biological components, determine their interactions, and generate genome-wide data sets has led to the emergence of a fundamental paradigm for systems biology [178]. It is composed of four principal steps (Figure 1.6):

- First, define and enumerate the list of biological components that participate in a cellular process.
- Second, the interactions between these components are studied, the 'wiring diagrams' of genetic circuits are reconstructed, and genome-scale maps are formed in a step-wise manner. This process is one of biochemical reaction network reconstruction.
- Third, reconstructed networks are converted into a mathematical format that formally describes the biological knowledge that underlies the reconstructed network. Computer models are then generated to analyze, interpret, and predict the biological functions that can arise from reconstructed networks.
- Fourth, the models are used in a prospective manner. Prediction entails generating specific hypotheses that can then be tested experimentally. These *in silico* models of reconstructed networks are then improved in an iterative fashion [311].

Much creative work has led to the development of high-throughput technologies (Step 1). Workflows now exist for network reconstruction (Step 2). Many different mathematical methods have been formulated for the analysis of biochemical reaction networks (Step 3) and the phenotypic space explored by experimentation (Step 4) is essentially infinite. In contrast, the reconstruction effort leads to one result.

The need for genome-scale models   It is a common experience of students of biochemistry to come across the same molecule in different chapters of their textbooks.

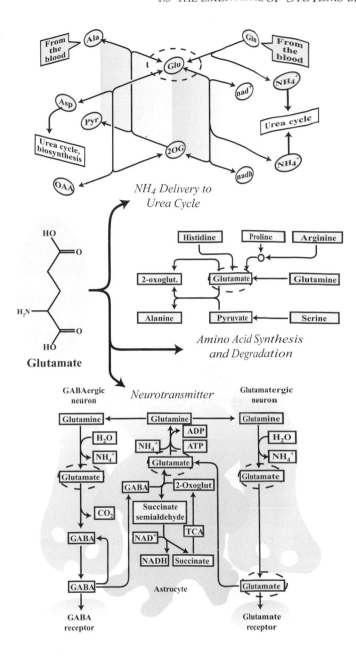

Figure 1.7 The many roles of glutamate. Images adapted from [216] (reprinted with permission). Prepared by Nathan Lewis.

After learning about one function of a molecule, another function arises in a different context. One example is given for glutamate in Figure 1.7. This figure shows the role of glutamate in the urea cycle, as a neurotransmitter, as well as its biosynthetic and degradative pathways. How can all of these functions be reconciled? Thoughtful students of biochemistry struggle with this question.

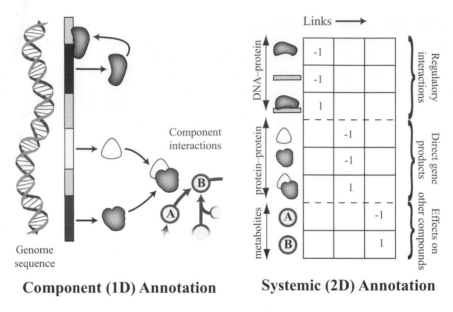

Component (1D) Annotation   **Systemic (2D) Annotation**

Figure 1.8   Systemic, or two-dimensional annotation of genomes: the origin of the stoichiometric matrix, **S**. Modified from [313].

Enter the power of systems science. It turns out that once a genome-scale network is built, all these functions can be reconciled simultaneously with a simple matrix equation. This ability may seem far-fetched at first, but it is achieved through relatively simple accounting principles. For this reason, bottom-up systems biology becomes a mathematical pursuit as its principal goal cannot be achieved without mathematics.

Two-dimensional genome annotation   The unity represented by Step 2 in Figure 1.6 leads to an effort to create a *two-dimensional annotation* of a genome (Figure 1.8). The classical annotation of a genome leads to the identification of open reading frames, their location, and often the corresponding DNA regulatory sequences; basically, a one-dimensional list of components. The open reading frames can then be assigned function based on homology searches of known genes. If the function of a gene product is known, one can describe its interactions with other known gene products, resulting in components and the links between them; fundamentally, a two-dimensional description.

A two-dimensional annotation therefore not only accounts for the components, but all their chemical states (represented as rows in the table in Figure 1.8) and the links between them. The links are represented as columns in the table in Figure 1.8 and ideally should represent the stoichiometric coefficients that correspond to the underlying chemical transformations that are possible between the components. In principle, this table represents a full genome-scale stoichiometric matrix (**S**) for a genome.

Calling for the formulation of this matrix may represent as bold an undertaking as asking for the full base pair sequence of the human genome over 20 years ago.

However, significant progress is being made and genome-scale networks are being reconstructed for metabolism, signaling networks, and transcriptional regulatory networks. Note that the construction of such a connectivity matrix brings us into the world of systems science, where networks are described by nodes and links, or by vertices and edges. This conceptual transition is important.

Building a complex relationship: workflows   The availability of detailed data sets and detailed literature information (the bibliome) about the function of the gene products has led to complex workflows to reconstruct networks. The need for error elimination, traceability of supporting information, data mapping needs, etc., has led to quality-controlled/quality-assured (QC/QA) workflows. For instance, after more than a decade of work on metabolic network reconstruction, a 96-step standard operating procedure (SOP) was developed [427]. Such workflows can be partially automated (Chapter 3), thereby minimizing the amount of detailed and laborious work required for complete workflow execution.

This development is important, as it leads to the introduction of industrial and operations engineering-type thinking into systems biology. We are faced with complex workflows that integrate various data types, various experimental procedures, various data analysis algorithms, and mathematical model building and computation. With some of the workflows now in place, the field can look forward to the development and establishment of similar workflows for the reconstruction of many different cellular functions.

1.4   Building Foundations

The construction of genome-scale models is a key outcome of such workflows, as such models represent mechanistic genotype–phenotype relationships. These models have to take into account dual causation (both proximal and distal). The generation of such models is now based on a few fundamental features that can be stated in an axiomatic fashion [315]:

Axiom #1:   All cellular functions are based on chemistry. A simple but consequential statement, it implies the fundamental events in a cell can be described by chemical equations. These equations, in turn, come with chemical information and physico-chemical principles.

Axiom #2:   Annotated genome sequences along with experimental data enable the reconstruction of genome-scale metabolic networks. The reconstruction process is a grand-scale systematic assembly of information in a QC/QA-based setting [427] that leads to a knowledge base, which is a collection of established biochemical, genetic, and genomic data represented by a network reconstruction.

Axiom #3:   Cells operate under a variety of constraints. Factors constraining cellular functions fall into four principal categories [314]: physico-chemical (see Axiom #5), topological (molecular crowding effects and steric hindrance), environmental (Axiom #4), and regulatory (basically self-imposed constraints, or restraints). These constraints cannot be violated and allow for the estimation of all functional (i.e., physiological) states that a genome-scale network reconstruction can achieve.

Axiom #4:   Cells function in a context-specific manner. When a cell is placed in a particular environment, it expresses a subset of its genes in response to environmental cues. The abundance of cellular components can be profiled using 'omics' methods (i.e., transcriptomics, proteomics, metabolomics). Such omic data can be mapped onto a network reconstruction to tailor it to the particular condition being considered. This 'relative' thinking is a departure from the 'absolute' characteristics of the physical laws, and it introduces the 'environment' into the genotype–phenotype relationship.

Axiom #5:   Mass (and energy) is conserved. This statement is one of basic physical laws. Because all proper chemical equations can be described by stoichiometric coefficients, and because a set of chemical equations can be described by the stoichiometric matrix, $\mathbf{S}$, this means that all steady states (normally close to the homeostatic states of interest) of a network can be described by a simple linear equation, $\mathbf{Sv} = \mathbf{0}$, where $\mathbf{v}$ is a vector of fluxes through chemical reactions [314]. Thus, the computation of functional, or physiological, states of a network is enabled based on the known underlying chemistry.

Axiom #6:   Cells evolve under a selection pressure in a given environment. This statement has implicit optimality principles built into it. Consequently, if we know the selection pressure, we can state a so-called *objective function* and determine optimal states given a network reconstruction and governing constraints, as well as study the 'evolution' of optimal states in a fixed or varying environment (i.e., distal causation).

Each one of these axiomatic statements by themselves may seem trivial and they are accepted in various scientific disciplines as being fundamental. Taken together, though, they combine to form the conceptual basis for constraint-based reconstruction and analysis (COBRA), and enable the development of the mechanistic genotype–phenotype relationship for metabolism and other cellular functions.

1.5   About This Book

Purpose   The availability of annotated genome sequences in the mid to late 1990s enabled the reconstruction of genome-scale metabolic networks [87]. Similar reconstructions of signaling and transcriptional regulatory networks are appearing [164, 402]. The topological structure and functional properties of these networks can now be studied through mathematical models. This requires the conversion of biochemical, genetic and genomic knowledge into a mathematical description. This mathematical description can then be used as a basis for computer models to compute biological functions. Thus, we can now analyze, interpret, and predict the phenotypic functions that such networks can produce using COBRA. The purpose of this book is to describe how biological knowledge is assembled to reconstruct the biochemical reaction networks, their conversion into a mathematical format, the formulation of computational models of biological functions, and how such computations can be used to address basic and applied questions in biology.

Approach   We will divide the material into five parts:

Part I   will describe the network reconstruction process. The efforts to reconstruct networks are intensive, combining the analysis of omic data sets and legacy data, sometimes referred to as bibliomic data. Reconstructions basically culminate in the formation of a biochemically, genetically, and genomically structured knowledge base (BiGG and k-base), as, once curated, a genome-scale reconstruction represents the structured integration of all available information about the target organism.

Part II   will describe the formulation of the stoichiometric matrix, **S**, including its function as a mathematical mapping operation, the chemical constraints on its structure, and its topological properties. Methods for the analysis of basic topological properties are described. We then explore the more subtle and intricate properties of the stoichiometric matrix. To do so, we need to study the fundamental spaces associated with **S**. The two null spaces of **S** contain systematically defined reaction pathways and concentration conservation quantities. An understanding of basic linear algebra will thus be essential to the reader.

Part III   will describe the mathematical methods that have been developed to interrogate the properties of reconstructed networks. The reconstructions and their associated information are not sufficient to completely define the state of a network. Flexibility in network function exists, leading to the development of COBRA methods. The COBRA approach is consistent with the biological reality of operating under governing constraints, but allowing for evolution within them to adapt and improve biological function.

Part IV   will describe the myriad applications of COBRA methods. We discuss the representation of environmental and genetic parameters and how to compute the consequences if they vary. Then we discuss applications that fall into four broad categories: (1) omics data mapping and analysis, (2) gap-filling or the identification of missing information in a reconstruction, (3) the understanding of complex biological processes, and (4) the use of genome-scale models for design or synthesis of biological functions.

Part V   will survey the educational values that underpin the field of systems biology. We finish the book with a chapter on future perspectives of the field.

1.6   Summary

- The genotype–phenotype relationship is fundamental to biology.
- Detailed biological part catalogs of cells have emerged and our ability to measure them is improving steadily due to the continued development of high-throughput technologies.

- The interactions of these parts are being documented. As a result, two-dimensional annotations of genomes are emerging.
- 'Wiring diagrams' representing genetic circuits and genome-scale networks are being reconstructed based on this information. They can be converted into a mathematical format.
- The stoichiometric matrix describes the underlying chemical nature of two-dimensional annotation and thus becomes foundational to the field of systems biology.
- The systems biology paradigm of:

  components $\rightarrow$ networks$\rightarrow$ *in silico* models$\rightarrow$ phenotype

  has arisen. Systems biology is thus inherently mathematical.
- Being able to implement this paradigm quantitatively has led to mechanistic and computable genotype–phenotype relationships.
- These relationships represent a fundamental advance in the life sciences.

PART I
Network Reconstruction

What I cannot create, I do not understand – Richard Feynman

Cellular functions rely on the interactions of their chemical constituents. Various high-throughput experimental methods now allow us to determine the chemical composition of cells on a genome-scale. These methods include whole-genome sequencing and annotation (genomics), the measurement of the messenger RNA molecules that are synthesized under a given condition (transcriptomics), the protein abundance, interactions, and functional states (proteomics), measurements of the presence and concentration of metabolites (metabolomics), and metabolic fluxes (fluxomics). In addition, methods now exist to determine the binding sites of proteins on the DNA (location analysis), to determine transcription start sites (TSSs), transcription breakage points, and the location of ribosomes on the transcripts. Furthermore, the physical location of protein products and segments of the DNA can be determined using various fluorescent reporting molecules. All these omic data types along with bibliomic data (primary literature information) enable the reconstruction of the biochemical reaction networks that operate in cells. Part I will discuss the process of network reconstruction, that effectively amounts to a 2D annotation of genomes.

2 Network Reconstruction: The Concept

The journey of a thousand miles begins with one step – Lao Tzu

Network reconstruction is a long and arduous process that involves building a large network in a step-by-step fashion by identifying one reaction at a time. It is foundational to the bottom-up approach to biology. A reconstruction collects all the available biochemical, genetic, and genomic (BiGG) information that is available on a cellular process of interest, and then organizes it in a formal, mathematical fashion that is consistent with the corresponding fundamental chemical and genetic properties. In this chapter we illustrate this process by looking at the familiar glycolytic pathway. Then we introduce the module-by-module nature of the network reconstruction process. We then show how detailed information about the enzymes, genetic information, and structural properties are incorporated in a reconstruction. Next, we will detail the reconstruction of the central metabolic pathways in *Escherichia coli* and how a knowledge base is formed from the reconstruction process. We close the chapter by discussing the features of genome-scale reconstructions and the computational models formed from them.

2.1 Many Reactions and Their Stoichiometry

Networks are composed of compounds (nodes) and reactions (links). When many reactions are known that share reactants and products, they can be graphically linked together. More and more reactions can be added to a graphical representation as one grows the scope of the network under consideration, and Figure 2.1A shows the first few steps in glycolysis as an example. Such reaction maps were formed historically as more metabolic reactions were discovered and they were eventually joined together.

Such a map can be represented mathematically (Figure 2.1B). Such mathematical representation is based on the stoichiometric coefficients that count the molecules that are consumed and produced by a biochemical reaction. All such coefficients for all the reactions in a network can be organized in a matrix format; a mathematical matrix akin to a table. This information is exact, quantitative, and forms the basis for mathematical characterization and assessment of integrative biochemical properties of a network

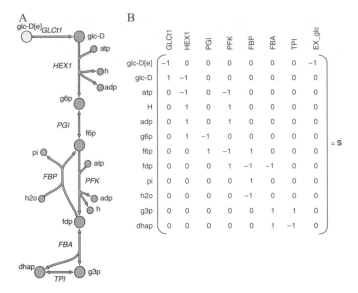

Figure 2.1   Stoichiometric representation of metabolic networks. Panel (A) shows the first few reactions of glycolysis in a graphical form as a network of interacting reactions that share substrates and products. Panel (B) shows the stoichiometric matrix (**S**) corresponding to the reactions in panel (A). As indicated, each column corresponds to a particular reaction and each row to a particular metabolite. The last column, labeled 'EX_glc,' is an exchange reaction for glucose that allows glucose to enter and leave the system. Modified from [34].

as a whole. It is important to note that a map for a network can be drawn in many different ways, but the mathematical representation is unique.

2.2   Reconstructing a Pathway

The glycolytic pathway degrades glucose to form pyruvate or lactate as end products. During this degradation process, the glycolytic pathway builds redox potential in the form of NADH and high-energy phosphate bonds in the form of ATP via substrate-level phosphorylation. Glycolysis also assimilates an inorganic phosphate group that is converted into a high-energy bond and then is hydrolyzed in the ATP use reaction (or the 'load' reaction).

The nodes (compounds) and links (reactions)   A network has nodes consisting of compounds and links that are made up of the reactions between them. In its simplest form, glycolysis has 12 primary metabolites, 5 co-factor molecules (ATP, ADP, AMP, NAD, NADH), inorganic phosphate (P_i), protons (H^+), and water (H_2O). There are thus 20 compounds participating in this pathway (Table 2.1).

The links formed between these compounds are the glycolytic reactions. The reactions and transport reactions are summarized in Table 2.2. There are 21 reactions, including all the transport reactions; i.e., a pathway has inputs and outputs. In this case, the primary input is glucose and the primary output is lactate (or pyruvate).

Table 2.1   The compounds in the glycolytic system and their abbreviations. The compounds have been divided into three groups: (i) the glycolytic intermediates (1–12), (ii) the co-factors (13–17), and (iii) inorganic components (18–20).

#	Abbreviation	Intermediates and co-factors
1	Gluc	Glucose
2	G6P	Glucose 6-phosphate
3	F6P	Fructose 6-phosphate
4	FDP	Fructose 1,6-diphosphate
5	DHAP	Dihydroxyacetone phosphate
6	GAP	Glyceraldehyde 3-phosphate
7	DPG13	1,3-Diphosphoglycerate
8	PG3	3-Phosphoglycerate
9	PG2	2-Phosphoglycerate
10	PEP	Phosphoenolpyruvate
11	PYR	Pyruvate
12	LAC	Lactate
13	NAD	Nicotinamide adenine dinucleotide (oxidized)
14	NADH	Nicotinamide adenine dinucleotide (reduced)
15	AMP	Adenosine mono-phosphate
16	ADP	Adenosine di-phosphate
17	ATP	Adenosine tri-phosphate
18	P_i	Inorganic phosphate
19	H^+	Proton
20	H_2O	Water

This pathway can be represented as a map (see Figure 2.2). This map can be drawn in many different ways, and different textbooks will have different graphical representations. Later we will see that a *system* is formed by drawing a boundary around the reconstructed network, with the inputs and outputs crossing this boundary. Boundaries can be either physical or virtual.

The stoichiometric matrix for glycolysis   If the reactions in Table 2.2 are proper chemical reactions, i.e., elementally and charge-balanced, they obey the basic laws of chemistry. The stoichiometric coefficients in these reactions are integers that are time- and condition-invariant; they are always the same. All these numbers can be organized into a table of stoichiometric coefficients. In fact, this table is a mathematical object that is key to characterizing the systems properties of a network.

The stoichiometric matrix, **S**, can be formed for the glycolytic system based on the reactions of which it is comprised. It is shown in Figure 2.2b. Its dimensions are

Table 2.2 The glycolytic enzymes and transporters, their abbreviations, and chemical reactions. These reactions can be grouped into several categories: (i) the glycolytic reactions (1–11), (ii) adenosine phosphate metabolism (12,13), (iii) primary exports (14,15), (iii) the load functions (16,17), (iv) the primary inputs, here fixed (18,19), and (v) the water and proton exchanges.

#	Abbrev.	Enzymes/transporter/load	Elementally balanced reaction
1	HK	Hexokinase	$Gluc + ATP \rightarrow G6P + ADP + H^+$
2	PGI	Glucose 6-phosphate isomerase	$G6P \leftrightarrow F6P$
3	PFK	Phosphofructokinase	$F6P + ATP \rightarrow FDP + ADP + H^+$
4	TPI	Triose-phosphate isomerase	$DHAP \leftrightarrow GAP$
5	ALD	Fructose 1,6-diphosphate aldolase	$FDP \leftrightarrow DHAP + GAP$
6	GAPDH	Glyceraldehyde 3-phosphate dehydrogenase	$GAP + NAD + P_i \leftrightarrow$ $DPG13 + NADH + H^+$
7	PGK	Phosphoglycerate kinase	$DPG13 + ADP \leftrightarrow PG3 + ATP$
8	PGLM	Phosphoglycerate mutase	$PG3 \leftrightarrow PG2$
9	ENO	Enolase	$PG2 \leftrightarrow PEP + H_2O$
10	PK	Pyruvate kinase	$PEP + ADP + H+ \rightarrow PYR + ATP$
11	LDH	Lactate dehydrogenase	$PYR + NADH + H+ \leftrightarrow LAC + NAD$
12	AMP	AMP export	$AMP \rightarrow$
13	APK	Adenylate kinase	$2ADP \leftrightarrow AMP + ATP$
14	PYR	Pyruvate exchange	$PYR \leftrightarrow$
15	LAC	Lactate exchange	$LAC \leftrightarrow$
16	ATP	ATP hydrolysis	$ATP + H_2O \rightarrow ADP + P_i + H^+$
17	NADH	NADH oxidation	$NADH \rightarrow NAD + H^+$
18	GLU_{in}	Glucose import	$\rightarrow Gluc$
19	AMP_{in}	AMP import	$\rightarrow AMP$
20	H	Proton exchange	$H^+ \leftrightarrow$
21	H_2O	Water exchange	$H_2O \leftrightarrow$

20×21, representing the 20 metabolites and the 21 fluxes given in Tables 2.1 and 2.2, respectively.

The stoichiometric matrix is remarkably important in systems biology. After learning how to reconstruct networks in Part I of this book, we will learn about the mathematical properties of the stoichiometric matrix in Part II. Then in Part III we will learn how it enables the formulation of mathematical models of network states. These states are called functional states and correspond to observable phenotypic properties.

For now, we will continue with the conceptual description of the reconstruction process. We will talk about the module-by-module approach to reconstruction, and

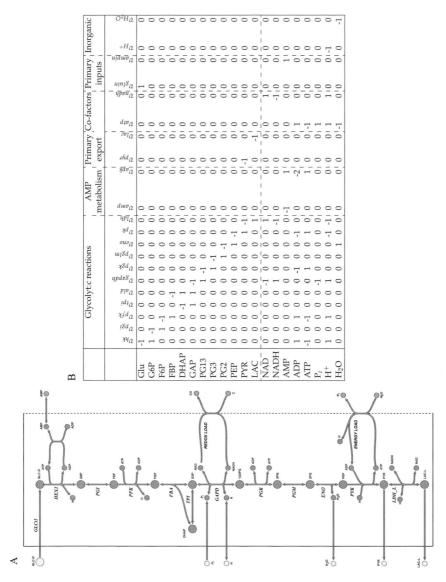

Figure 2.2 Glycolysis as a system, with a systems boundary and inputs and outputs. (A) The reaction schema, co-factor interactions, and environmental exchanges. (B) The stoichiometric matrix for this glycolytic system. The matrix is partitioned to show the co-factors separate from the glycolytic intermediates and to separate the exchange reactions and co-factor loads. Prepared by Hooman Hefzi.

then about how to parse out the properties of the participating macromolecules and their functional states. The important notion of the *scope* of a reconstruction arises. Ultimately, we would like to include in a reconstruction all the relevant information found in the target organism, leading to the formulation of a genome-scale network.

2.3   Module-by-module Reconstruction

Glycolysis is commonly the first pathway that one learns about in biochemistry. It is normally followed by a discussion of the pentose phosphate pathway. We learn that there are molecules that are found in both pathways. Glycolysis and the pentose pathway are thus connected and form a larger network.

The pentose pathway   The pentose pathway originates from G6P in glycolysis. G6P undergoes two oxidation steps including decarboxylation, releasing CO_2, leading to the formation of one pentose and two NADPH molecules. These reactions are called the *oxidative branch* of the pentose pathway. The branch forms two NADPH molecules that are used to form glutathione (GSH) from an oxidized dimeric state, GSSG, by breaking a di-sulfite bond. GSH and GSSG are present in high concentrations, and thus buffer the NADPH redox charge. The pentose formed, R5P, can be used for biosynthesis.

If the pentose formed by the oxidative branch is not used for biosynthetic purposes, it undergoes a number of isomerization, epimerization, transaldolation, and transketolation reactions that lead to the formation of F6P and GAP. Specifically, two F6P and one GAP that return to glycolysis come from three pentose molecules (i.e., $3 \times 5 = 15$ carbon atoms go to $2 \times 6 + 3 = 15$ carbon atoms). This part of the pathway is the *non-oxidative branch*, and it is composed of a series of reversible reactions, while the oxidative branch is irreversible.

The pentose pathway has 12 compounds (Table 2.3) and 11 reactions (Table 2.4). These compounds and reactions are indexed so that they are a continuation of the corresponding list for glycolysis. These two lists can be concatenated to form a list of all the compounds and all the reactions in glycolysis and the pentose pathway to form a reconstructed network that contains both of these classical biochemical pathways.

Coupling to glycolysis: forming a unified reaction map   Because the inputs to and outputs from the pentose pathway are from glycolysis, the pentose pathway and glycolysis are readily interfaced to form a single reaction map (see Figure 2.3). The dashed arrows in the reaction map represent the return of F6P and GAP from the pentose pathway to glycolysis and do not represent actual reactions. The additional exchanges with the environment, over those in glycolysis alone, are CO_2 secretion and redox load on the NADPH pool, which are shown in Figure 2.3 as a load on GSH. We will not consider R5P production here, but it will appear in the next chapter as it is involved in the nucleotide salvage pathways.

Structure of the combined stoichiometric matrix   The stoichiometric matrix for glycolysis (from Figure 2.2b) can be appended with the reactions in the pentose pathway (Table 2.4). The resulting stoichiometric matrix is shown in Figure 2.3B. This matrix

Table 2.3   Pentose pathway intermediates, their abbreviations, and steady-state concentrations for the parameter values used in this chapter. The index on the compounds is added to that for glycolysis (Table 2.1).

#	Abbreviation	Intermediates/co-factors
21	GL6P	6-Phosphogluconolactone
22	GO6P	6-Phosphoglyconate
23	Ru5P	Ribulose 5-phosphate
24	X5P	Xylose 5-phosphate
25	R5P	Ribose 5-phosphate
26	S7P	Sedoheptulose 7-phosphate
27	E4P	Erythrose 4-phosphate
28	NADP	Nicotinamide adenine dinucleotide phosphate (oxidized)
29	NADPH	Nicotinamide adenine dinucleotide phosphate (reduced)
30	GSH	Glutathione (reduced)
31	GSSG	Glutathione (oxidized)
32	CO_2	Carbon dioxide

Table 2.4   Pentose pathway enzymes and transporters, their abbreviations, and chemical reactions. The index on the reactions is added to that for glycolysis, Table 2.2. The reactions of the oxidative branch are irreversible, while those of the non-oxidative branch are reversible.

#	Abbreviation	Enzymes/transporter/load	Elementally balanced reaction
22	G6PDH	G6P dehydrogenase	$G6P+NADP \rightarrow GL6P+NADPH + H$
23	PGLase	6-Phosphogluconolactonase	$GL6P+H_2O \rightarrow GO6P + H$
24	GL6PDH	GO6P dehydrogenase	$GO6P+NADP \rightarrow RU5P+NADPH+CO_2$
25	R5PE	X5P epimerase	$RU5P \rightleftharpoons X5P$
26	R5PI	R5P isomerase	$RU5P \rightleftharpoons R5P$
27	TKI	Transketolase I	$X5P+R5P \rightleftharpoons S7P + GAP$
28	TKII	Transketolase II	$X5P+E4P \rightleftharpoons F6P+GAP$
29	TALA	Transaldolase	$S7P+GAP \rightleftharpoons E4P+ F6P$
30	GSSGR	Glutathione reductase	$GSSG+NADPH+H \rightleftharpoons 2GSH + NADP$
31	GSHR	Glutathione oxidase	$2GSH \rightleftharpoons GSSG + 2H$
32	CO_2 exch	freely exchanging CO_2	$CO_{2,in} \rightarrow CO_{2,out}$

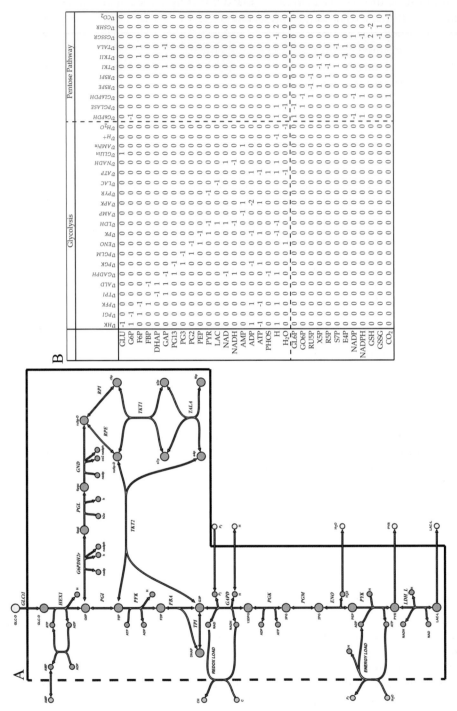

Figure 2.3 (A) Coupling glycolysis and the pentose pathway. The reaction schema, co-factor interactions, and environmental exchanges. (B) The stoichiometric matrix for the coupled glycolytic and pentose pathways. Prepared by Hooman Hefzi.

has dimensions of 32×32. We have put partitioning dashed lines into the matrix to illustrate its structure. The two blocks of matrices on the diagonal are those for each pathway. The lower left block is filled with zero elements showing that the pentose pathway intermediates do not appear in glycolysis. Conversely, the upper right-hand block shows that three of the glycolytic intermediates either leave (GAP and F6P) or enter (G6P) the pentose pathway. Both glycolysis and the pentose pathway produce and/or consume protons and water.

Building networks alters the notion of a pathway The stoichiometric matrix combines the two pathways into a seamless whole. There are no longer two pathways, just a single network, which begs the question, "what actually is a pathway?" When one analyzes the functions of integrated network reconstructions through the methods of systems biology (see Chapter 12) pathways actually become properties of the stoichiometric matrix and they describe the state of the entire reconstructed network.

2.4 Proteins and Their Many States

Macromolecules and their properties can be included in a reconstruction. The proteins that are of particular interest are those that regulate fluxes. Regulatory enzymes bind to many ligands. Their functional states can be described through all the reactions that they undergo. Glycolysis is composed of many well-studied enzymes. We chose human phosphofructokinase (PFK) as an example for reconstruction. PFK was historically considered to be the key regulatory enzyme in glycolysis and is referred to as the 'pacemaker' of glycolysis in older literature. We now know that many of the glycolytic enzymes play key regulatory roles and affect cell functions in a notable fashion.

The enzyme PFK is a tetrameric enzyme. There are isoforms of the subunits of the enzyme, meaning that there is more than one gene for each subunit and the genes are not identical. The isoforms are expressed differentially in various tissues, and therefore different versions of the enzyme are active in different tissues. The regulation of PFK is quite complicated [293]. Here, for illustrative purposes, we will consider a homotetrameric form of PFK (Figure 2.4) with one activator (AMP) and one inhibitor (ATP).

The reaction catalyzed PFK is a major regulatory enzyme in glycolysis. It catalyzes the reaction

$$F6P + ATP \xrightarrow{\text{PFK}} FDP + ADP + H^+ \tag{2.1}$$

This reaction is a part of the glycolytic pathway (Table 2.2).

The detailed reaction mechanism of PFK is shown in Figure 2.4. The reactants (F6P and ATP) and products (FDP and ADP) enter and leave the PFK subnetwork. These exchanges will reach a steady state. The regulators – AMP is an activator and ATP is an inhibitor – also enter and leave the PFK subnetwork. They reach an equilibrium in the steady state, as they have no flow through the subnetwork. The bound states of PFK thus equilibrate with its regulators while the reactants and products will flow through the PFK subnetwork.

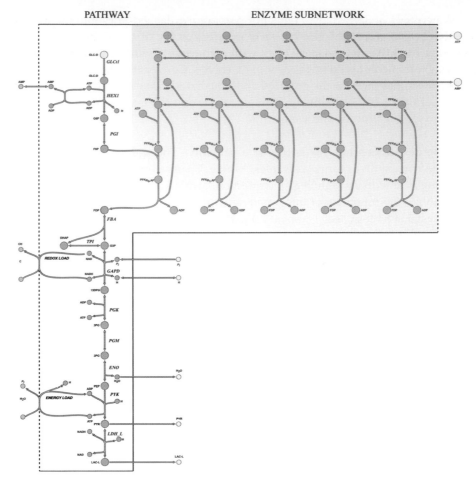

Figure 2.4   The phosphofructokinase subnetwork and its integration with glycolysis. The subnetwork shows the reaction schema for catalysis, regulation, and the exchanges with the rest of the metabolic network. Prepared by Hooman Hefzi.

Bound states of PFK   PFK stays within the boundaries of the subnetwork, where it can have various states. The enzyme exists in two natural forms, T (tight) and R (relaxed).

$$R_0 \rightleftharpoons T_0 \tag{2.2}$$

The R form is catalytically active and the T form is inactive. The two forms are in equilibrium as

$$L = \frac{T_0}{R_0} \tag{2.3}$$

where L is the *allosteric constant* for PFK.

There are many ligands that can bind to PFK to modulate its activity by altering the balance of the T and R forms. Here we will consider AMP as an activator and ATP as an inhibitor. AMP will bind serially to all the subunits of the tetramer in the R state:

$$R_0 + AMP \rightleftharpoons R_1 \tag{2.4}$$
$$R_1 + AMP \rightleftharpoons R_2 \tag{2.5}$$
$$R_2 + AMP \rightleftharpoons R_3 \tag{2.6}$$
$$R_3 + AMP \rightleftharpoons R_4 \tag{2.7}$$

and ATP will similarly bind serially to the T state:

$$T_0 + ATP \rightleftharpoons T_1 \tag{2.8}$$
$$T_1 + ATP \rightleftharpoons T_2 \tag{2.9}$$
$$T_2 + ATP \rightleftharpoons T_3 \tag{2.10}$$
$$T_3 + ATP \rightleftharpoons T_4 \tag{2.11}$$

The chemical reaction, see Equation (2.1), will proceed in three steps:

1   the binding of the co-factor ATP

$$\xrightarrow{\text{substrate in}} ATP + R_i \rightleftharpoons R_{i,A} \tag{2.12}$$

2   the binding of the substrate F6P

$$\xrightarrow{\text{substrate in}} F6P + R_{i,A} \rightleftharpoons R_{i,AF} \tag{2.13}$$

3   the catalytic conversion to the products and their simultaneous release from the enzyme

$$R_{i,AF} \xrightarrow{\text{transformation}} R_i + FDP + ADP + H \xrightarrow{\text{products out}} \tag{2.14}$$

The simultaneous release of the products is an assumption used here to simplify the reconstruction. The subscript i ($i = 0, 1, 2, 3, 4$) indicates how many activator molecules (AMP) are bound to the R form of the enzyme. The free form ($i = 0$) and all the bound forms ($i = 1, \ldots, 4$) are catalytically active.

The stoichiometric matrix for glycolysis with PFK   The list of reactions that describe the bound states of PFK can be added to the reaction list for glycolysis. This leads to an integrated stoichiometric matrix (Table 2.5). Here, the column for the PFK reaction is replaced with the PFK subnetwork. This results in a stoichiometric matrix that is of dimension 40×44.

These two extensions (the PPP and PFK) to the original glycolytic pathway reconstruction illustrate the step-by-step and module-by-module nature of the reconstruction process. A boundary is drawn around the network of interest and all the

Table 2.5 The stoichiometric matrix for glycolysis with PFK shown in Figure 2.4. The matrix is partitioned to show the addition of PFK subnetwork to the stoichiometric matrix for glycolysis.

	v_{hk}	v_{pgi}	$v_{R_{0,1}}$	$v_{R_{0,2}}$	$v_{R_{0,3}}$	$v_{R_{1,0}}$	$v_{R_{1,1}}$	$v_{R_{1,2}}$	$v_{R_{1,3}}$	$v_{R_{2,0}}$	$v_{R_{2,1}}$	$v_{R_{2,2}}$	$v_{R_{2,3}}$	$v_{R_{3,0}}$	$v_{R_{3,1}}$	$v_{R_{3,2}}$	$v_{R_{3,3}}$	$v_{R_{4,0}}$	$v_{R_{4,1}}$	$v_{R_{4,2}}$	$v_{R_{4,3}}$	v_T	v_{T_1}	v_{T_2}	v_{T_3}	v_{T_4}	v_{tpi}	v_{ald}	v_{gapdh}	v_{pgk}	v_{pglm}	v_{eno}	v_{pk}	v_{ldh}	v_{amp}	v_{apk}	v_{pyr}	v_{lac}	v_{atp}	v_{nadh}	v_{gltin}	v_{amptin}	v_{H^+}	v_{H_2O}
GLU	-1	0	0	0	0	0	0	0	0	0	0	0	0	0	0	0	0	0	0	0	0	0	0	0	0	0	0	0	0	0	0	0	0	0	0	0	0	0	0	0	1	0	0	0
G6P	1	-1	0	0	0	0	0	0	0	0	0	0	0	0	0	0	0	0	0	0	0	0	0	0	0	0	0	0	0	0	0	0	0	0	0	0	0	0	0	0	0	0	0	0
F6P	0	1	-1	0	0	0	0	0	0	0	0	0	0	0	0	0	0	0	0	0	0	0	0	0	0	0	0	0	0	0	0	0	0	0	0	0	0	0	0	0	0	0	0	0
R0	0	0	-1	-1	1	1	0	0	0	0	0	0	0	0	0	0	0	0	0	0	0	-1	0	0	0	0	0	0	0	0	0	0	0	0	0	0	0	0	0	0	0	0	0	0
R0,A	0	0	1	-1	0	0	0	0	0	0	0	0	0	0	0	0	0	0	0	0	0	0	0	0	0	0	0	0	0	0	0	0	0	0	0	0	0	0	0	0	0	0	0	0
R0,AF	0	0	0	1	-1	0	0	0	0	0	0	0	0	0	0	0	0	0	0	0	0	0	0	0	0	0	0	0	0	0	0	0	0	0	0	0	0	0	0	0	0	0	0	0
R1	0	0	0	0	0	-1	-1	1	0	1	0	0	0	0	0	0	0	0	0	0	0	0	-1	0	0	0	0	0	0	0	0	0	0	0	0	0	0	0	0	0	0	0	0	0
R1,A	0	0	0	0	0	0	1	-1	0	0	0	0	0	0	0	0	0	0	0	0	0	0	0	0	0	0	0	0	0	0	0	0	0	0	0	0	0	0	0	0	0	0	0	0
R1,AF	0	0	0	0	0	0	0	1	-1	0	0	0	0	0	0	0	0	0	0	0	0	0	0	0	0	0	0	0	0	0	0	0	0	0	0	0	0	0	0	0	0	0	0	0
R2	0	0	0	0	0	0	0	0	0	-1	-1	1	0	1	0	0	0	0	0	0	0	0	0	-1	0	0	0	0	0	0	0	0	0	0	0	0	0	0	0	0	0	0	0	0
R2,A	0	0	0	0	0	0	0	0	0	0	1	-1	0	0	0	0	0	0	0	0	0	0	0	0	0	0	0	0	0	0	0	0	0	0	0	0	0	0	0	0	0	0	0	0
R2,AF	0	0	0	0	0	0	0	0	0	0	0	1	-1	0	0	0	0	0	0	0	0	0	0	0	0	0	0	0	0	0	0	0	0	0	0	0	0	0	0	0	0	0	0	0
R3	0	0	0	0	0	0	0	0	0	0	0	0	0	-1	-1	1	0	1	0	0	0	0	0	0	-1	0	0	0	0	0	0	0	0	0	0	0	0	0	0	0	0	0	0	0
R3,A	0	0	0	0	0	0	0	0	0	0	0	0	0	0	1	-1	0	0	0	0	0	0	0	0	0	0	0	0	0	0	0	0	0	0	0	0	0	0	0	0	0	0	0	0
R3,AF	0	0	0	0	0	0	0	0	0	0	0	0	0	0	0	1	-1	0	0	0	0	0	0	0	0	0	0	0	0	0	0	0	0	0	0	0	0	0	0	0	0	0	0	0
R4	0	0	0	0	0	0	0	0	0	0	0	0	0	0	0	0	0	-1	-1	1	0	0	0	0	0	-1	0	0	0	0	0	0	0	0	0	0	0	0	0	0	0	0	0	0
R4,A	0	0	0	0	0	0	0	0	0	0	0	0	0	0	0	0	0	0	1	-1	0	0	0	0	0	0	0	0	0	0	0	0	0	0	0	0	0	0	0	0	0	0	0	0
R4,AF	0	0	0	0	0	0	0	0	0	0	0	0	0	0	0	0	0	0	0	1	-1	0	0	0	0	0	0	0	0	0	0	0	0	0	0	0	0	0	0	0	0	0	0	0
T0	0	0	0	0	0	0	0	0	0	0	0	0	0	0	0	0	0	0	0	0	0	1	-1	0	0	0	0	0	0	0	0	0	0	0	0	0	0	0	0	0	0	0	0	0
T1	0	0	0	0	0	0	0	0	0	0	0	0	0	0	0	0	0	0	0	0	0	0	1	-1	0	0	0	0	0	0	0	0	0	0	0	0	0	0	0	0	0	0	0	0
T2	0	0	0	0	0	0	0	0	0	0	0	0	0	0	0	0	0	0	0	0	0	0	0	1	-1	0	0	0	0	0	0	0	0	0	0	0	0	0	0	0	0	0	0	0
T3	0	0	0	0	0	0	0	0	0	0	0	0	0	0	0	0	0	0	0	0	0	0	0	0	1	-1	0	0	0	0	0	0	0	0	0	0	0	0	0	0	0	0	0	0
T4	0	0	0	0	0	0	0	0	0	0	0	0	0	0	0	0	0	0	0	0	0	0	0	0	0	1	0	0	0	0	0	0	0	0	0	0	0	0	0	0	0	0	0	0
FDP	0	0	0	0	1	0	0	0	1	0	0	0	1	0	0	0	1	0	0	0	1	0	0	0	0	0	-1	0	0	0	0	0	0	0	0	0	0	0	0	0	0	0	0	0
DHAP	0	0	0	0	0	0	0	0	0	0	0	0	0	0	0	0	0	0	0	0	0	0	0	0	0	0	1	-1	0	0	0	0	0	0	0	0	0	0	0	0	0	0	0	0
GAP	0	0	0	0	0	0	0	0	0	0	0	0	0	0	0	0	0	0	0	0	0	0	0	0	0	0	1	1	-1	0	0	0	0	0	0	0	0	0	0	0	0	0	0	0
PG13	0	0	0	0	0	0	0	0	0	0	0	0	0	0	0	0	0	0	0	0	0	0	0	0	0	0	0	0	1	-1	0	0	0	0	0	0	0	0	0	0	0	0	0	0
PG3	0	0	0	0	0	0	0	0	0	0	0	0	0	0	0	0	0	0	0	0	0	0	0	0	0	0	0	0	0	1	-1	0	0	0	0	0	0	0	0	0	0	0	0	0
PG2	0	0	0	0	0	0	0	0	0	0	0	0	0	0	0	0	0	0	0	0	0	0	0	0	0	0	0	0	0	0	1	-1	0	0	0	0	0	0	0	0	0	0	0	0
PEP	0	0	0	0	0	0	0	0	0	0	0	0	0	0	0	0	0	0	0	0	0	0	0	0	0	0	0	0	0	0	0	1	-1	0	0	0	0	0	0	0	0	0	0	0
PYR	0	0	0	0	0	0	0	0	0	0	0	0	0	0	0	0	0	0	0	0	0	0	0	0	0	0	0	0	0	0	0	0	1	-1	0	0	-1	0	0	0	0	0	0	0
LAC	0	0	0	0	0	0	0	0	0	0	0	0	0	0	0	0	0	0	0	0	0	0	0	0	0	0	0	0	0	0	0	0	0	1	0	0	0	-1	0	0	0	0	0	0
NAD	0	0	0	0	0	0	0	0	0	0	0	0	0	0	0	0	0	0	0	0	0	0	0	0	0	0	0	0	-1	0	0	0	0	1	0	0	0	0	0	1	0	0	0	0
NADH	0	0	0	0	0	0	0	0	0	0	0	0	0	0	0	0	0	0	0	0	0	0	0	0	0	0	0	0	1	0	0	0	0	-1	0	0	0	0	0	-1	0	0	0	0
AMP	0	0	0	0	0	0	0	0	0	0	0	0	0	0	0	0	0	0	0	0	0	0	0	0	0	0	0	0	0	0	0	0	0	0	1	0	0	0	-1	0	0	1	0	0
ADP	-1	0	0	-1	0	0	0	-1	0	0	0	-1	0	0	0	-1	0	0	0	-1	0	0	0	0	0	0	0	0	0	1	0	0	-1	0	0	-2	0	0	1	0	0	0	0	0
ATP	1	0	0	1	0	0	0	1	0	0	0	1	0	0	0	1	0	0	0	1	0	0	0	0	0	0	0	0	0	-1	0	0	1	0	0	1	0	0	-1	0	0	0	0	0
P_i	0	0	0	0	0	0	0	0	0	0	0	0	0	0	0	0	0	0	0	0	0	0	0	0	0	0	0	0	-1	1	0	0	0	0	0	0	0	0	1	0	0	-1	1	0
H	1	0	0	0	0	0	0	0	0	0	0	0	0	0	0	0	0	0	0	0	0	0	0	0	0	0	0	0	1	0	0	0	1	-1	0	0	0	0	-1	1	0	0	-1	0
H_2O	0	0	0	0	0	0	0	0	0	0	0	0	0	0	0	0	0	0	0	0	0	0	0	0	0	0	0	0	0	0	0	1	0	0	0	0	0	0	0	0	0	0	0	-1

reactions inside are defined and described. Best practices in manual reconstruction call for a module-by-module reconstruction where each addition is tested for accuracy and effect on the functions of the network. In the rest of Part I, we learn how to systematically expand the scope of a reconstruction.

2.5   Central *E. coli* Energy Metabolism

The reconstruction process introduced will quickly grow into biologically meaningful networks. As knowledge about the central metabolism of enterobacteria is well characterized, a realistic but small-scale network reconstruction of this process is used for educational purposes. We will use this network as an example for computations throughout this book.

This core metabolic network, shown in Figure 2.5, was built in a module-by-module fashion. It contains the most important reactions of glycolysis, the pentose phosphate pathway, the TCA cycle, and oxidative phosphorylation. It also contains basic nitrogen metabolism in the form of L-glutamate and L-glutamine synthesis from α-ketoglutarate, as L-glutamate is a precursor to many other amino acids. It accounts for the 54 unique metabolites, 20 of which are extracellular and can be transported from the cell. The reactions list has 95 chemical reactions and 20 exchange reactions. A combined removal of the so-called 12 biosynthetic precursors and the co-factor needed for biomass synthesis can be used as a proxy for growth requirements (see Chapter 21).

In addition, a full genetic basis for the enzymes that catalyze the reactions is included in the reconstruction. The association between the genes, the proteins, and the reactions (abbreviated as GPR associations) are illustrated for two enzymes in Figure 2.5. This figure illustrates the enormous amount of information that is included in a reconstruction. A reconstruction thus is a knowledge base that represents the curated integration of disparate data types and sources of information. Detailed information about this core network reconstruction is found in [296].

2.6   Genome-scale Networks

The scope of a reconstruction can grow with each additional biochemical process that is included. Ultimately, a reconstruction can reach a genome-scale when all the known biochemical and genetic processes that are found on a genome are accounted for. Such a reconstruction process is clearly quite laborious and detailed. The scope and content of reconstructions is steadily growing and they now include the details of both protein synthesis and structure, transcriptional regulation, and other processes.

Vision   A genome-scale reconstruction aims to be comprehensive, ideally accounting for all relevant genetic elements found in the genome of the target organism. It should be able to put everything found in the target organism in context, systematically integrate all knowledge about the target organism, and in principle should be able to address every phenotypic trait. In a way, such genome-scale reconstruction efforts represent a grand unification theory for biology and constitute the basis for

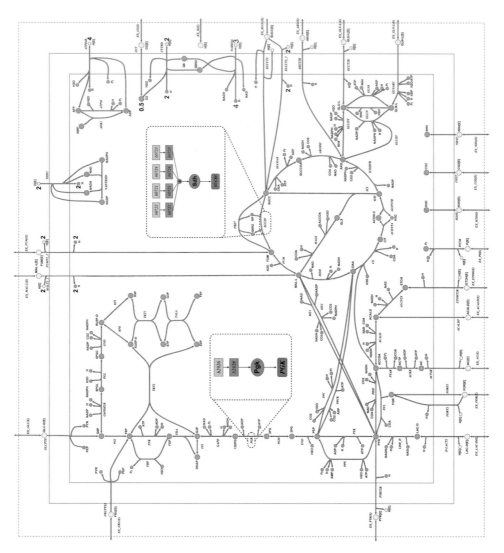

Figure 2.5 The reaction map of the central metabolic pathways in *E. coli* that illustrates a couple of GPRs. Prepared by Hooman Hefzi.

formulating mechanistic genotype–phenotype relationships. Such formulation requires the conversion of a reconstruction into a computational genome-scale model (GEM).

Today, such a grand-scale vision of a comprehensive *in silico* model of a target cell is still conceptual. However, a 'network-by-network' or cellular 'process-by-process' effort is underway. Today, we have integrated models of metabolism, GPRs, and genome location that, once put into a genome-scale model, are able to compute some genotype–phenotype relationships.

Illustration   An example of the development of a genotype–phenotype relationship is given in Figure 2.6 where the consequences of a gene knock-out is to be assessed. The genomic location of the gene to be removed is shown. It comes with a variety of information such as the promoter structure and flanking sequence information that can be used to design the knock-out procedure. The consequence of the removal of the gene on metabolism can be traced through the GPR. The corresponding reactions in the network are rendered inoperative. The alteration in the functional capabilities of the network can be assessed using a GEM. One can address a question such as whether this is a lethal gene deletion in a certain environment. We will detail answers to questions like this one later in the book, but this example serves to illustrate that a comprehensive reconstruction that contains diverse data types and

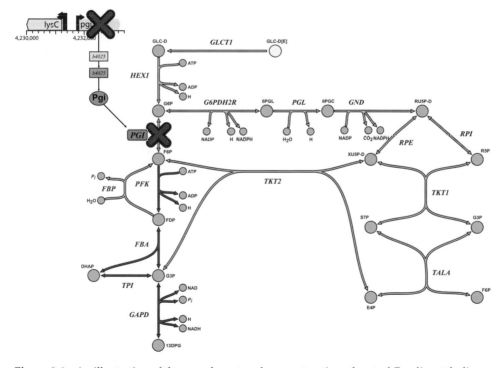

Figure 2.6   An illustration of the use of a network reconstruction of central *E. coli* metabolism to compute the consequences of a gene knock-out . The removal of the gene is traced to the reactions that are consequently rendered inoperative. The reconstruction accounts for detailed biochemistry, GPRs, and genome location. Prepared by Hooman Hefzi.

processes can be used as the basis for the computation of phenotypes that, in turn, can be measured.

Recap In this chapter, we illustrated the concept of a network reconstruction, the characteristics of the reconstruction process, and the goal of reaching the genome scale. In the following chapters, we will illustrate the detailed reconstruction process and what it involves. Part II then discusses the mathematical representation of a network reconstruction and the assessment of its topological properties. A network reconstruction is the basis for a computational model for network states (i.e., phenotypic states), and these are discussed in Part III.

2.7 Summary

- The process of *network reconstruction* is fundamental to systems biology.
- A reconstruction involves the systematic acquisition of information about the small molecules, proteins, and genes that participate in a cellular process of interest and ultimately the representation of this information in a chemically self-consistent form.
- The reconstruction process leads to a *knowledge base* (k-base); sometimes called a BiGG (biochemically, genetically, and genomically structured) k-base.
- A k-base can be described mathematically and thus converted into a computational model. There are several ways to evaluate the content of a k-base computationally.
- If a well-annotated genome sequence is available for the target organism for study, then a genome-scale reconstruction can be performed and a *genome-scale model* (GEM) formulated.
- A GEM essentially represents a *mechanistic genotype–phenotype relationship* as it can be used to trace the effect of genetic parameters on a computed phenotypic state.

3 Network Reconstruction: The Process

A model that proves very inadequate will be quickly rejected, without contributing much to the genesis and progression of knowledge, while a succession of adjustments to a model that is useful, though not perfect, will lead to an increasingly detailed representation of the phenomenon – Antoine Danchin

In this chapter we describe the process of network reconstruction in general terms and outline the basic underlying concepts. This process is fundamental to systems biology and culminates in a structured *knowledge base*, or *k-base*, that contains comprehensive curated biochemical, genetic, and genomic (BiGG) information on the target organism. A k-base can be updated continually as more is learned about the target organism. It is fair to say that once one has performed such a reconstruction, one knows the target organism in great detail.

3.1 Building Knowledge Bases

2D genome annotation A bottom-up network reconstruction is effectively a two-dimensional annotation of a genome (Figure 1.8). It includes a list of components (represented as rows in a table) and the known links between them (represented by columns). If all these links are actual chemical interactions, this table is filled with stoichiometric coefficients. The table basically becomes a connectivity map for the network and it is based on the known chemical components of the network and the chemical interactions (covalent transformations or associations) between the components. The network grows as more and more information about the organism becomes available, or as the scope of the reconstruction grows.

Similarity to the DNA sequence assembly process The process of generating the full DNA sequence of a genome is a familiar one (see top part of Figure 3.1). The irreducible unit of a DNA sequence is a base pair (bp). A *read* is a number of base pairs sequenced as a group. The read length can vary from 30 to 800 bp. Overlap between reads is used to build *contigs*. These can be quite large, reaching hundreds of thousands of base pairs. Finally, targeted sequencing is then performed to fill in the gaps between the contigs to obtain the full contiguous genomic sequence.

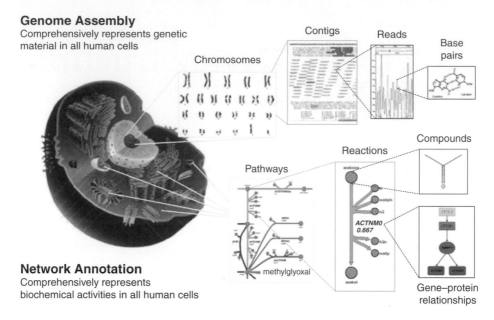

Figure 3.1   Comparing the process of establishing the 2D genome annotation to that of sequence assembly. Both are bottom-up processes where the whole is systematically built from the known parts.

The network reconstruction procedure, in principle, follows a similar process (see lower part of Figure 3.1). The smallest element of a network reconstruction is a chemical reaction. To define this, we need to know the participating compounds and determine the genetic basis for the transformation. This defines the chemical transaction. A series of these are then strung together to form a pathway. All pathways then come together to form a map of the network. A network assembled in this manner will have gaps that require curation and gap-filling.

Network reconstruction, like sequence assembly, is an organized process that starts with fundamental elements that are systematically assembled and built into larger and larger constructs. This process is hierarchical and it culminates with genome-scale knowledge. The properties of the resulting k-base are thus subject to multi-scale analysis.

Reconstruction jamborees   1D annotation (i.e., the list of genes and their location) of genomic sequences are of common interest in genomic sciences. Similarly, reconstruction, or a 2D annotation, is of common interest to all that study the systems biology of a target organism. This common interest has led to the formation of 2D annotation jamborees where the research community that focuses on a particular organism comes together to form a community consensus network reconstruction that will be used by all who are focused on that particular target organism [97, 165, 422].

3.2   Reconstruction is a Four-step Process

The bottom-up reconstruction process at the genome-scale can be broken down into four basic steps (Figure 3.2). First, an automated draft to the reconstruction can be obtained based on the annotated sequence of the target organism and associated information. Second, detailed biochemical, genetic, and physiological data are used to complete and curate the reconstruction. Third, the reconstruction is converted into a mathematical format forming a queryable knowledge base for integrated cellular functions. Fourth, the reconstruction is validated against various data sets that can lead to an iterative loop, and often includes a module-by-module approach as discussed in Chapter 2. Each one of these steps relies on different data types (see left side of Figure 3.2). The incorporation of each data type requires a different computational and analytical approach (see right side of Figure 3.2). We now briefly describe these procedures. More detail is found in [117, 427].

1. Draft reconstruction   The starting point for a bottom-up reconstruction is the enumeration of components in the network. A list of components can be obtained from an annotated genomic sequence. For metabolism, one collects a list of all the open reading frames (ORFs) that are associated with enzymes or enzyme complexes that

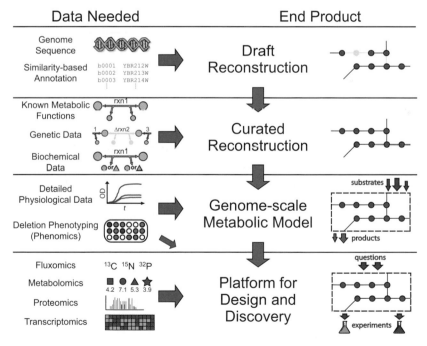

Figure 3.2   The four basic steps of the bottom-up reconstruction procedure: draft reconstruction, curated reconstruction, network models, validation and iteration. Modified from [117].

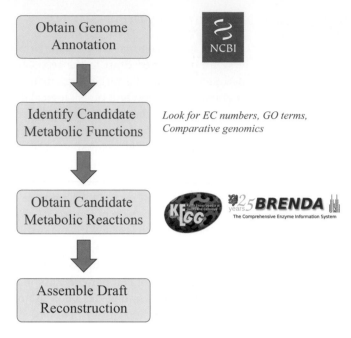

Figure 3.3   The principal steps in generating an automated reconstruction.

catalyze metabolic transformations. This initial step can now be automated, leading to a first draft of a genome-scale network reconstruction.

The automated process involves the retrieval of disparate data types and inter-linking them (Figure 3.3). The process begins with obtaining the annotated genome sequence of the target organism. The genome sequence is then used to subset the ORFs based on the information in the functional annotation of the corresponding gene products. In the case of metabolism, this would involve selecting all ORFs that encode functions relevant to metabolism, and similar subsetting would apply to other networks that one wants to reconstruct. This list is then compared to entries in a database that has information about the detailed biochemical functions of the gene product. In the case of metabolism, this involves the identification of the metabolic reactions the gene product is involved in catalyzing.

The list of all such reactions is then assembled and the genetic basis for each is documented. The full relationship between an ORF and a metabolic reaction can be complicated. The transcripts that come from an ORF are defined, and, if needed, expression profiling data can then be associated with this transcript based on its sequence. The protein that is formed from the translation associated with the transcript is identified, and one can obtain an identifier that can be used to map proteomic data onto this protein. Finally, the assembly of the enzyme complex is described. This completes the gene-to-protein-to-reaction (GPR) association that describes the genetic basis for a metabolic reaction and the results are shown in Figure 3.4. There are several challenges that are faced when one attempts to automate the network reconstruction process (see Table 3.1 and [356]).

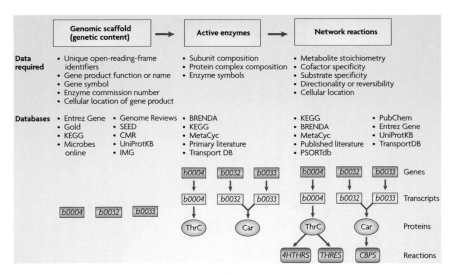

Figure 3.4 The gene-to-protein-to-reaction (GPR) associations and associated data types. In the GPR scheme, the first level (purple) corresponds to genetic loci, the second level (orange) to transcripts, the third level (green) to functional proteins, and the fourth level (blue) to reactions. From [117].

2. Manual curation The enzymes identified on the genomic sequence are either identified from previous cloning experiments in the target organism, or, more likely, from sequence similarity analysis with a known gene from another organism. A 'standard' biochemical reaction will have been associated with this enzyme during the formation of the draft sequence because genomic annotation only contains generic information about a metabolic function.

Much more information is now needed to validate a reaction in a target organism and the information that has been automatically retrieved needs to be manually curated. This process requires the retrieval of primary bibliomic data on the enzyme and its functions in the target organism. A 'standard' reaction from textbooks or some general source is the starting point. This information needs to be established for the target organism, as the reconstruction will be organism-specific.

The manual curation step includes the examination of the draft reconstruction in great detail. The additional information that is incorporated at this stage falls into many categories, including known metabolic functions, information about the co-factor usage of the enzyme, any relevant genetic and physiological data that are available, and information about cellular location where the reaction takes place. Thus, different levels of information are needed to obtain a detailed description of a biochemical transformation. Adding a new reaction requires a five-step process (Figure 3.5). Biochemical accuracy is especially important if the mathematical representation of the reconstruction is to be used for subsequent computations, otherwise the calculated network properties are likely to be incorrect.

   1   The first level defines the metabolite specificity of the reaction carried out by the gene product. Although primary metabolites are often the same for homologous enzymes across organisms, the use of coenzymes might vary.

Table 3.1 Common issues encountered during metabolic network reconstruction. Prepared by Adam Feist.

Issue	Common solutions	Example
Reaction directionality is unknown	Compute thermodynamic reversibility estimates	Thermodynamic consistency analysis performed for *E. coli* reconstruction iAF1260 [115]
Proton translocation stoichiometry information missing for an enzyme or pathway	Determine the translocation efficiencies for individual reactions or the overall pathway through primary literature search or experimental investigation	The apparent P/O ratio for *E. coli* and individual enzyme translocation efficiencies were determined by simultaneously examining oxygen and phosphate consumption rates of respiring cells [288]
Missing conversion in a pathway between gene-associated reactions	1. Biochemically determine if the reaction occurs in the organism from either primary literature or experimental assay 2. Utilize a gap-filling algorithm to find the most likely candidate reaction and/or encoding gene	Network topology-based gap-filling algorithms [71, 201, 203] were used to generate tentative ORF assignments that were further investigated through biochemical characterization studies utilizing genetic mutants to characterize and assign function to an encoding gene [135]
Maintenance energies necessary to account for non-metabolic activity is unknown	Compute maintenance values through simulation by examining uptake, production, and growth rates under a desired growth condition	Maintenance costs (ATP equivalents) were determined for *Lactobacillus plantarum* from fermentation data at different growth rates [417]
Co-factor specificity for an enzyme	1. Utilize co-factor specificity from a closely related organism 2. Biochemically characterize the gene product in your strain through experimental analysis	The first enzyme in the *de novo* biosynthesis of NAD uses oxygen or fumarate as a co-substrate in most bacteria and many archaea (a FAD-dependent enzyme, e.g., NadB in *E. coli*). In *Thermotoga maritima* and methanogenic archaea, this same step converting aspartate to iminoaspartate is replaced by a non-homologous NAD-dependent enzyme [463]

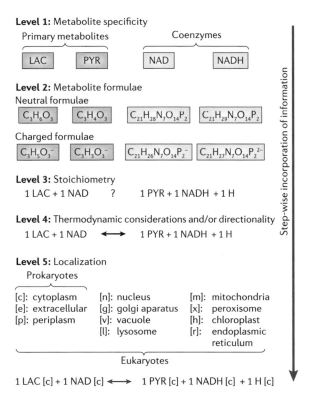

Figure 3.5 The five levels of information required to add a biochemical reaction into a bottom-up reconstruction. From [355].

In the case of lactate dehydrogenase in *E. coli* (see Figure 3.5), NAD serves as an electron acceptor for lactate (LAC) resulting in the formation of pyruvate (PYR) and NADH.

2 The second level of detail accounts for the charged molecular formula of each metabolite at a physiological pH. This requires the elemental composition of a compound.

3 The knowledge of the elemental composition leads to the third level of detail; the stoichiometric coefficients of the reaction. By balancing out the elements and charge in the reaction, the overall stoichiometry of the reaction is defined. It is here where protons and water molecules are often added to balance the chemical equation.

4 The directionality of the reaction represents the fourth level, at which biochemical studies and thermodynamic properties define the *in vivo* reaction directionality.

5 At the fifth level, the cellular compartment in which the reaction takes place has to be determined.

Most of this information is not found in the genome annotation.

The assembled reaction network is likened to the genome of the organism. GPRs formally connect reactions in the metabolic network to proteins and genes in the

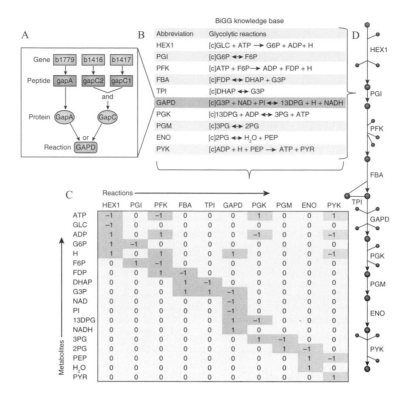

Figure 3.6   The conversion of BiGG data into a mathematical format. (A) Connections between genes and reactions can be represented as GPR associations by using Boolean rules or visualized using graphic images. (B) A list of charge and elementally balanced metabolic reactions can be represented in a stoichiometric matrix (**S**) (C). (D) The resulting pathway is then shown as a traditional reaction map. From [375].

organism. These GPR associations indicate which genes encode which proteins and which enzymatic reactions these proteins catalyze (Figure 3.6).

Once constructed, GPR associations can be used to relate various data types, including genomic, transcriptomic, proteomic, and flux data. GPR associations need to distinguish between isozymes, enzyme complexes, enzyme subunits, and single and multifunctional enzymes so that they capture the complexity and diversity of the biological relationships. GPR associations are available for several reconstructed organisms.

Following manual curation, a variety of data and information has now been assembled for the target organism. If you have not gone through this process personally, it is hard to appreciate the depth of the knowledge that is represented in a reconstruction. When the process is completed, visual representations of the results often take the form of tables of reactions, pathway maps, and GPRs (see Figure 3.6).

3. Conversion to a computational format   The BiGG database that the manually curated reconstruction represents can be converted into a mathematical format. The chemical reactions identified are shown in the table in Figure 3.6B. These are a set of elementally and charge-balanced chemical reactions. The stoichiometric coefficients in this list of chemical reactions can be organized as a table (Figure 3.6C). This table is a mathematical matrix (the *stoichiometric matrix*) whose properties are detailed in Part II of this text. The genetic basis for these reactions is then shown as GPRs (Figure 3.6A). This information is described with logical statements. Finally, the information can be shown in the familiar format of a network map (Figure 3.6D); a simple linear pathway in this case.

A curated reconstruction that is mathematically represented can be used to compute integrated cellular functions that result from the simultaneous activity of multiple gene products. The mathematical form of the curated reconstruction is used to compute basic cellular states, such as the provision of all the needed compounds to produce growth. Initially, a metabolic reconstruction would just have some basic features. At this stage, one can perform various tests on the functionality of the reconstruction (see Figure 3.7):

- Examine the ability to generate all the precursor metabolites that are required for the synthesis of biomass components in order to enable the model to compute a growth state.
- Determine the presence of biosynthetic pathways that are required for the formation of biomass components.
- Evaluate whether the reconstructed network can support growth on all the primary substrates on which the organism is known to grow.
- Determine if the reconstruction is able to secrete the known secretion products under the particular growth condition.

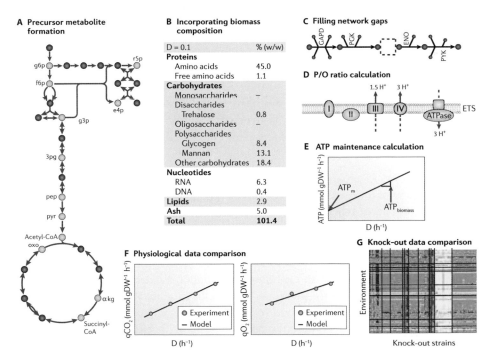

Figure 3.7 The validation of a computational model based on a bottom-up reconstruction. From [355].

- Ensure that the reconstruction uses a logical set of metabolic pathways, not any unorthodox routes that might occur due to any flux loops.
- Check for loops that might allow the generation of infinite ATP. Loops can be detected by shutting off all exchanges and maximize for ATP maintenance.

The mathematical methods used for this evaluation are described in Part III.

The initial curated reconstruction may not be able to reproduce known organism functions and would need to be further expanded. Once this has been accomplished, we can move to the final stage of the reconstruction process that involves comprehensive evaluation and validation of network properties .

4. Network evaluation and validation   With a computational model of a network that has the ability to meet growth and basic physiological demands, we incorporate and examine other organism-specific network properties. This ability then leads to the evaluation of the properties and capabilities of the reconstructed network and validation against data sets available for the target organism. This validation process can be quite comprehensive if much data is available for the target organism.

Examples of the steps that are taken during the validation process are as follows (see Figure 3.7):

- If experimental data are available for the P/O ratio and the stoichiometry of the electron-transport system (ETS) in the cell, the efficiency of the network for making ATP through oxidative phosphorylation can be calculated.
- The energy maintenance that is required for growth-associated and non-growth-associated activities must also be incorporated into the network reconstruction.
- ATP-maintenance values can be extrapolated from growth data. Once the energy maintenance is determined, the ability of the network to establish uptake and secretion rates for molecules such as CO_2 or O_2 (q_{CO_2} or q_{O_2}) can be calculated and compared to experimental measurements.
- Evaluating the inconsistency between model and experimental knock-out results can lead to experimentation and biological discovery and increase network accuracy.

Although network evaluation is more or less a sequential procedure in the order described, many of the steps might need to be repeated iteratively following changes to the network that arise from its evaluation. Some commonly encountered issues are summarized in Table 3.1.

Recapitulation   The result of this four-step process is a mathematical representation of all the biochemical, genetic, and genomic information available for the target organism. This network underlies the integrated metabolic functions of the organism. These integrated functions can be computed from the reconstruction that is derived from a BiGG k-base and one can compare the computations with measured organism functions. The mathematical methods that are deployed for this purpose are detailed in Part III of this text.

3.3 Reconstruction is Iterative and Labor-intensive

Timelines As the reader can probably already appreciate, the bottom-up network reconstruction process is very labor-intensive. The collection, evaluation, and presentation of all the data types discussed above takes hard work and requires much patience. A reasonable draft reconstruction can now be obtained for a prokaryotic organism relatively rapidly. The manual curation phase may take months of work, depending on how many data are available for the target organism.

The first genome-scale reconstruction of a metabolic network in a mammal was successfully completed in 2007 [100]. The reconstruction proceeded in two phases; first, the use of the sequence annotation of build-35 of the human genome, followed by extensive evaluation of literature data. The first phase took six people six months to accomplish. This process involved the examination of many different sources of genome annotation. The second phase took 6 people almost 18 months to complete. This process involved the examination of the biochemical literature available on human metabolism and iterative curation against 288 known metabolic functions in humans. The resulting Recon 1 reconstruction is discussed in Chapter 6.

One by-product of the laborious bottom-up reconstruction process is that the participants in the process get to know the target organism very well. One has to learn about its genome, its content, the biochemical functions of the gene products, data on the described physiological functions, and so on. One thus becomes an expert in the target organism, and all this knowledge is now formally represented in one place. Once the reconstruction is done, its many uses make the difficult process worthwhile.

Confidence scores and QC/QA procedures The information that is included in a reconstruction needs to be scored based on the quality and extent of information on which it is based. Genome annotation has some ambiguities associated with it and bibliomic data often need expert evaluation. Thus a *confidence score* needs to be established.

As illustrated above, there are several sources of data that are used to add components and links to a reconstruction. Some data types provide stronger evidence than others. These differences lead to the formulation of a confidence score associated with additions to a reconstruction. A schema for metabolic reactions is shown in Table 3.2; it has a five-level scoring schema based on data of increasing reliability. This score can be associated with individual reactions, pathways, or subsystems in a reconstruction depending on the detail of the information desired. An example of subsystem-level confidence scoring is given for the *Pseudomonas pudita* metabolic reconstruction (see Figure 3.8).

Automating the procedure A reconstruction comes with confidence scores, unique chemical specifiers, and QC/QA steps, such as elemental and charge balancing of reactions. These well-defined criteria enable the generation of a *standard operating procedure*, or a SOP [427]. A SOP is a workflow that can be graphically illustrated, as in Figure 3.9.

A SOP, in turn, leads to the definition of the steps within it that can be automated, thus vastly reducing the amount of manual labor that goes into a reconstruction.

Table 3.2 Currently employed scheme for assigning confidence level for metabolic reactions. Prepared by Ines Thiele.

Evidence type	Confidence score	Examples
Biochemical data	4	Protein purification and biochemical assays, experimentally solved protein structures
Genetic data	3	Knock-out characterization, over-expression experiments, gene deletions
Physiological data	2	Physiological evidence for existence of reaction, e.g., known secretion products imply indirectly the existence of transporter as well as metabolic reactions. However, no enzyme/gene has been directly identified in target organism; thus, the actual reaction mechanism may differ from the reaction(s) included in the reconstruction
Sequence data	2	Genome annotation, SEED annotation, significant sequence similarity to another gene with known function
Modeling data	1	No supporting evidence available but reaction is required for modeling, e.g., growth or known by-product secretion. These reactions represent hypotheses, and the actual reaction mechanism may differ from the included reactions
Not evaluated	0	None

With better definition of SOPs and their workflows, an increasing number of automation approaches are being developed (Table 3.3). The extent to which and the reliability of the automation that can be achieved is unknown at present, but progress continues to be made. Table 3.1 illustrates some of the issues that can be hard to resolve automatically. Reconstructions are likely to always require manual intervention and curation to some extent.

The iterative process The one-dimensional DNA sequence annotation for an organism is constantly updated as new open reading frames and regulatory regions are discovered and described. Similarly, a network reconstruction (a two-dimensional annotation) is constantly updated. Network reconstruction is thus an iterative process (Figure 3.10). As new open reading frames are described, as new biochemical

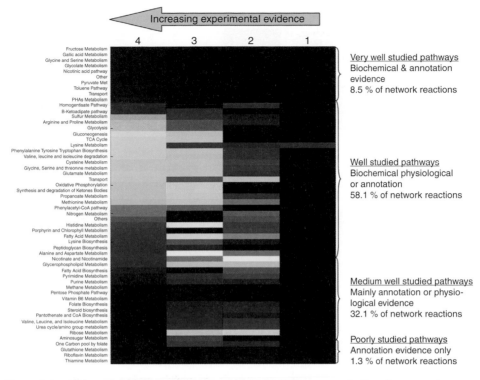

Figure 3.8 A heat map of the confidence scores of the different metabolic subsystems in iJN746, a genome-scale reconstruction of *Pseudomonas pudita* metabolism [287]. The four columns in the map represent the different confidence score (from left to right: 4, 3, 2, 1). The various colors correspond to the percentage of subsystems reactions that have the corresponding confidence score (red = 100%, blue = 0%).

properties are discovered, and as model-driven experiments are performed for the target organism, new information becomes available and needs to be incorporated into the reconstruction. Network reconstructions thus grow in scope and completeness over time. This process is illustrated in the following three chapters.

Available reconstructions   There are now many different reconstructions published and publicly available (Figure 3.11). Most of these are for prokaryotic microorganisms. These reconstructions represent model organisms, pathogenic organisms, environmentally important organisms, and those of bioprocessing importance. Reconstructions for eukaryotic organisms, such as yeast and human, are beginning to emerge. The available reconstructions now span many of the branches of the phylogenetic tree. Some of these reconstructions will be highlighted in Chapters 5 and 6.

3.4   The Many Uses of Reconstructions

Retrospective analysis vs. prospective uses   The reconstruction of a network is based on what is *known*. In other words, it requires a retrospective evaluation of all

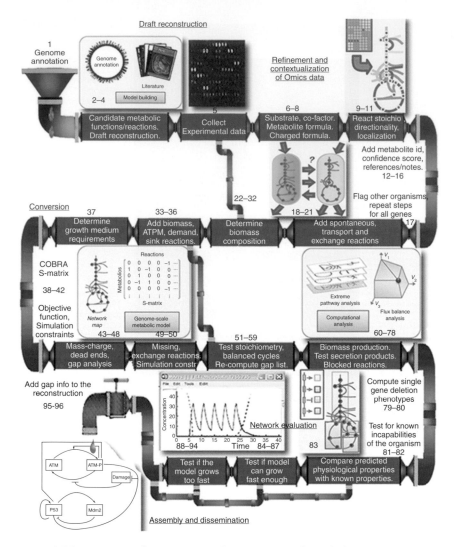

Figure 3.9   A 96-step process for automating the generation of metabolic reconstructions. Prepared by Daniel Hyduke.

the knowledge that is available about the target organism. This knowledge is then converted to a mathematical format. Once a computational model has been formed based on the k-base, then prospective uses of the network reconstruction emerge. As is shown in Figure 1.6, a curated reconstruction that has been converted into a mathematical format can be used for at least four different types of studies. These applications will be detailed in Part IV of this text, but briefly mentioned here.

Mapping omics data: content for context   A reconstruction represents a k-base against which comprehensive omics data can be mapped. For instance, the genome-scale metabolic reconstruction for human can be tailored to particular cell and tissue

Table 3.3 Global automated reconstruction tools.

Name	Model SEED	MetaCyc pathway tools	Subliminal Toolbox	Grow match
Citation	[163]	[61]	[411]	[217]
Description	Independent gene calling and annotation based on RAST server, incorporates biolog phenotype arrays and gene essentiality data (if available), gap-filling	Gap-filling, dead-end metabolite finder, choke-point finder, based on GenBank, genome annotation	Predicts protonation state for elemental/ charge balancing, compartment assignment, allows model merging, relies on 'pre-draft' models from KEGG or EcoCyc	Reconciles growth/no growth prediction of *existing* models

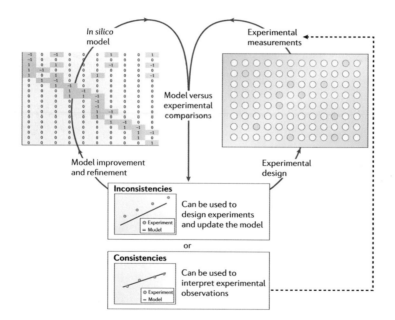

Figure 3.10 Network reconstruction is an iterative process involving experiment and computation. From [355].

types based on expression profiling [6, 396]. Proteomic data can also be used to build models of metabolism, in mitochondria [446], in the human red cell [51], or in CHO cells [30]. Several methods have been developed to map omics data onto reconstructions (see Table 6.2).

Finding missing parts: gap-filling The incompleteness of a bottom-up reconstruction brings out the holes or gaps in our knowledge base about the target organism.

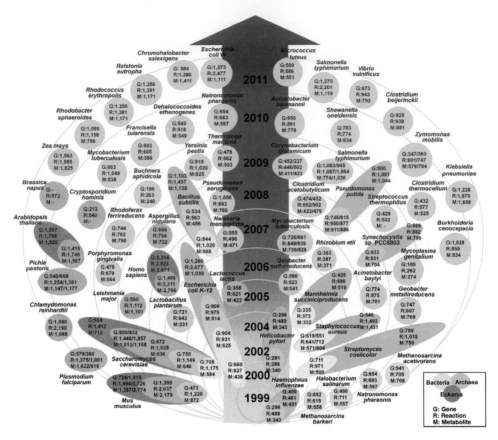

Figure 3.11 The growth in available genome-scale network reconstructions. Some organisms have gone through multiple rounds of iterative reconstruction. From [208].

Algorithms have been developed to automatically fill in such gaps [297]. Such computations basically generate hypotheses for what is missing and thus systematize the discovery process. The computation of all possible pathways to and from a novel metabolite represents another example [158].

Understanding complex processes Genome-scale models can be used to compute, simulate, and study complex cellular processes. For instance, the process of adaptive evolution in bacteria has been successfully studied using genome-scale models [80]. Other unexplored examples would be a differentiation process to a mature cell type, and studies of the disease progression process.

Synthetic biology Genome-scale models have been used for the purposes of design. In fact, 'there is nothing more practical than a good theory.' Several synthetic biology approaches to the design of optimally performing metabolic networks have appeared [58, 334]. Some of these have been reduced to practice [126].

3.5   Summary

- The bottom-up reconstruction of metabolism is now a four-step process.

  1   A draft genome-scale reconstruction is enabled by well-annotated sequences and bibliomic data.
  2   Curated reconstructions are then obtained by auditing specific biochemical, genetic, and physiological data.
  3   A biochemically, genetically, and genomically structured reconstruction is a k-base that can be converted to a mathematical format that enables computational interrogation of its properties.
  4   The computational interrogation leads to an iterative validation and curation process.

- The reconstruction process is an iterative, data- and labor-intensive process representing growing knowledge about the target organism. Automated procedures to perform reconstructions are appearing.
- Network reconstruction represents a k-base, and a number of reconstructions for important organisms now exist. They are being used for an increasing number of basic and applied uses.

4 Metabolism in *Escherichia coli*

Although not everyone is mindful of it all cell biologists have two cells of interest: the one they are studying and Escherichia coli – Frederick Neidhardt

We now turn our attention to examining how the genome-scale reconstruction process for metabolic networks has been applied to particular organisms. The reconstruction process is enabled by genome sequencing. The first full genomic sequence appeared for *Haemophilus influenzae* in 1995. Four years later, the genome-scale metabolic reconstruction for *H. influenzae* appeared, the first of its kind [105]. The second genome-scale reconstruction to appear was for *E. coli* K-12 MG1655 in 2000 [106]. Due to a wealth of bibliomic data available for *E. coli*, there have been several subsequent updates and expansions of this reconstruction. This chapter discusses the history of the reconstruction of the genome-scale metabolic network in *E. coli*. We then discuss how this network reconstruction can be converted to a computational model and illustrate the range of basic scientific questions that can be addressed and describe its practical uses.

4.1 Some Basic Facts about *E. coli*

E. coli is a Gram-negative bacterium, a prokaryote belonging to the Enterobacteria family. It is perhaps the best-characterized organism as evidenced by its high Species Knowledge Index (SKI) [184, 295]. Non-pathogenic strains have proven to be the workhorse of the biotechnology industry. *E. coli* is normally harmless, but needs only to acquire a combination of a few mobile genetic elements to become a highly adapted pathogen [92]. Pathogenic strains cause a range of diseases, from gastroenteritis to extra-intestinal infections of the urinary tract, bloodstream, and central nervous system (see Figure 4.1). The worldwide burden of these diseases is high. With the low cost of sequencing, there are now many strains of *E. coli* that have been sequenced [251].

  E. coli is a remarkable and diverse organism. It can grow on a wide variety of different carbon and nitrogen sources both aerobically and anaerobically. Fundamentally, its life cycle consists of five key shifts in growth conditions: a heat shock (i.e., from ambient temperature to 37°C), a pH shock (from neutral to gastric pH of about 2), oxygen deprivation, nutritional richness and community competition, and finally a cold shock back to ambient temperature.

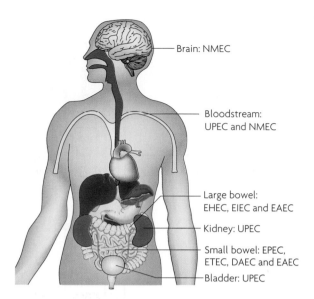

Figure 4.1   Sites of pathogenic *E. coli* colonization. Enteropathogenic *E. coli* (EPEC), enterotoxigenic *E. coli* (ETEC) and diffusely adherent *E. coli* (DAEC) colonize the small bowel and cause diarrhea, whereas enterohemorrhagic *E. coli* (EHEC) and enteroinvasive *E. coli* (EIEC) cause disease in the large bowel; enteroaggegrative *E. coli* (EAEC) can colonize both the small and large bowels. Uropathogenic *E. coli* (UPEC) enters the urinary tract and travels to the bladder to cause cystitis and, if left untreated, can ascend further into the kidneys to cause pyelonephritis. Septicemia can occur with both UPEC and neonatal meningitis *E. coli* (NMEC), and NMEC can cross the blood–brain barrier into the central nervous system, causing meningitis. Taken from [92].

From a systems biology standpoint, *E. coli* has a number of advantages. It continues to be studied extensively and thus many research groups are producing large amounts of high-throughput data, including genome sequence, expression profiling, metabolomic, flux, and phenotyping data. Furthermore, *E. coli* is relatively easy to grow and its genomic content easy to manipulate, making dual perturbation experiments possible. These advantages have led to *E. coli* being a model organism in systems biology.

4.2   History

The reconstruction of the *E. coli* metabolic network, its mathematical representation, and the analysis of its properties was initiated in 1990. It has grown continually in scope and content since then (Figure 4.2), with a noticeable acceleration when the genome sequence was published in 1997. The history of the reconstruction of the *E. coli* metabolic network illustrates the iterative nature of the reconstruction process.

A continual growth in scope   Functional reconstruction of the *E. coli* K-12 MG1655 metabolic network was possible before its genome sequence was available. Prior to the annotation of its genome, there was enough known about the biochemistry of *E. coli*

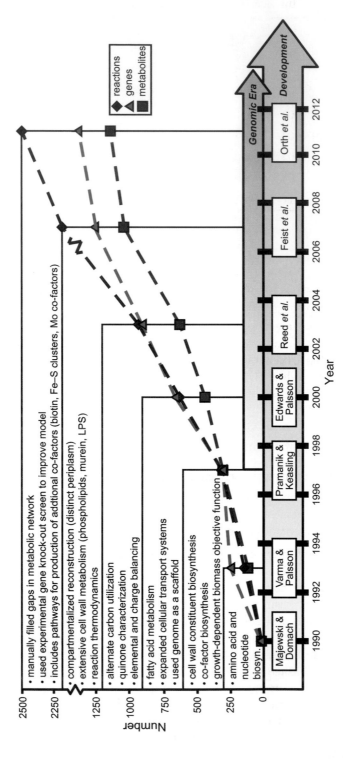

Figure 4.2 History of the reconstruction of the *E. coli* metabolic network. For each of the seven reconstructions indicated, the number of included reactions (diamonds), genes (triangles), and metabolites (squares) are displayed. See Table 4.1 for details. The expansion in scope with subsequent iterations is indicated with the inserted text. Figure modified from [118].

Table 4.1   A summary of the size of the *E. coli* metabolic reconstructions.

Study	Year	Genes	Reactions	Metabolites	Citation
Majewski & Domach	1990	24	14	17	[258]
Varma *et al.*	1993–95	250	146	118	[436,437,439,440,442]
Pramanik & Keasling	1997–98	306	317	305	[340]
Edwards & Palsson	2000	660	627	438	[106]
Covert *et al.* (regulated)	2002	149	113	63	[85]
Reed *et al.*	2003	904	931	625	[360]
Feist *et al.*	2007	1260	2077	1039	[115]
Orth *et al.*	2011	1366	2251	1136	[295]

metabolism to reconstruct the functional model of its basic metabolic pathways. Since the publication of the sequenced *E. coli* genome, there have been three major updates to the sequence annotation [362]. Every time the genome annotation is updated, one can update the metabolic network reconstruction to incorporate the new information. Updates to a reconstruction based on direct biochemical or physiological data can also be carried out. Note that every time the reconstruction is updated and expanded, it grows in coverage and utility. Thus, network reconstructions are knowledge bases that grow in value with time. The characteristics of the *E. coli* reconstructions are summarized in Table 4.1.

The *E. coli* network reconstruction has reached a stage where it is a comprehensive representation of the known biochemistry, genetics, and genomics that relate to *E. coli*'s metabolism. It has now entered a prospective discovery phase based on gap-filling procedures and newly discovered metabolic capabilities published in the literature.

4.2.1   Pre-genome era reconstructions

Metabolic reconstruction and constraint-based modeling for *E. coli* has a history dating back to 1990 (see Figure 4.2). There were three notable reconstructions formulated prior to the publication of the genome sequence. These three pre-genome era reconstructions of *E. coli* metabolism were based solely on biochemical information and provided a foundation for subsequent work at the genome scale.

Core energy metabolism   Beginning in 1990, a small metabolic reconstruction was generated, consisting of 14 reactions, the TCA cycle, and, partially, glycolysis. This simple network included 14 reactions and 17 metabolites in the central metabolism of *E. coli*. This network was reconstructed to analyze the production and secretion of acetate during aerobic growth on glucose [258]. Although this reconstruction was basically a pathway model, it did illustrate the importance of co-factor use and coupling in *E. coli* metabolism.

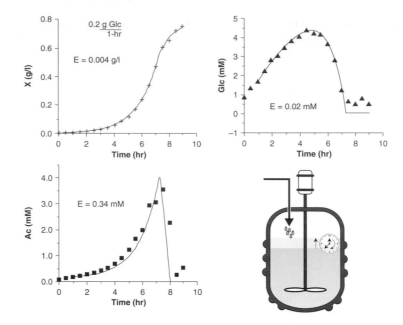

Figure 4.3   Prediction of acetate secretion and re-consumption with a pre-genome era model. Shown is an aerobic fed-batch culture with continuous glucose injection at the rates indicated. The culture was not limited for minerals. The time profiles of cell density (X), glucose concentration (Glc), and acetate (Ac) concentration are shown, with predictions of the model represented as solid lines. The quantity E indicates average deviation between predictions of the model and experimental measurements. Figure modified from [438].

Basic catabolism and anabolism   In 1993, a metabolic network reconstruction consisting of 118 metabolites and 146 reactions was generated representing key catabolic and anabolic pathways in *E. coli* [439]. This biochemically based reconstruction was used for computing a number of network properties. First, the optimal yield of charged co-factors (i.e., ATP, NADH, NADPH) and the 12 biosynthetic precursors were computed from various substrates. Such computations were used to characterize integrated functions of the core metabolic pathways. Second, the properties of the biosynthetic pathways were studied by computing molar yields of all the amino acids and the nucleic acids on select substrates. Third, optimal growth properties were computed in a given environment. The computational predictions resulting from these studies were compared to experimental data and found to be consistent with measurements under both aerobic and anaerobic glucose minimal media conditions.

The growth rate and by-product secretion patterns could be predicted. The outcome of fed-batch experiments could be quantitatively predicted by considering the metabolic network of the cell to be in a steady state relative to a time varying growth environment (Figure 4.3); a process now called *dynamic flux balance analysis*. In this figure, we see how growth increases the biomass over time. In the early stages of the experiment, glucose enters faster than it is metabolized. After about 4-5 hours, these rates reverse and all the glucose is consumed as the amount of biomass increases. Remarkably, acetate is predicted to increase and is then re-consumed after glucose is exhausted. This early reconstruction could thus quantitatively recapitulate the basic growth patterns of *E. coli* grown clonally in the lab in minimal media. Some of these computations are repeated in Part III of the text using the core *E. coli* metabolic network.

Detailing cell wall and co-factor biosynthesis   Subsequently, an expanded reconstruction consisting of 305 metabolites and 317 reactions was generated [339, 340]. Major updates in this reconstruction were in coenzyme, nucleotide, and lipid metabolism. This expanded reconstruction was used to study three key issues. First, the computational characterization of cell wall and co-factor synthesis was performed. Second, the growth rate dependency of biomass composition, and thus metabolic demands, was evaluated. Third, the reconstruction was used for computations that incorporated measured metabolite uptake and secretion rates to predict central metabolic fluxes consistent with enzymatic flux values determined from isotopomer-based measurements.

Recapitulation   As shown above, even before the full genome sequence was determined, a variety of integrated metabolic functions in *E. coli* could be studied using comprehensive metabolic network reconstructions. These networks were based on all the relevant biochemical and physiological data available at the time. These achievements provided a strong impetus for building genome-scale models once genomic sequences began to appear.

4.2.2   Genome era reconstructions

With the publication of the genome sequence of *E. coli* K-12 MG1655 in 1997 [40] and its annotation, it was possible to get a genome-scale view of its metabolic capabilities. This ability led to the formulation of a genome-scale reconstruction of its metabolic network. Reconstructions have subsequently grown in scope with the re-annotation of the K-12 MG1655 genome. There are now several full-length sequences available for additional hundreds of *E. coli* strains.

Naming conventions for genome-scale network reconstructions   A naming convention was developed for reconstructions that was modeled after those used to name bacterial plasmids. Names of plasmids start with lower case 'p' followed by letters and numbers. pBR322 was a workhorse plasmid early in the history of biotechnology. Its name starts with a 'p', for plasmid, and then BR representing the initials of the person who made it (Bolivar Rodriguez) and a number, 322, that characterizes the plasmid.

A similar naming convention for network reconstructions has developed. This nomenclature uses an '*i*' for *in silico* followed by the initials of the person (or the organization) primarily responsible for the reconstruction, and a number that designates the number of open reading frames (ORFs) accounted for in the reconstruction. Additional specifiers can also follow.

*i***JE660**   By utilizing the annotated genome sequence published in 1997 [40], the first genome-scale metabolic reconstruction was generated for *E. coli*, consisting of 627 unique reactions produced by 660 gene products [106]. This reconstruction, titled

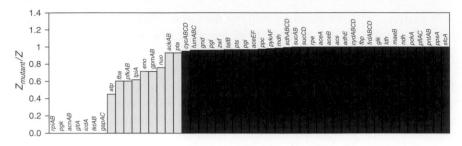

Figure 4.4   Prediction of growth phenotypes of gene knock-out strains. The growth rates on the *y*-axis are normalized to the predicted growth of the wild type. The genes knocked out are indicated above the corresponding bar. Taken from [106].

*i*JE660, was used as a computational platform to study various *E. coli* metabolic capabilities. First, the consequences of gene deletions could be determined because reactions in the reconstruction had an explicit genetic basis. The viability of knock-out strains was computed (see Figure 4.4) and it was found that the growth characteristics of 68 of 79 (86%) examined strains were computed accurately. Second, a broad characterization of the robustness characteristics of the network was performed using parameter variation and phenotypic phase plane (PhPP) analysis (see Chapter 22). Third, based on PhPP analysis, genome-scale prediction of the outcome of adaptive evolution was computed [177].

These results demonstrated the utility of a genome-scale metabolic reconstruction for understanding growth characteristics of gene knock-out strains of *E. coli*. The consequence of the loss of a gene function could be predicted in a given growth environment. In other words, a quantitative genotype–phenotype relationship had been formulated for microbial metabolism. Furthermore, optimal growth patterns could be studied quantitatively using the genome-scale reconstruction as an underlying framework, and the outcome of adaptive laboratory evolutions (ALEs) could be predicted.

***i*JR904**   An updated annotation of the K-12 MG1655 genome appearing in 2000 [389] and continual functional characterization of *E. coli* ORFs enabled an expanded metabolic network reconstruction that appeared in 2003 [360]. It consisted of 931 reactions catalyzed by 904 gene products. This metabolic reconstruction, titled *i*JR904, was a significant improvement over previous reconstructions.

In terms of advances in reconstruction technology, *i*JR904 introduced charge and elementally balanced reactions. Such balancing is important to avoid the computation of chemically unrealistic states, such as net protein production, that can occur with unbalanced chemical reactions. This balancing enabled the prediction of the pH effect of growth on different substrates (Figure 4.5). Another significant addition in *i*JR904 was the detailed relationship between given genes, proteins, and reactions contained in the reconstruction (the GPR associations). *i*JR904 presented the first genome-scale assembly of such associations and it enabled more accurate and automatic tracing of the consequences of gene deletions.

In terms of content, a larger number of characterized transport systems and their encoding genes were added that expanded carbon source utilization pathways in the

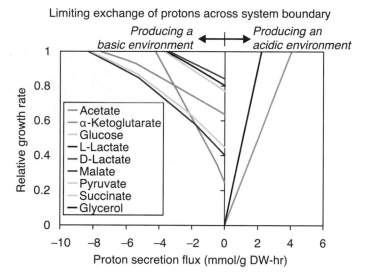

Figure 4.5 Prediction of acidification of medium based on primary substrate. Figure taken from [360]. Some of these predictions were subsequently verified experimentally [403].

reconstruction. These additions, in turn, allowed the computational interpretation of large phenotypic growth screens that allowed a broad validation of the growth capabilities of *E. coli in silico*. In addition, *i*JR904 included a detailed accounting of quinone usage in the electron transport chain. The detailing of transmembrane processes proves to be one of the most difficult aspects of metabolic reconstruction.

Subsequently, three versions of *i*JR904 were developed.

- Utilizing the *i*JR904 reconstruction, an expanded reconstruction of *E. coli* was generated (containing 979 reactions and titled MBEL979) for the purpose of designing over-producing strains within a fluxomic computational framework [227].
- A transcriptional regulatory network was reconstructed based on 106 gene products to generate the first genome-scale integrated metabolic-regulatory model [83]. Remarkably, this model could predict the outcome of a very comprehensive phenotypic screen with 78.7% agreement with experimental data (see Figure 4.6).
- Finally, a subset of *i*JR904 reactions was used for designing an isotopomer model for *E. coli* [376].

*i*AF1260 The next iteration of the metabolic reconstruction for *E. coli* that appeared incorporated data from an updated *E. coli* K-12 MG1655 genome annotation [362] and other legacy data. *i*AF1260 consists of 2077 reactions and accounts for proteins encoded in 1260 ORFs [115].

The expansion in scope represented by *i*AF1260 over *i*JR904 was in five categories: (i) an increased scope with the inclusion of 356 additional ORFs; (ii) compartmentalization into three distinct compartments (cytoplasmic, periplasmic, and extracellular);

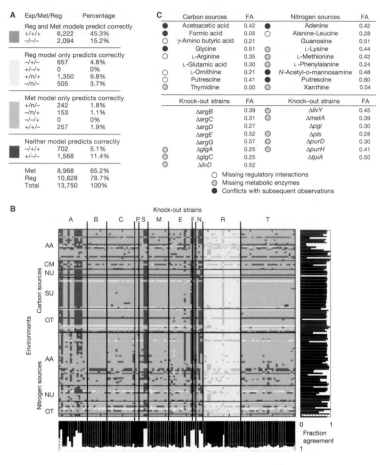

Figure 4.6   Growth phenotype study using *i*MC1010. (A) Comparison of high-throughput phenotyping array data (Exp) with predictions for the *E. coli* network, both considering regulatory constraints (Reg) and ignoring such constraints as a control (Met). Each case is categorized by comparison type (Exp/Met/Reg), and results are listed as '+' (predicted or observed growth), '-' (no growth) or 'n' (for cases involving a regulatory gene knock-out not predictable by the Met model). The comparisons are further divided into four subgroups represented by different colors. (B) Chart showing individual results for each knock-out under each environmental condition, with results categorized and colored as in A. The environments involve variation of a carbon or nitrogen source and are further divided into subgroups: AA, amino acid or derivative; CM, central metabolic intermediate; NU, nucleotide or nucleoside; SU, sugar; OT, other. The knock-out strains are also divided by functional group: A, amino acid biosynthesis and metabolism; B, biosynthesis of co-factors, prosthetic groups and carriers; C, carbon compound catabolism; P, cell processes (including adaptation and protection); S, cell structure; M, central intermediary metabolism; E, energy metabolism; F, fatty acid and phospholipid metabolism; N, nucleotide biosynthesis and metabolism; R, regulatory function; T, transport and binding proteins; U, unassigned. Each environment and knock-out strain is associated with a fraction of agreement (FA) between regulatory model predictions and observed phenotypes, as shown in the bar charts to the right and below. (C) Table showing all environments or knock-out strains for which FA, 0.60. Of these substrates or knock-out strains, 18 point to uncharacterized metabolic or regulatory capabilities in this organism, as indicated. From [83].

(iii) the removal of all grouped, or lumped, reactions (most often associated with lipid and lipopolysaccharide biosynthesis); (iv) the incorporation of reaction thermodynamics; calculated Gibbs free energy values for 950 metabolites and 1935 reactions; and (v) alignment with the well-known EcoCyc database [1].

During the formulation of *i*AF1260 the number of grouped, or lumped, reactions in the network reconstruction was minimized. The predecessors of *i*AF1260, discussed above, included many lumped reactions, which simply represent a summation of two or more discrete enzymatically catalyzed reactions in metabolic processes such as membrane lipid and lipopolysaccharide (LPS) biosynthesis. The detailing of these reactions allowed a fine-grained resolution description of the biosynthetic requirements for growth (Table 4.2). This detailed view of the molecular composition of the cell notably increases the scope of applications of a whole-cell model. Although *i*AF1260 includes a smaller total number of lumped reactions than previous reconstructions, some cases remain in which the reaction mechanism(s) have yet to be fully characterized in *E. coli* (e.g., biotin synthase [243]). The *i*AF1260 reconstruction has been the basis for a large number of studies [118,263].

***i*JO1366**   An update to *i*AF1260 was built in 2011, named *i*JO1366 [295]. It accounts for 1366 genes, 2251 metabolic reactions, and 1136 unique metabolites. *i*JO1366 (i) was updated in part using a new experimental screen of 1075 gene knock-out strains, illuminating cases where alternative pathways and isozymes are yet to be discovered; (ii) continues to make improved phenotypic predictions of growth on different substrates and for gene knock-out strains compared to *i*AF1260; and (iii) mapped to the genomes of all available sequenced *E. coli* strains, including pathogens, leading to the identification of hundreds of unannotated genes in these strains.

From a strain to a species   Although *i*JO1366 is a metabolic network reconstruction for the *E. coli* K-12 MG1655 strain, gene homology mapping can be used to create models of other *E. coli* and closely related *Shigella* strains (see Figure 4.8). Thus, the highly curated *i*JO1366 reconstruction can prove useful for the study of other recently sequenced *E. coli* strains. While it is known that equivalent function is not guaranteed by gene homology, it is still one of the most commonly used and effective methods of genome annotation [102]. In addition, the combination of sequence homology with constraint-based analysis of metabolic networks can be used to determine the most likely metabolic gene content of an organism by generating models that match known biology.

Using this approach, a set of 55 *E. coli* genome-scale reconstructions was built and used to compare gene, reaction, and metabolite content between strains [268]. The reactome shared among all reconstructions defines the *core* metabolic capabilities among all the strains. Similarly, the metabolic capabilities of all the strains were combined to define the full reactome that encompasses all models and thereby defines the *pan* metabolic capabilities among all the strains. The core metabolic content is the intersection of the gene, reaction, and metabolite content of all 55 models, while the pan metabolic content is the union of these features among the models (Figure 4.8A).

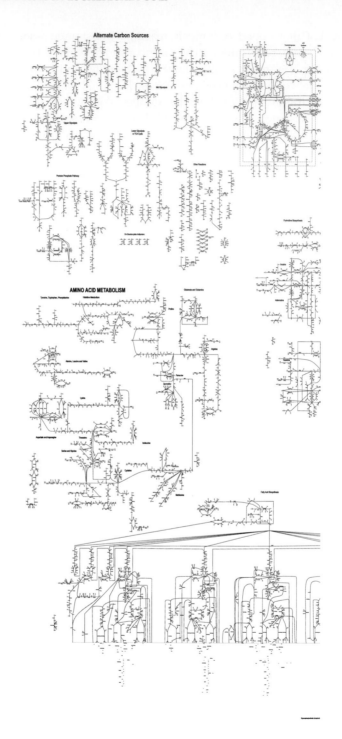

Figure 4.7   Visual representation of the *i*JO1366 metabolic reconstruction.

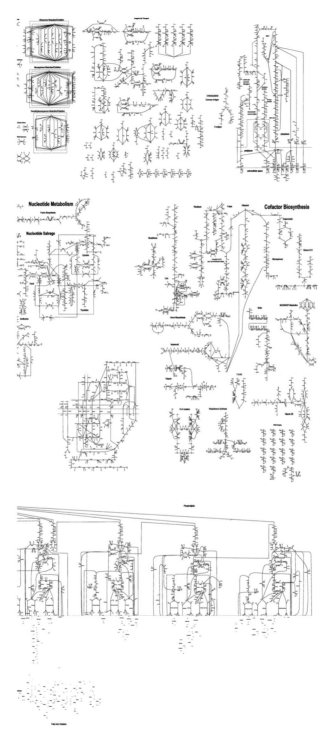

Figure 4.7 *(cont.)*

Table 4.2 The biomass composition of the average wild-type *E. coli* cell. From [115]. The average *E. coli* wild-type macromolecules (and the weight percentage for each) are listed along with their corresponding network metabolites or metabolic precursors. The non-essential wild-type metabolites were determined using gene essentiality data [23,194] and are shown in red. Metabolites listed in blue were determined to have a reduced 'core' structure different from the wild-type metabolite(s) and these are listed in the 'core' biomass composition substitutes. [a]Was determined to be non-essential from [204]. [b]Determined to be essential under minimal media conditions and was not essential under the rich media condition examined.

Typical 'wild-type' composition

Protein (55%)

L-alanine	L-arginine	L-asparagine
L-aspartate	L-cysteine	L-glutamine
L-glutamate	glycine	L-histidine
L-isoleucine	L-leucine	L-lysine
L-methionine	L-phenylalanine	L-proline
L-serine	L-threonine	L-tryptophan
L-tyrosine	L-valine	

RNA (20.5%)

ATP	CTP	GTP
UTP		

DNA (3.1%)

dATP	dCTP
dTTP	dGTP

Inorganic ions (1.0%)

ammonium	calcium	chlorine
cobalt	copper	iron
magnesium	manganese	molybdate
phosphorous	potassium	sulfate
zinc		

Lipid (9.1%)

structure

phosphatidylethanolamine	phosphatidylglycerol[a]	cardiolipin

acyl chain length:# of unsaturated bonds

16:0	16:1	18:1

LPS (3.4%)

inner/outer core KDO_2 lipid A

Co-factors, Prosthetic Groups and Other (<2.9%)

S-adenosylmethionine	FAD	coenzyme A	NAD(P)
thiamine diphosphate	riboflavin	undecaprenyl pyrophosphate	
pyridoxal 5'-phosphate[b]	folates	quinones	hemes
chorismate	enterobactin	glutathione	putrescine
spermidine	vitamin B$_{12}$		

Murein (2.5%)

structure

murein disaccharide

peptide chain length

pentapeptide	tetrapeptide	tripeptide

Glycogen (2.5%)

glycogen

'Core' biomass composition substitutes
inner/outer core KDO_2 lipid A: substituted with KDO_2 lipid (IV) A
quinones: substituted with 2-octaprenyl-6-hydroxyphenol
hemes: protoheme; siroheme included
folates: tetrahydrofolate; 10-formyltetrahydrofolate; 5,10-methylenetetrahydrofolate included

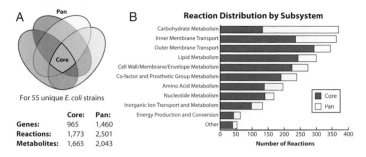

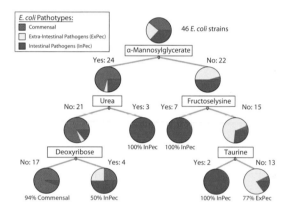

Figure 4.8   Core and pan metabolic capabilities of the *E. coli* species. The core and pan metabolic content was determined for genome-scale metabolic models of 55 unique *E. coli* strains. (A) The core reactome, illustrated by the intersection of the Venn diagram, is shared among all the strains. The pan content consists of all reactions in any model and includes the core reactome. The Venn diagram is not to scale. (B) Classification of reactions in the core and pan reactomes by metabolic subsystem. From [268].

Figure 4.9   Classification of *E. coli* pathotypes based on growth-supporting conditions. Growth-supporting nutrients were used to create a classification tree. This tree can be used to determine if an *E. coli* strain is commensal, an intestinal pathogen or an extra-intestinal pathogen. For example, following the tree to the right shows that 77% of *E. coli* strains that cannot grow on α-mannosylglycerate, fructoselysine, or taurine as sole carbon sources are expected to be extra-intestinal pathogens. Thus, a small number of nutrient sources can be used to classify *E. coli* strains of different types. From [268].

The commonalities and differences between the strains can be analyzed by the metabolic functions of the gene products (Figure 4.8B). The most pronounced differences are found in catabolism. About 64% of reactions in carbohydrate metabolism were part of the pan reactome, the largest group. A majority of these reactions are involved in alternate carbon source metabolism. These differences in catabolic capabilities lead to a set of substrates that differentiate pathogenic strains from commensal (non-pathogenic) strains (Figure 4.9). Based on simulated growth phenotypes, a general separation of commensal strains from both Extra-intestinal Pathogenic *E. coli* (ExPec) and Intestinal Pathogenic *E. coli* (InPec) strains of *E. coli* is observed, suggesting that a classification schema of strains based on metabolic capabilities is possible. Growth screens were used to validate predicted differences between the strains, and the differentiating substrates map onto the colonization sites of the pathogenic strains [268].

Going forward As a result of this 20-year history, the reconstruction of the *E. coli* metabolic network represents one of the two best-developed genome-scale networks to date, the other being yeast (see Chapter 6). Reconstruction of the *E. coli* network is thus approaching exhaustion of known metabolic gene functions and is now being used in a prospective fashion to discover new metabolic capabilities [275, 297, 478]. Information about *E. coli* metabolism continues to grow; for instance, there are 58 annotated ORFs appearing in the literature after the 2007 publication of *i*AF1260 that are included in *i*JO1366.

Thus, the content of the *E. coli* metabolic reconstruction is still expected to grow as new gene functions are discovered. There are many online resources that are

Table 4.3 Online resources that contain a wealth of information about *E. coli*.

Name	Description
ASAP	A Systematic Annotation Package for community analysis of genomes
coliBase	An online database for *E. coli*, *Salmonella* and *Shigella* comparative genomics
Colibri	France's *E. coli* database of DNA and protein sequences
CyberCell Database	A comprehensive collection of detailed enzymatic, biological, chemical, genetic, and molecular biological data about *E. coli*
EchoBASE	An integrated post-genomic database for *E. coli*
EcoCyc	Encyclopedia of *E. coli* K-12 genes and metabolism
EcoGene	A collection of information about the genes, proteins, and intergenic regions of the *E. coli* K-12 genome and proteome
EcoliWiki	Community-contributed content about *E. coli*
EcoReg	The *E. coli* Regulation Consortium
E. coli Consortium	The *E. coli* Model Cell Consortium
Genobase	Japan's *E. coli* K-12 W3110 database of sequence information, proteome, transcriptome, bioinformatics, and literature knowledge
GenProtEC	*E. coli* genome and proteome database and classification
PortEco	Aggregated search results from 14 different *E. coli* data resources
RegulonDB	Database on *E. coli* transcriptional regulation and operon organization
TransportDB	Transporter protein analysis database
Transport Links	Transport protein overview

continually updated that contain information about *E. coli* (Table 4.3). As this available information grows, it will find its way into future reconstructions of the metabolic network in *E. coli*. The reconstruction process is thus an ongoing one, although for *E. coli* the progress curve is beginning to 'bend' as the new content incorporated is slowing down with subsequent reconstructions.

4.3 Content of the *i*JO1366 Reconstruction

*i*JO1366 is the result of 20 years of painstaking, systematic, and detailed reconstruction work. It is thus important to pause and review what has been compiled in terms of the scope and content of this reconstruction. This section summarizes the contents of *i*JO1366.

Overall features The *i*JO1366 reconstruction of the *E. coli* metabolic network is a knowledge base (k-base) that contains a structured representation of biochemical, genetic, and genomic (BiGG) information. The reconstruction can be visualized as a genome-scale metabolic map (Figure 4.7.) A breakdown of genes, reactions, and metabolites included in *i*JO1366 are given in Table 4.4, where it is compared to the contents of its two immediate predecessors. The 1366 ORFs included account for 32% of all the identified ORFs in *E. coli* and about 40% of currently functionally annotated ORFs on the genome.

   Given the fact that *E. coli* was a workhorse organism during the era that elucidated the biochemistry of metabolic reactions, there is extensive biochemical information about the actual functional properties of the corresponding gene products. *i*JO1366 has direct information about the biochemical properties of 1161 ORFs. Additionally, there are 133 transport and metabolic processes included based on biochemical or physiological information only, i.e., the corresponding ORFs have not been identified. Many of the proteins included in *i*JO1366 have either experimental or computational information about their structure. In addition, there is extensive structural information available on these enzymes. *i*JO1366 thus represents a comprehensive and curated BiGG k-base on *E. coli* metabolism.

Detailing *i*JO1366 The detailed classification of the contents of *i*JO1366 is presented in Figure 4.10. The figure was generated using the functional categories assigned through manual curation to classify the reactions included in the *E. coli* metabolic reconstruction.

- The number of reactions from each functional class that are included in *i*JO1366 are detailed in Figure 4.10A. The highest number of reactions are associated with the two classes of transporters, in the inner and outer membrane, followed by lipid and carbohydrate metabolism. The transporters also represent the highest number of non-gene-associated reactions.
- Overall, transport and carbohydrate metabolism has the highest number of ORFs (Figure 4.10B). All classes of genes are highly specific to the category into which they fall.
- The classification of metabolites in *i*JO1366 tied to each functional class are detailed in Figure 4.10C. The largest number of metabolites associated to

Table 4.4 Summary of the contents of the *iJO1366*, *iAF1260* and *iJR904* metabolic reconstruction for *E. coli* K-12 MG1655. Adapted from [115, 295].

Category	iJR904	iAF1260	iJO1366
Included genes	904 (20%)	1260 (28%)	1366 (32%)
Experimentally based function	838 (93%)	1161 (92%)	1328 (97%)
Computationally predicted function	58 (6%)	99 (8%)	38 (3%)
Unique functional proteins	817	1148	1254
Multigene complexes	105	167	185
Genes involved in complexes	289	415	483
Instances of isozymes	149	346	380
Reactions	931	2077	2251
Metabolic reactions	747	1387	1473
Unique metabolic reactions	745	1339	1424
Cytoplasmic	745	1187	1272
Periplasmic	0	192	193
Extracellular	2	8	8
Transport reactions	184	690	778
Cytoplasm to periplasm	0	390	447
Periplasm to extracellular	0	298	329
Cytoplasm to extracellular	184	2	2
Gene–protein–reaction associations			
Gene-associated (metabolic/transport)	706/166	1294/625	1382/706
Spontaneous/diffusion reactions	2/9	16/9	21/14
Total (gene-associated and no association needed)	708/175 (95%)	1310/634 (94%)	1403/720 (94%)
No gene association (metabolic/transport)	37/9 (5%)	77/56 (6%)	70/58 (6%)
Exchange reactions	143	304	330
Metabolites			
Unique metabolites	625	1039	1136
Cytoplasmic	618	951	1039
Periplasmic	0	418	442
Extracellular	143	299	324

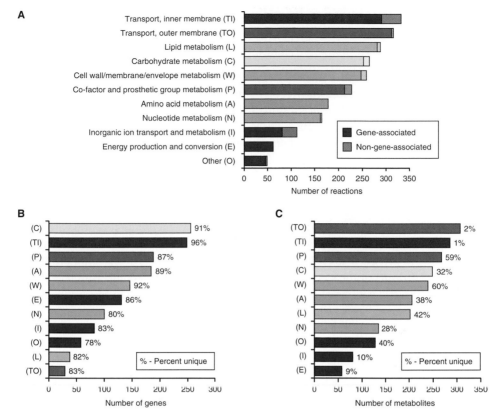

Figure 4.10 The reaction, gene, and metabolite content of *i*JO1366. (A) The number of reactions in each of 11 functional categories. Non-gene-associated (orphan) reactions are indicated by the lighter portion at the far right of each bar. (B) The number of genes accounted for with associated reactions in each category. The number of genes unique to each category (i.e., associated only with reactions in one category) is given as a percentage. (C) The number of unique metabolites that participate in at least one reaction in each category, with the number of metabolites unique to each category indicated. From [295].

ORFs are in two functional classes of transport and co-factor and prosthetic group metabolism. Furthermore, the highest number of unique metabolites were found in the cell wall/membrane/envelope metabolism.

These few observations are indicative of the detailed knowledge contained in *i*JO1366. The interested reader can browse its contents and extract information that he or she is particularly interested in.

4.4 From a Reconstruction to a Computational Model

Although this process is detailed in Part III of the book, it is helpful to preview the model formulation process at this point. As previously stated, a reconstruction needs to be converted to a mathematical model to enable the computation of biological functions. A genome-scale network reconstruction can be converted through a defined series of steps into a computational model that is more complex than those discussed above for a simple core model. The following steps are necessary to convert a genome-scale reconstruction to a predictive computational model.

1   *Explicit assignment of the metabolites participating in a reaction.* Some enzymes are promiscuous and can act on a number of different metabolites. Each of these potential substrates for an enzyme need to be explicitly defined as participating in a separate reaction. For example, an enzyme acting on a phospholipid needs to be defined explicitly to act on a specific structure(s) with defined chemical composition. This basically represents a validated network reconstruction.

2   *Definition of a systems boundary.* A boundary is necessary to define the system that is going to be modeled. For a whole-cell model, the systems boundary is defined around the entire reaction network and exchange reactions are generated to exchange metabolites through this boundary. Constraints can then be assigned to each of these exchange reactions to restrict the inputs and outputs of the system, depending on the chemical composition of the growth environment. Thus, if a substrate is not available in the medium, the uptake rate is zero.

3   *Conversion of the defined system into a mathematical format that forms the basis for a computational model.* After detailing the contents of the network and defining the systems boundary, a reconstruction can be represented in mathematical terms (see Chapter 9). A variety of available software can then be deployed to perform the computations, such as the COBRA toolbox [34, 375].

4   *Initial curation: filling gaps.* In order to produce essential biomass components (amino acids, deoxy-nucleotides, etc.) from minimal media components, there need to be continuous pathways from media substrates to the required metabolites for biosynthesis. COBRA tools, see Chapter 18, in conjunction with a biomass composition, are used to aid in filling essential gaps in the network.

5   *Determining the strain-specific parameters.* In order to examine the network's ability to fulfill the biomass requirements needed for cellular growth, a biomass objective function (BOF) is needed. The BOFs are formulated as linear combinations of experimentally measured biomass requirements.

After the conversion of the reconstructed network into a genome-scale model (GEM), constraint-based approaches can be applied to predict cellular phenotypes under different genetic and environmental conditions.

4.5   Validation of *i*JO1366

The GEMs built from the *E. coli* metabolic reconstructions have been extensively compared to various data types. Here we illustrate how *i*JO1366 was validated against growth data in various nutritional environments and against information in gene deletion (knock-out) strains.

Table 4.5 Tabulation of the number of growth-supporting carbon, nitrogen, phosphorus, and sulfur sources for *i*JO1366 and *i*AF1260. Adapted from [295].

Source	*i*JO1366		*i*AF1260	
	Potential substrates	Growth-supporting	Potential substrates	Growth-supporting
Carbon	285	180	262	174
Nitrogen	178	94	163	78
Phosphorus	64	49	63	49
Sulfur	28	11	25	11

Table 4.6 Gene essentiality predictions from *i*JO1366 on glucose and glycerol minimal media. Adapted from [295].

	Experimental	
	Essential	**Non-essential**
Growth on glucose		
Computational		
Essential	168 (12.3%)	39 (2.8%)
Non-essential	80 (5.9%)	1079 (79.0%)
Growth on glycero		
Computational		
Essential	161 (11.8%)	45 (3.3%)
Non-essential	87 (6.4%)	1073 (78.5%)

Growth capabilities High-throughput growth data are available for various organisms. Growth rate computed capabilities based on metabolic reconstructions are described above. The growth phenotypes of *E. coli* on all possible carbon, nitrogen, phosphorus, and sulfur sources were computed. Such computations are shown for *E. coli* based on GEMs derived from *i*AF1260 and *i*JO1366 (Table 4.5). Such growth predictions can be compared to the experimentally determined growth ability of *E. coli* K-12 MG1655.

Gene essentiality Growth predictions can be further tested by computing the effects of gene knock-out under a given growth condition and comparing the growth ability to the actual strain. For *i*JO1366, growth phenotypes were predicted on both glucose and glycerol minimal media, and the results were compared with experimental data sets (Table 4.6). The results of the comparison fall into four categories.

1   Computationally essential/experimentally essential (true positive): for growth on glucose, 12.3% of these cases fell into this category. These are real predictions of the growth consequences of a gene knock-out as a phenotype is produced.

2   Computationally non-essential/experimentally non-essential (true negative): for growth on glucose, 79% of the cases fell into this category. These are weaker predictions, and might be thought of as more of a consistency check.

3   Computationally essential/experimentally non-essential (false negative): for growth on glucose, 2.8% of the cases fell into this category. Here the reconstruction is missing a component that enables growth under these conditions. This outcome offers the potential to systematically discover a missing component.

4   Computationally non-essential/experimentally essential (false positive): for growth on glucose, 5.9% of the cases fell into this category. In this case, the model fails to predict correctly. There can be several reasons for this, including the down-regulation of genes that are actually used in the computed solution. These outcomes also offer an opportunity for discovery.

The results for glycerol are similar.

Assessment of gaps in the network   A more comprehensive assessment of the growth predictive ability of *i*JO1366 has been performed. It has been used to predict growth phenotypes for 13,470 growth conditions [298]. A total of 11,855 true positives, 639 true negatives, 711 false positives, and 265 false negatives were identified (see Figure 4.11A). The prediction can be classified into the same 11 functional categories of metabolic reactions as seen in Figure 4.10B (see Figure 4.11B).

The prediction failures can be addressed by the gap-filling methods described in Chapter 25. Gap-filling methods are computational procedures that are used to produce hypotheses that explain the mismatch between computation and experiment. One should keep in mind that the experimental data can be incorrect, as some validation studies have shown [39].

The initial gap-filling of *i*JO1366 has been performed, and perhaps the most interesting results include the ability to get experimental evidence supporting the prediction that the gene *yhiJ* is a mhyo-inositol:oxygen oxidoreductase [298]. Gap-filling also predicted the functions of over a dozen other genes which have not yet been experimentally tested. These results suggest experiments to perform in the future to continue to fill in the missing components in the reconstruction. Thus, the reconstructions exhaustively account for known metabolic functions, but can now also be used prospectively to conduct experimental studies. More detailed discussion of the resolution of false predictions is found in in Chapter 25.

4.6   Uses of the *E. coli* GEM

Ask not what you can do for a reconstruction, but what a reconstruction can do for you   Microbial life mostly revolves around metabolism and growth. Thus, not

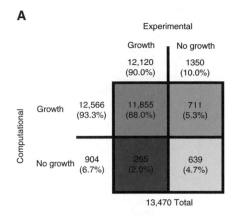

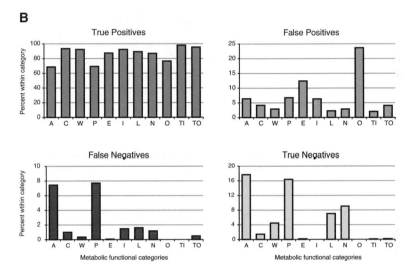

Figure 4.11 Comparison of *i*JO1366 GEM predicted growth phenotypes to experimental data. (A) The overall comparison, indicating numbers of true positives, true negatives, false positives, and false negatives. (B) The numbers of each type of prediction within 11 functional categories of metabolic reactions. The categories are: amino acid metabolism (A), carbohydrate metabolism (C), cell wall/membrane/envelope metabolism (W), co-factor and prosthetic group metabolism (P), energy production and conversion (E), inorganic ion transport and metabolism (I), lipid metabolism (L), nucleotide metabolism (N), other (O), inner membrane transport (TI), and outer membrane transport (TO). From [298].

surprisingly, the reconstructed metabolic network in *E. coli* has found a wide variety of uses. The process of formulating a network reconstruction is labor intensive, lengthy, and requires patience. The conversion to a computational model requires knowledge of mathematics, programming, and numerical methods. Over the past 10 years, the types and numbers of applications for GEMs has grown steadily [118, 263]. Applications of the *E. coli* metabolic model range from pragmatic to theoretical studies, and have been classified into six general categories (Figure 4.12). We will briefly describe some of them here.

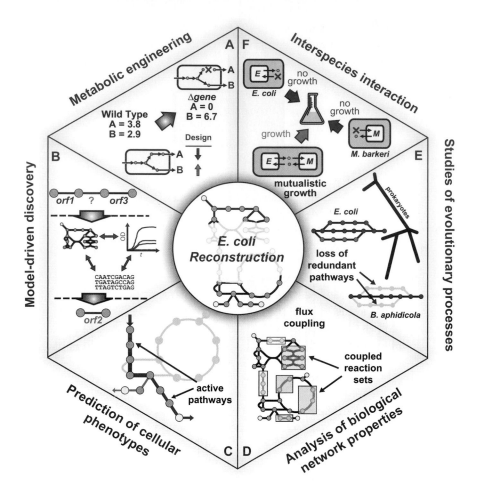

Figure 4.12   Six broad categories of use of the *E. coli* metabolic reconstruction. Modified from [118].

Metabolic engineering   There is significant interest in redirecting metabolic fluxes in bacteria, primarily for the purposes of production of certain compounds of interest. This field of study is known as metabolic engineering. In such studies, the *E. coli* GEM was principally used to analyze the metabolite production potential of *E. coli* and identify metabolic interventions needed to enable the production of the product of interest. Thus, *E. coli* strains have been systematically designed through *in silico* analysis to over-produce target metabolites such as ethanol, lycopene, lactic acid, succinic acid, amino acids, as well as diverse products from hydrogen to vanillin.

Bacterial adaptation   The GEMs of *E. coli* have been used to examine the process of bacterial adaptation and evolution. Specifically, the network reconstructions have been used to (1) interpret adaptive evolution events, (2) examine horizontal gene transfer, (3) simulate evolution to minimal metabolic networks, and (4) predict the endpoints of short-term laboratory adaptive evolution. Such studies, which utilize the *E. coli* reconstruction as an organism-specific genetic and metabolic content database, and the corresponding GEM, have been able to provide insight into evolutionary events by combining known physiological data. These issues are detailed in Chapter 26.

Essentiality of the components of a reconstruction A range of computational studies have sought to understand phenotypes through determining the essential genes, metabolites, and reactions in the *E. coli* metabolic network. A benchmark used for examining GEM predictive ability is to determine the agreement with growth phenotype data from knock-out collections of *E. coli*. Such studies will be further enabled by the recent availability of a comprehensive single-gene knock-out library for *E. coli* [23]. Implications for examining essentiality of components in *E. coli* metabolism include: (1) determining network essentiality in similar organisms; (2) deciphering network makeup and enzyme dispensability; and (3) aiding in metabolic network annotation, validation, and refinement.

Toward the fulfillment of the promise of molecular biology As the molecular basis for life began to emerge in the mid twentieth century, the field of molecular biology arose, which, for the rest of the century, had a broad and fundamental impact on basic biology and the biomedical sciences. The growth, acceptance, and influence of molecular biology was based on the premise that if you understood the molecules of life you could understand the form and function of the living processes that they produced. Then, with the emergence of high-throughput technologies and large-scale omics data sets, the enormity of this process became clear as the number of molecular components of cells discovered continued to grow.

Now, with use of genome-scale measurements and bibliomic data, the reconstruction of the interactions of all these components into validated networks has become possible. In addition, functional models that compute phenotypes based on known underlying data organized in the form of BiGG k-bases have become possible. Thus, in a way, the validated genome-scale models that have become available since 2000 not only represent significant steps towards the fulfillment of molecular biology as a field, but also the formalization of the quest to generate mechanistic genotype–phenotype relationships.

After over 10 years of development, such models are beginning to show their applicability and utility, and their scope, resolution, and predictive ability is expected to grow over the coming decades, forming the foundations for (molecular) systems biology. In some way, this achievement represents partial completion of *Project K: the complete solution of E. coli*, a remarkable vision articulated by Francis Crick 40 years ago [90].

4.7 Summary

- The *E. coli* metabolic network reconstruction represents the best-developed network of its kind.
- It was generated in an iterative fashion over a 20-year period, with three notable versions before the genome sequence was made available, and four

iterations following the availability of the genome sequence and its updated annotations.

- The *E. coli* metabolic reconstruction now comprehensively represents known metabolic functions in *E. coli*. This reconstruction has been validated against a number of known cellular functions.
- A wide variety of applications of the genome-scale reconstruction have been developed.

5 Prokaryotes

If an alien visited earth, they would likely take some note of humans, but probably spend most of their time trying to understand the dominant form of life on our planet – microorganisms like bacteria and viruses – Nathan Wolfe

The *E. coli* metabolic reconstruction and its GEM are highly developed. There are now a number of other microorganisms for which GEMs have been built and put to use. A significant feature of microbial life and biology revolves around metabolism and growth, and thus GEMs of microbial metabolism have been used to address a number of important issues in industrial, environmental, and medical microbiology. We will describe some of these accomplishments in this chapter. It is likely that the number of such reconstructions and their applications will grow significantly over the coming years given the rapidly growing amount of sequenced genomes and semi-automated reconstruction tools.

5.1 State of The Field

Phylogenetic coverage Bacteria display an astonishing spectrum of metabolic capabilities that reflect the wide spectrum of microenvironmental niches in which they grow. Metabolism in some of the branches of the phylogenetic tree, such as enterobacteria, is generally well characterized, whereas little molecular detail is known about members of other branches. In fact, a large number of microorganisms cannot even be cultured in the laboratory, but their genomes can be sequenced and a metabolic reconstruction can be performed. Interestingly, metabolic network reconstructions can then be used to formulate growth media [441].

The phylogenetic coverage of existing genome-scale metabolic reconstructions is shown in Figure 5.1A. It shows that the coverage of the tree is biased by well-understood model species. There is clearly a need to get better uniform coverage of the phylogenetic tree so that we can have a broad view of the metabolic capabilities of microbes that inhabit the various microenvironments on this planet.

The number of manually curated reconstructions has grown steadily since the publication of the first GEM in 1999 (Figure 5.1B). The iterative nature of the process also shows up in this summary, as metabolic reconstructions for some organisms have gone through multiple iterations (see Figure 3.11 for more details). The number of

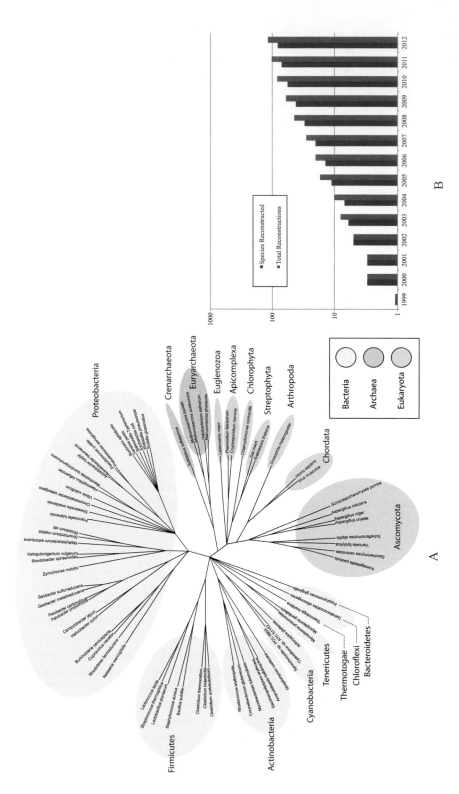

Figure 5.1 Metabolic reconstructions of microbial metabolism. (A) Phylogenetic tree for microorganisms highlighting the phylogenetic distribution of existing metabolic reconstructions. (B) The number of genome-scale metabolic network reconstructions that have appeared over time. The interactive tree of life tool (http://itol.embl.de/) was used to generate the tree based on the NCBI taxonomy database (http://www.ncbi.nlm.nih.gov/taxonomy). Prepared by Jon Monk.

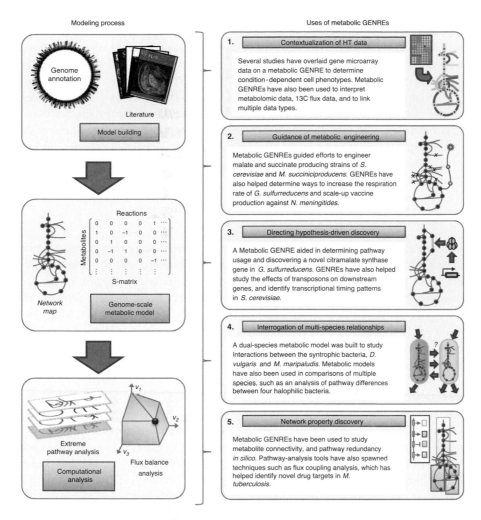

Figure 5.2   Uses of metabolic network genome-scale reconstructions (GENREs). The building and analysis of metabolic GENREs are shown in the left panels, and the five categories of uses for metabolic GENREs are illustrated in the right set of panels. From [290].

GEMs is likely to grow faster going forward as automated reconstruction methods are developed (Table 3.3) and validated.

Snapshot of uses of genome-scale metabolic reconstructions   A survey of uses of genome-scale metabolic reconstructions, from the very first appearing around 2000, to about 2010, showed that their uses fell into five general thematic categories [290]: (1) contextualization of omics data, (2) applications to metabolic engineering, (3) directing hypothesis-driven discovery, (4) interrogation of multi-species relationships, and (5) network property discovery (see Figure 5.2). These categories are similar to those previously shown for *E. coli* (Figure 4.12). The applications have been numerous and varied.

Scope We can only describe the metabolic reconstruction efforts for bacteria at a high level in this chapter. We begin by surveying the number of reconstructions of pathogenic bacteria and their uses. Photosynthetic metabolic reconstructions have focused on *Synechocystis*, and we discuss the latest in a series of its metabolic reconstructions. Considerable interest has developed in microbial communities in the wake of the progress with metagenomics in recent years. Surprisingly, there are several examples showing that reconstruction efforts for single organisms scale to models of simple communities. Finally, we will showcase a particularly detailed story of what used to be a poorly characterized organism of environmental importance at the end of the chapter.

5.2   Metabolism in Pathogens

In addition to their application in environmental microbiology and industrial microbiology, GEMs have been applied to medical microbiology and the analysis of the functions of pathogens [69]. Among the GEMs summarized in Figure 5.1 are many pathogenic organisms. Table 5.1 summarizes the target pathogens that have been studied, the analysis performed using their GEMs, and the results obtained. Many of the analysis methods shown as columns in this table are described in Part III of the book. Some of the results from applying these analysis methods to the GEMs for pathogens are now described.

Gene essentiality analysis The analysis of the consequences of gene knock-outs can readily be performed using GEMs (recall Figure 2.6). Such analysis can be used to represent the metabolic intervention of a drug that inhibits the corresponding enzyme. If an *in silico* gene knock-out leads to the loss of ability of the GEM to produce growth in the environment considered, the corresponding enzyme can be considered to be a candidate target for drug development.

Another easy application of GEMs is to compute synthetically lethal pairs of genes or reactions. A synthetic lethal is a double gene knock-out that is lethal, while the knock-out of each of the genes individually is not. Confirmed synthetic lethals can be used for two-drug treatment strategies and other applications described elsewhere in the text.

Enzyme robustness and flux variability The analysis of gene essentiality can be complemented by assessing robustness analysis. To determine the robustness of a metabolic network to the inhibition of an enzyme-catalyzed reaction, its flux is constrained in a step-wise fashion and the effects on the growth rate or other critical functions is evaluated. This complements gene essentiality by assessing the effects of different levels of partial enzyme inhibition.

Metabolite essentiality Drugs have strong structural similarity to natural metabolites, thus they can compete for and/or inhibit normal enzymatic activity. This observation has led to another approach for identifying drug targets in metabolic networks. This is called the identification of essential metabolites. GEMs can be used to find

Table 5.1 Drug targeting-related analysis of pathogen metabolic networks. From [69].

Pathogen	Disease	Drug targeting-related *in silico* analysis								Validation		Novel compounds identified	Refs
		Gene/reaction essentiality	Minimal media prediction	Conditional essentiality	Synthetic lethality	Flux variability analysis	Enzyme robustness	Metabolite essentiality	Correlated reaction sets	Literature-derived	Novel experimental validation		
Acinetobacter baumannii	Opportunistic; nosocomial infection	X						X		X			[205]
Burkholderia cenocepacia	Opportunistic; cepacia syndrome	X		X						X			[112]
Francisella tularensis	Tularemia	X		X		X	X			X	X		[352]
Haemophilus influenzae	Otitis media and respiratory infections	X	X	X	X					X			[105,381]
Helicobacter pylori	Gastritis; peptic ulceration; gastric cancer	X	X	X	X					X	X		[380,419]
Klebsiella pneumoniae	pneumonia; urinary tract infection	X								X			[237]
Mycobacterium tuberculosis	Tuberculosis	X		X	X	X			X	X	X		[39,113,182]
Neisseria meningitidis	Meningitis; meningococcal septicemia		X							X	X		[22]
Porphyromonas gingivalis	Periodontal disease	X	X							X			[262]
Pseudomonas aeruginosa	Opportunistic; nosocomial infection	X								X			[289]
Salmonella typhimurium	Gastroenteritis; diarrhea	X			X	X				X	X		[5,351,422]
Staphylococcus aureus	Opportunistic; nosocomial infection	X	X		X					X	X		[35,161,222]
Vibrio vulnificus	Cellulitis; septicemia							X		X	X	X	[206]
Yersinia pestis	Bubonic; pneumonic; and septicemic plague	X	X		X					X			[279]
Cryptosporidium hominis	Cryptosporidiosis	X								X			[434]
Leishmania major	Leishmaniasis	X	X		X					X			[68]
Plasmodium falciparum	Malaria	X			X		X			X	X	X	[174,336]
Trypanosoma cruzi	Chagas disease	X				X				X			[363]

such essential metabolites by removing all the reactions in which a given metabolite participates.

Multiple targets of a single compound   Many existing anti-infectives act on multiple targets. This multiplicity of targets was exemplified in a network analysis of 890 FDA-approved drugs showing that approximately 38% of these drug molecules were associated with more than one target [466]. A drug discovery strategy incorporating compounds known to act simultaneously on multiple targets can be readily applied using COBRA methods.

Groups of targets and network topology   GEMs can be used to determine subtle dependencies among the components of a model. Sets of correlated reactions (or co-sets, which are discussed in Chapter 13) are groups of reactions whose fluxes are linked in such a way that they are always used the same way to produce a physiological state. Thus, the removal of any one of them is predicted to have the same effect on the phenotype. More intricate dependencies among the function of the gene products have also been defined [56]. Co-sets can be used to suggest alternative and equivalent drug targets by identifying reactions that are functionally related to each other.

Prospects   Numerous initial studies have been performed that deploy COBRA methods for the analysis of pathogens as summarized in Table 5.1. With the development of GEMs that are larger in scope we will, in the future, be able to analyze in much more detail the effects of antibiotics on pathogens. Such detailed analysis relies on our improved understanding of the molecular mechanisms underlying the action of antibiotics and how they kill bacteria [215, 464]. Such molecular information and understanding can now be put in the context of the cells as an integrated system.

As will be discussed in the following chapter, there are now a series of metabolic reconstructions available for human cells and tissues. Having metabolic models of a host cell and a pathogenic organism opens up the possibility of building integrated-host pathogen models. Metabolic network reconstructions exist for a number of pathogens and the tissues that they invade or populate, as summarized in Figure 5.3. Such efforts have been initiated by various investigators.

The more detailed treatment of the interactions between the host and the pathogen and its disruption brings up the issues of communities in which many organisms or cell types interact. We will discuss reconstruction of multiple interacting organisms later in the chapter.

5.3   Metabolism in Blue-Green Algae

Synechocystis: **a model organism for biosynthetic prokaryotes**   The cyanobacterium *Synechocystis* sp. PCC6803 has been studied as a model photosynthetic organism because it is capable of carrying out oxygenic photosynthesis with higher efficiency than vascular plants with its simpler photosystem structures. In addition, its cultivation is simple and inexpensive, it is a source of natural high-value products

Porphyromonas gingivalis W83
G: 478
M: 564
R: 679

Leishmania major Friedlin
G: 560
M: 657
R: 1112

Neisseria meningitis serogroup B.
G: 555
M: 471
R: 496

Haemophilus influenzae Rd
G: 400
M: 367
R: 461

Staphylococcus aureus N315
G: 619/ 551/ 546
M: 571/ 604/ 1431
R: 640/ 712/ 1493

Francisella tularensis LVS
G: 683
M: 586
R: 605

Pseudomonas aeruginosa PAO1
G: 1056
M: 760
R: 883

Yersinia pestis 91001
G: 818
M: 825
R: 1020

Acinetobacter baumanni AYE
G: 650
M: 778
R: 891

Mycobacterium tuberculosis H37Rv
G: 661/ 721/ 663
M: 740/ 739/ 742
R: 939/ 849/ 1049

Klebsiella pneumoniae MGH 78578
G: 1228
M: 1055
R: 1970

Plasmodium falciparum 3D7
G: 579/ 366
M: 1622/ 616
R: 1375/ 1001

Burkholderia cenocepacia J2315
G: 1028
M: 748
R: 859

Vibrio vulnificus CMCP6
G: 673
M: 765
R: 943

Helicobacter pylori 26695
G: 341
M: 411
R: 476

Cryptosporidium hominis
G: 213
M: NR
R: 540

Streptococcus thermophilus LMG18311
G: 429
M: NR
R: 522

Salmonella enterica ssp. typhimurium LT-2
G: 945/ 1270
M: 1036/ 1119
R: 1964/ 2201

Lactobacillus plantarum WCFS1
G: 721
M: 554
R: 761

Mycoplasma genitalium G-37
G: 187
M: 276
R: 264

Escherichia coli W (ATC 9637)
G: 1273
M: 1111
R: 2477

Escherichia coli K12 MG-1655
G: 1366
M: 1136
R: 2251

Bacteroides thetaoio-taomicron VPI-5482
G: 853
M: 914
R: 1305

Lactococcus lactis ssp. lactis IL1403
G: 358
M: 422
R: 621

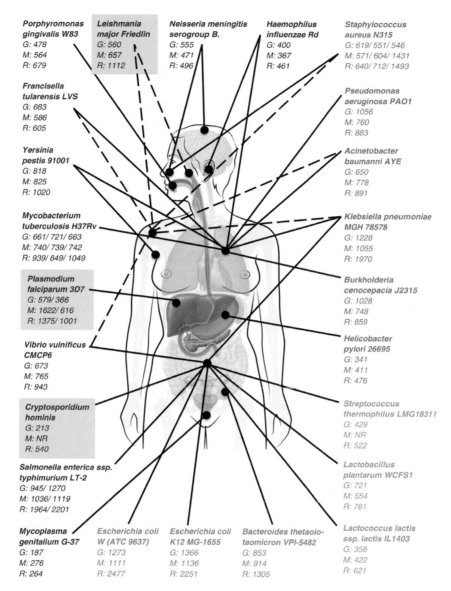

Figure 5.3   A summary of microbes that live on or in man for which genome-scale metabolic reconstructions have been published. Their predominant sites of colonization are shown. In red are highlighted pathogens, orange are opportunistic pathogens, green are commensals, and blue are probiotic bacteria. Dotted lines represent skin as body site. White background represents prokaryotic organisms, while eukaryotes are shaded gray. G; number of genes included in the metabolic reconstruction. R; number of reactions. M; number of metabolites. NR; not reported in the original reconstruction paper. Taken from [425].

such as carotenoids, lipids, and vitamins, and it is amenable to genetic manipulations. Therefore, it is not surprising that the reconstruction and modeling of the metabolism of this cyanobacterium at the genome-scale has received great attention in the last few years.

Genome-scale metabolic network reconstructions Nine different metabolic reconstructions of *Synechocystis* are currently available. Their content and properties are
summarized in Table 5.2. The content of these reconstructions and the level of
modeling details vary among these studies; however, it has been found that the
complete mass and charge balancing as well as the inclusion of well-compart-
mentalized reactions considerably improved the prediction accuracy, irrespective
of the model's size [286, 469]. Light-driven metabolism is the most distinct feature
of phototrophs compared to other non-photosynthetic microorganisms. Thus, while
reconstruction and modeling of photosynthesis represents a notable challenge, at
the same time it provides an opportunity to expand the current scope of metabolic
reconstructions and understanding of microbial metabolism.

*i*JN678 Among the currently available *Synechocystis* reconstructions, *i*JN678 is the
most complete and detailed reconstruction of its photosynthetic metabolism (see
Table 5.2). Instead of using a lumped reaction, in *i*JN678 the photosynthetic linear
electron flow pathway was reconstructed by including individual reactions for
each component, while several alternate electron flow pathways were included as
well. A GEM including photosynthesis was completed by placing the oxidative
phosphorylation reactions both in the periplasmic and thylakoid membranes, enabling
the interaction between respiration and photosynthesis [286].

Studying photosynthesis using GEMs The detailed GEM built from *i*JN678 allowed
the analysis of the light-driven metabolism at the systems level, highlighting the complexity and versatility of the photosynthetic process in *Synechocystis* under different
light and inorganic carbon conditions. Two main states of the photosynthetic apparatus, a light-limited state and a CO_2-limited state, were identified. By computing
the main bioenergetic parameters produced by the linear electron flow by itself as
well as when it's assisted by alternate electron flow pathways, the first functional
characterization of these latter pathways in the context of the whole network became
possible (Figure 5.4).

Genome-scale analysis showed that while alternative electron flow pathways with
high photosynthetic yield provide extra ATP levels under light-limited conditions,
those pathways with lower photosynthetic yield participated in redox balancing under
high-light conditions. The GEM-based analysis also showed that a high degree of
cooperativity between complementary alternate electron flow pathways was required
for optimal autotrophic metabolism.

Under genetic perturbations, the *Synechocystis* metabolic network showed poor
robustness characteristics, suggesting that the high robustness of photosynthesis
required for autotrophic metabolism comes at a cost of reduced metabolic robustness.
Thus, the systems analysis of *Synechocystis* metabolism using a GEM not only
promoted a better understanding of the photosynthetic process, but also enabled
the identification of an inherent tradeoff in key metabolic properties of phototrophs.

5.4   Metabolism in Microbial Communities

Microbial interactions are foundational to all global geochemical cycles and have
an important role in human health and disease. A comprehensive understanding

Table 5.2 Comparison of *i*JN678 with previous *Synechocystis* metabolic reconstructions. *Biomass objective function (BOF) level definition was taken from [119]. Compartment symbols were taken from [427]. [e] extracellular space; [p] periplasm; [c] cytoplasm; [u] thylakoid. From [286].

Model	Genes	Reactions	Metabolites	BOF level*	Photosynthesis modeling	Lipid modeling	Complete mass/charge balancing	Compartments	Ref
*i*JN678	678	863	795	Advanced	Complete	Complete	Yes	[e],[p],[c],[u]	[286]
Synechocystis	Nd	93	Nd	Basic	Lumped	No	No	[e],[c]	[392]
Synechocystis	78	56	72	Basic	Lumped	No	No	[e],[c]	[171]
Synechocystis	505	652	701	Basic	Lumped	No	No	[e],[c]	[134]
Synechocystis	Nd	46	29	Basic	Lumped	No	No	[e],[c]	[278]
Synechocystis	343	380	291	Intermediate	Lumped	Partial	No	[e],[c]	[212]
*i*Syn669	669	882	790	Intermediate	Partial	Partial	No	[e],[c]	[270]
*i*Syn811	811	965	911	Intermediate	Partial	Partial	No	[e],[c]	[271]
Synechocystis	376	493	465	Intermediate	Partial	Partial	Yes	[e],[c],[u]	[469]

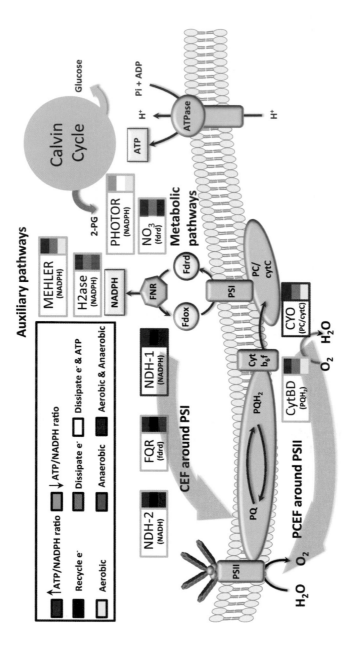

Figure 5.4 The linear electron flow pathway (PSII, PQ pool, CytBF, PSI, and FNR), the ATPase, and the Calvin cycle are shown in green. The final products from linear electron flow and photophosphorylation (reduced ferredoxin (Fdrd), NADPH, and ATP) are shown in blue. The high and low light-yield pathways and their electron donors are shown with red and yellow boxes, respectively. The alternative pathways participating in the cyclic electron flow around photosystem I (CEF) and pseudocyclic electron flow around the photosystem II (PCEF) as well as those auxiliary and metabolic pathways involved in photosynthesis optimization were color-coded based as shown on their contribution to the ATP/NADPH ratio, the mechanism of energy dissipation, and by dependence on O_2 availability. Taken from [286].

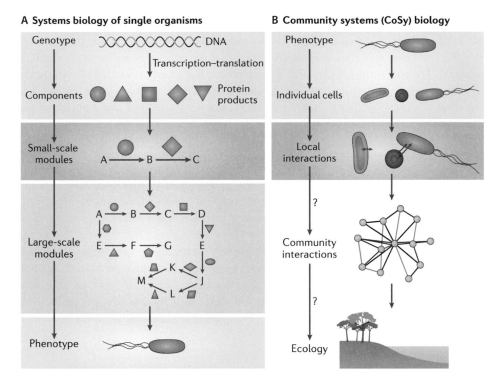

Figure 5.5 Systems biology approaches for modeling single organisms and communities. (A) Reconstructing the genotype–phenotype relationship for single organisms. The colors depict the current status of the process: green is for feasible and well-established processes, yellow indicates a feasible but not as clearly defined process, and orange shows a process that is currently very challenging. (B) Just as a workflow has been established for genome-scale modeling approaches for single organisms, an analogous path forward for community systems (CoSy) biology is proposed, with status colors as for part A. First, phenotypes for multiple individual cells have to be established; individual cells can represent different organisms, species or strains. The definition of local and community interactions would allow ecological questions to be addressed. Question marks indicate areas for which fundamental principles and concepts have to be established if we are to understand community interactions mechanistically. Taken from [471].

of microorganisms, their interaction with other species in a community, and their interplay with their environment is one major goal in microbial ecology. With the rapid development of metagenomics, development of systems biology for microbial communities has become not only plausible, but pressing.

5.4.1   Systems biology of communities

Community systems (CoSy) biology is emerging and one can envision that a bottom-up approach to this new field will build on the success of single-organism biology to follow an analogous path forward. As described above for the development of systems biology of single organisms, we now need to develop a new set of fundamental principles and concepts that will allow us to understand microbial communities (Figure 5.5).

Although metagenomic data is useful, by itself it is inadequate for this purpose. As for GEMs of single species, information is needed about the biochemical properties of the gene products and chemical mechanisms that describe species interactions, both metabolically and through physical and chemical communication [29]. Community reconstruction processes are needed. Progress towards analyzing the various types of species interactions using COBRA methods is being achieved and have been discussed [236].

5.4.2   Model-based analysis of microbial communities

Integration of metabolic capabilities and resource allocation in a community is one of its chief characteristics [236]. Organisms compete for scarce resources, as demonstrated for *Geobacter sulfurreducens* above, or depend on the metabolic capabilities of their neighbors.

Evolution should select for cells that exploit community structure [307]. Thus, reconstruction and statement of objectives represents an approach that is likely to be fruitful. COBRA methods have been used to characterize the role of metabolism in a microbial community structure [211, 406, 413], and these studies are providing insight into mutualism [456], competition [476], parasitism [48, 174], and community evolution [307, 468].

Mutualism Mutualism is the way two organisms of different species biologically interact such that each individual derives a fitness benefit from the interaction. Synthetic mutualism between a pair of auxotrophic *E. coli* mutants has been constructed experimentally, grown in co-culture, and the results have been analyzed using COBRA methods [456]. Their coupled metabolism was described and analyzed to identify mutant pairs that exchange essential metabolites to improve growth (Figure 5.6A). Shadow prices, a measure of the value of a metabolite to the growth of a cell (Chapter 18), demonstrated the balance between the cost (from metabolite loss) and the benefit (from receiving missing essential metabolites) to each rescued auxotroph. The cooperative efficiency (the ratio of uptake benefit to production cost) recapitulated the observed growth of the co-cultures. Substantial increases in growth (Figure 5.6B) were witnessed in co-cultures that exchanged beneficial but less costly metabolites (in co-cultures that had higher cooperative efficiency). Thus, optimality principles and COBRA methods were successfully used to describe a system of two mutualistic organisms.

Competition Metabolic competition for scarce nutrients can also be assessed using COBRA methods. One COBRA method, called dynamic multi-species metabolic modeling (DMMM), was used to characterize the competition for acetate, Fe(III), and ammonia between two bacterial species, *G. sulfurreducens* and *R. ferrireducens* (Figure 5.6C) [476]. DMMM was used to simulate the simultaneous growth rate of these two organisms as well as the rates of change of external metabolites to dynamically predict community composition. The community composition was predicted under geochemically distinct conditions of low, medium, and high acetate flux.

The prediction was that *Rhodoferax ferrireducens* dominates the community when sufficient ammonia is available under low acetate flux. Conversely, *G. sulfurreducens* is

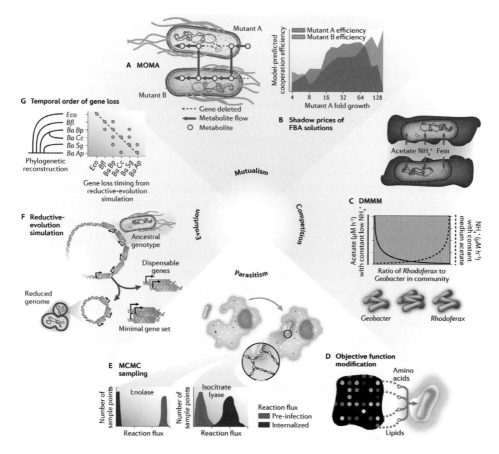

Figure 5.6 Integrating COBRA methods to study community interactions. COBRA methods are providing insight into the metabolic interactions in various types of microbial communities as described in the text. Taken from [236].

predicted to dominate under low ammonia concentrations and high acetate flux. This competitive difference under these two environments was attributed to the nitrogen fixation abilities of *G. sulfurreducens* as well as to its high acetate uptake rate relative to *R. ferrireducens*. Under nitrogen-fixing conditions, it was predicted that *G. sulfurreducens* increases its respiration at the expense of biomass production, thus showing how balancing community structure can affect the efficacy of uranium bioremediation in low-ammonium zones.

Parasitism   Host–pathogen interactions have been studied using COBRA methods [174]. The metabolic interactions between a human alveolar macrophage and *Mycobacterium tuberculosis* have been studied [48]. Context-specific models of the infection process were built using methods that map transcriptomic data onto the network reconstructions. Three different states of the *M. tuberculosis* infection could be analyzed. The objective function for *M. tuberculosis* was derived from the infection-specific gene expression (Figure 5.6D). Gene deletion analysis was compared with *in vivo* gene essentiality data.

The metabolic changes in the pathogen during infection could be assessed. Monte Carlo randomized sampling (see Chapter 14) was used to demonstrate a substantial alteration in metabolic pathway use by *M. tuberculosis* during macrophage infection. These alterations included a suppression of glycolysis and an increased dependency on glyoxylate metabolism (Figure 5.6E). This constraint of central metabolism during *M. tuberculosis* infection was also suggested by differential producibility analysis that identifies genes that affect the production of each metabolite in the metabolic network [44]. Such analyses could give clues to antibiotic development.

Community evolution   Genetic drift and selective pressures during evolution lead organisms to optimize their function for a particular niche [276]. The assumption of optimal cellular states in a given environment makes COBRA methods appropriate tools for addressing hypotheses about evolution [326]. In natural settings, the optimization of microbial metabolism is a multi-species affair, as demonstrated by the aphid endosymbiont *Buchnera aphidicola*. It has lost a substantial portion of its genome as it adapted to the nutrient-rich environment in its host. *B. aphidicola* is a relative of *E. coli*. Thus, a reductive-evolution simulation [307] was carried out on the *E. coli* GEM to generate predictions of a minimal metabolic gene set for *B. aphidicola* based on its lifestyle. Such computations predicted minimal gene sets that are highly consistent with the metabolic gene content of *B. aphidicola* (Figure 5.6F). Remarkably, the predicted temporal order of gene losses is consistent with the phylogenetically reconstructed gene loss timing among the genomes of five *B. aphidicola* strains (Figure 5.6G) [468]. Interestingly, metabolic pathways that are retained in the computed minimal gene sets shed light on symbiotic evolution. The retained pathways include riboflavin and essential amino acids synthesis that are lacking from the aphid diet [307].

Recapitulation   There are many studies of optimal functions of pure cultures. It is unclear how far we can progress with identifying and deploying objective functions to analyze the interactions of many organisms (see Figure 5.5). The initial use of COBRA methods describes how communities function as detailed in this section and suggest that organism interactions in simple settings are likely to succumb to constraint-based analysis approaches.

5.5   An Environmentally Important Organism

An illustration of the use of GEMs for gaining detailed understanding of an organism and to perform predictive discovery necessarily requires a very detailed description. We finish this chapter with some detailed studies of a poorly characterized organism called *G. sulfurreducens*.

5.5.1   *Geobacter sulfurreducens*

Geobacteraceae are a family of dissimilatory metal-reducing bacteria with the ability to reduce Fe(III) oxides as well as soluble heavy metal contaminants such as uranium into insoluble forms (Figure 5.7). Given the important environmental impact of this

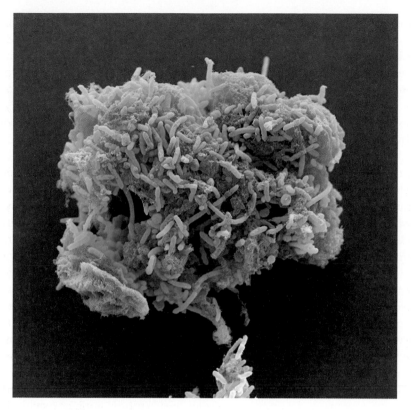

Figure 5.7 A microelectrogram of *Geobacter sulfurreducens* (false-colored in green) growing on iron as the electron donor (false-colored as pink). © Eye of Science/Science Photo Library.

capability, the metabolic characteristics of the Geobacteraceae are of special interest. The metabolic capabilities of the *Geobacter* species have only been recently investigated as these organisms were only discovered about 20 years ago [247,355].

In this section, we summarize over 10 years of work on the formulation of the GEM for *G. sulfurreducens* and touch on its many uses [254]. This history illustrates the use of GEMs and the importance of taking a systems biology point of view when studying the adaptations of metabolic capabilities to a particular micro-environment. In some ways, this material illustrates how the detailed mechanistic studies of the function of individual molecules in molecular biology is now mirrored by mechanistic studies of networks using methods of (molecular) systems biology.

A poorly characterized organism Compared to other organisms whose physiology has been studied for decades, the physiology of Geobacteraceae was poorly characterized in the early 2000s. Fortunately, with the availability of high-throughput experimental technologies such as genome sequencing, gene and protein expression profiling, as well as computational tools such as bioinformatics and metabolic modeling, the characterization of the physiology of Geobacteraceae has accelerated.

Genome sequencing of this class of organisms began in the 1990s and the first member of Geobacteraceae to be sequenced was *Geobacter sulfurreducens* [248].

Subsequently, several members of the genus have been sequenced. Many interesting features of its physiology have been unraveled based on the systematic study of its metabolism, including components of its unusual electron transport chain, metabolic pathways, and other adaptations such as chemotaxis to Fe(II), and its ability to store electrons (capacitance), several of which are highly unique to Geobacteraceae.

Metabolic challenges   The common physiological characteristic of this family is their ability to reduce Fe(III) oxides coupled to acetate oxidation [246]. Acetate is a common by-product of the fermentative metabolism and is thus a 'poor' substrate relative to highly reduced compounds, such as glucose. In addition, Fe(III) oxide, the predominant electron acceptor for Geobacteraceae, is insoluble compared to other electron acceptors such as sulfate, nitrate, or even oxygen. Consequently, the metal reducers face a significant challenge in generating metabolic energy from coupling acetate oxidation to Fe(III) oxide reduction.

Some of the key metabolic challenges faced by Geobacteraceae in their natural environments include: (1) low energy bearing acetate as the electron donor, (2) the number of carbon atoms in acetate necessitating two CO_2 fixation steps to synthesize oxaloacetate, (3) the need to access insoluble Fe(III) oxides as the electron acceptor, (4) the need to completely oxidize the electron donor to access all of the electrons, (5) the requirement to adapt to Fe(III) oxide exhaustion in the local microenvironment, and (6) the need for an enhanced rate of respiration to sustain cellular processes due to lower net energy of the acetate/Fe(III) oxide coupling. An understanding of how Geobacteraceae have met these metabolic challenges can be developed using GEMs.

5.5.2   Genome-scale science for *Geobacter*

A genome-scale *in silico* metabolic model of *G. sulfurreducens* has been developed [249]. The *G. sulfurreducens* metabolic reconstruction was an early effort in the field and one of first genome-scale reconstructions performed for a poorly characterized organism. It was thus initially unclear how much would be learned from this metabolic reconstruction. The workflow that followed is shown in Figure 5.8. Experience with the *G. sulfurreducens* GEM that accumulated over the ensuing years demonstrated that a GEM for a poorly characterized organism proved to be surprisingly useful.

The metabolic network reconstruction of Geobacteraceae and its GEMs have been used to understand the metabolic physiology of the members of this species and their role in environmental processes. We will describe the improved understanding of acetate and amino acid metabolism resulting from the GEMs. These and other lessons and uses are summarized in detail in [254].

Acetate uptake   Acetate is the primary electron donor and carbon source for Geobacteraceae and thus members of this species have developed strategies to compete effectively for it. One strategy involves the presence of multiple genes (*aplA–D*) that encode the acetate permease transporter responsible for acetate uptake. These genes were identified in the initial reconstruction of the metabolic network of *G. sulfurreducens* based on homology with the characterized acetate permease gene *actP* (formerly *yjcG*) in *E. coli*.

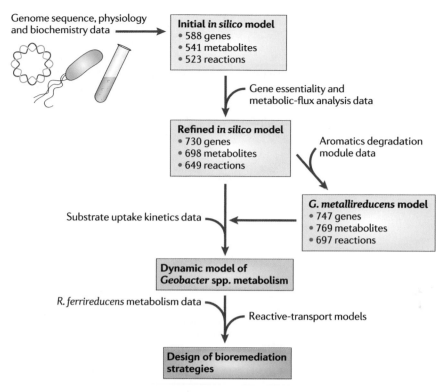

Figure 5.8 Iterative and integrated experimental and computational methodology for studying the physiology and metabolism of *Geobacter* spp. and their environmental impact. The initial model of *G. sulfurreducens* was refined through the comparison with data from gene essentiality and metabolic-flux analyses. This refined model was used for metabolic engineering and physiology studies. Subsequently, the aromatics degradation module was added to obtain a model of *G. metallireducens*. Then, data on the kinetics of substrate uptake were incorporated into the *Geobacter* spp. model to develop dynamic models of microbial ecology, which were used to predict the outcome of microbial competition. The dynamic model of *Geobacter* spp. metabolism can then be integrated with reactive-transport models in order to design bioremediation strategies. Taken from [254].

These genes were subsequently characterized using a combination of genetics, biochemistry, and sequence analysis and their role in acetate uptake was confirmed [18]. Further, these genes were also found to be conserved among the subsurface clades of Geobacteraceae identified from environmental samples [18,165]. The biochemical studies showed that there were three different uptake mechanisms with increasing values for K_M and V_{max}, suggesting that three transporters were active at different acetate concentrations and allowing for optimal acetate transport in a variety of conditions.

Acetate activation In addition to the versatility in acetate uptake systems, the Geobacteraceae are found to have multiple mechanisms for acetate activation for its metabolic incorporation (Figure 5.9A). These mechanisms include: (1) the typical activation pathway involving acetate kinase and phosphotransacetylase, and

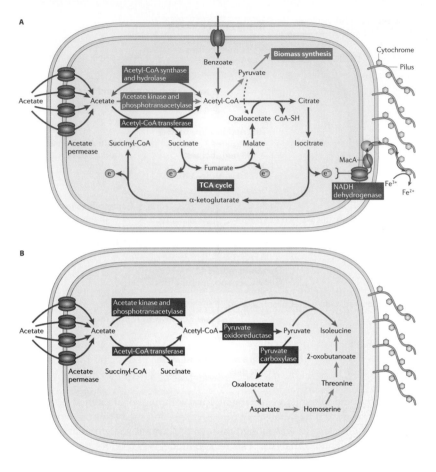

Figure 5.9   GEM-derived insights into acetate and isoleucine metabolism in *Geobacter* spp. Panel A: redundancy in acetate uptake and activation in *Geobacter* spp. There are four acetate transporters and three acetate activation pathways, along with the TCA cycle. Acetyl-CoA for biomass synthesis is obtained through the acetate kinase pathway (green arrows), whereas the acetyl-CoA that is synthesized through the acetyl-CoA transferase is consumed in the citrate synthase reaction in the TCA cycle (gray arrows). In contrast, acetyl-CoA hydrolase (blue arrows) was found to be upregulated during benzoate oxidation in *G. metallireducens*. The current model of extracellular electron transfer through cytochromes associated with pili is also shown. Panel B: most isoleucine is synthesized through the citramalate pathway (blue arrows), which was identified based on ^{13}C isotope-labeling data and model predictions. An alternative isoleucine pathway (green arrows) also exists. The roles of the reactions catalyzed by pyruvate oxidoreductase and pyruvate carboxylase were investigated using the model-based approach. Taken from [254].

(2) acetyl-CoA transferase pathway which has a dual role in the tricarboxylic acid (TCA) cycle. Specific members of Geobacteraceae that utilize aromatics and organics other than acetate, such as *G. metallireducens*, have additional reactions involved in acetate metabolism including the acetyl-CoA synthase and acetyl-CoA hydrolase. These additional reactions are energy-inefficient and if active during growth on

acetate, would retard growth. Thus, the role of the different pathways was not clear initially, but could be elucidated with the guidance of the GEMs.

Elucidation of pathway roles. In order to understand the function of these pathways, the *G. sulfurreducens* GEM was used to predict the metabolic functions of mutants defective in the genes encoding these pathways. These computations were compared with the experimentally observed phenotypes from such mutants [366]. GEM computations showed that the acetate kinase pathway could not substitute for acetyl-CoA transferase defect, suggesting that these two pathways had different roles. Acetyl-CoA transferase had a dual role in acetate activation and the TCA cycle, thereby coupling acetate activation to the TCA cycle flux or the respiration rate. Hence, any increase in the TCA cycle flux is accompanied by an increase in acetate activation and uptake. In contrast, the acetate kinase pathway is required to synthesize acetyl-CoA for pyruvate production via the pyruvate oxidoreductase reaction, as any acetyl-CoA generated by the transferase reaction (ATO) is consumed at the citrate synthase step. Hence, the primary role of acetate kinase pathways appears to be related to pyruvate generation for biomass synthesis. The distinct roles of these seemingly redundant pathways can be confirmed by observations that the gene expression of the acetate kinase pathway is down-regulated in conditions associated with increased respiration, such as during the expression of a futile cycle relative to wild-type and during Fe(III) reduction relative to fumarate reduction.

Analyzing omics data. The analysis of the *G. metallireducens* metabolic network revealed the presence of a multitude of energy-inefficient reactions in the acetate activation pathways. Integration of these pathways with the gene expression data revealed that the energy-inefficient reactions are not expressed during growth with low energy containing acetate, but are active during growth with higher-energy substrates, such as aromatic compounds. This suggests that, during growth with higher-energy carbon substrates, the activities of these reactions are required to generate acetate and to balance the acetyl-CoA and acetate pools for efficient function of the TCA cycle.

Therefore, based on these studies, it is quite clear that Geobacteraceae have evolved a multitude of redundant pathways for acetate uptake and activation, with different functional roles, in order to compete effectively for growth on acetate. The diversity in these metabolic functions is consistent with the lifestyle and growth challenges that this species faces.

Amino acid metabolism In addition to specific metabolic adaptations for growth on acetate, the Geobacteraceae have evolved distinct metabolic pathways that allow increased efficiency of amino acid synthesis that are advantageous for growth on acetate. GEM-guided discovery of two additional metabolic adaptations for the Geobacteraceae growing on acetate have been made.

Role of pyruvate oxidoreductase. One of the challenges in the utilization of acetate, a two-carbon substrate, is that two CO_2 fixation steps are necessary to generate four carbon intermediates for the anapleurotic reactions. In *E. coli*, this problem is solved through the use of the glyoxylate by-pass [115], where two moles of acetyl-CoA are utilized to generate one mole of malate. Geobacteraceae genomes do not have the genes for the glyoxylate by-pass [249]. Instead, they utilize pyruvate oxidoreductase (POR) in the reverse direction to generate pyruvate from acetyl-CoA and CO_2. Reduced ferredoxin generated in the TCA cycle provides the reducing power to provide the thermodynamic driving force to drive the POR reaction in reverse. In addition to utilizing only one mole of acetyl-CoA per mole of pyruvate formed, this step is a CO_2 fixation step and serves to increase the efficiency of amino acid synthesis. GEM-based computations of the amino acid yields with POR resulted in higher yields for almost all of the amino acids as compared to the situation when POR was absent (Figure 5.9B and [249]). These results illustrate an additional critical metabolic adaptation of Geobacteraceae to growth on acetate, without which biomass yields would be significantly reduced.

Citramalate pathway for isoleucine synthesis. Validation of the GEM predictions of metabolic flux distributions can be achieved using ^{13}C isotope labeled substrates (see Chapter 17). Measurements of ^{13}C label incorporation into metabolites were compared with predicted ^{13}C label incorporation and the GEM-derived predictions were found to be consistent with experimental data for all the amino acids, with the exception of isoleucine [283]. The source of this discrepancy was analyzed further and it was determined subsequently that the *G. sulfurreducens* uses the citramalate pathway to synthesize isoleucine. This pathway uses pyruvate and acetyl-CoA as the precursors instead of oxaloacetate that is used as a precursor in the isoleucine synthesis pathway in other organisms such as *E. coli*. Further genetic and biochemical analysis resulted in the elucidation of the genes encoding the citramalate pathway [283].

GEM computations showed that the yield of isoleucine per mole of acetate was significantly higher for the citramalate pathway as compared to the traditional oxaloacetate-based pathway (Figure 5.9B). Interestingly, in contrast, GEM computations also showed very little yield improvement of the citramalate pathway if glucose was the primary substrate. Thus, GEM computations provided detailed insights on the evolution of citramalate pathway in Geobacteraceae to optimally support their lifestyle and growth on acetate.

Recapitulation   Taken together, these detailed examples show how GEMs are useful to study the biochemical adaptation of metabolism to the particular nutritional challenges that an organism faces. They show that straight extrapolation from what is known about model organisms like *E. coli* could lead to incorrect conclusions. The combination of network reconstruction, GEM formulation, computations with GEMs, detailed biochemical and genetic studies, and GEM-enabled interpretation of data leads to the development of genome-scale science and an advancement about the detailed functioning of an organism from a systems perspective.

Geobacteraceae are a class of environmentally relevant bacteria with practical applications in biological energy generation and bioremediation that are linked to their growth and respiration. A GEM of microbial metabolism has been used to elucidate key metabolic features of the microorganisms involved. In addition to the physiological insights, the metabolic modeling approach has also been shown to help with the description of the fate of heavy metal contaminants at the field-scale and other issues as detailed in [254]. The history of development of GEMs for the Geobacteraceae demonstrates that basic and applied results can be obtained through genome-scale reconstruction even for poorly characterized bacteria.

5.6   Summary

- Genome-scale metabolic reconstructions are now available for many prokaryotic organisms, including those of industrial, environmental, and medical importance.
- GEMs built from these reconstructions can be used for many purposes and have been demonstrated to have both basic and applied uses.
- Metabolism in environmental, photosynthetic, and pathogenic organisms has been studied, leading to improved understanding of the role of the metabolic network in the functioning of the organism.
- Initial applications to simple communities suggest that genome-scale reconstructions and GEMs can help to elucidate properties of metabolically interacting microbes.
- Applications of prokaryotic reconstructions are found throughout the text.

6 Eukaryotes

Anything found to be true of E. coli *must also be true of elephants*
— Jacques Monod

The previous chapter described the procedures developed for studying the systems biology of metabolism in bacteria. In parallel, similar efforts have been undertaken for unicellular eukaryotes. The main challenge that arises is the presence of multiple intracellular compartments (organelles), that, in principle, can be dealt with during a reconstruction process, but in practice is difficult due to the scarcity of data of transporters that move metabolites in and out of organelles. The yeast *Saccharomyces cerevisiae* was the first eukaryote to undergo a genome-scale metabolic reconstruction in 2003. This achievement was followed by the reconstruction of other fungal species. A detailed reconstruction of photosynthetic green algae appeared in 2011. Reconstruction of metabolic networks in multicellular organisms have also appeared. The first version of the genome-scale human metabolic map was published in 2007, followed by parallel reconstruction efforts for other mammals. This global human map has since been customized for various cell and tissue types. Interacting models of multiple tissue types have appeared that, in principle, should be able to study systemic metabolism in humans. Therefore, it appears that the network reconstruction procedures that have developed for *E. coli* will extend to multicellular organisms; however, no model of metabolism in the elephant has yet appeared.

6.1 Metabolism in *Saccharomyces cerevisiae*

6.1.1 Reconstruction and its uses

Besides being an industrial workhorse for a variety of biotechnological products, *S. cerevisiae* (baker's yeast) is a well-developed model organism for biochemical, genetic, pharmacological, and post-genomic studies. Several attempts at reconstructing its metabolic network from genomic and literature data have been made, as summarized in Figure 6.1.

History   Shortly after the first pre-genome era *E. coli* models were published, a similar effort was undertaken for yeast. This undertaking is more difficult than for *E. coli*

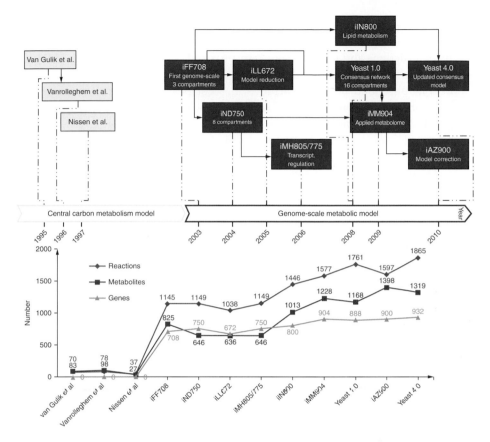

Figure 6.1 Metabolic network reconstruction in *Saccharomyces cerevisiae* over the past 16 years. Each box represents a metabolic model, the larger text is the name of the model and the small text in each box summarizes its scope. The arrows between the boxes show the historical relationship between the models. The graph shows the number of reactions, metabolites, and genes accounted for in the different models. Taken from [301].

because the organism is more complicated, and unlike for *E. coli*, much of the yeast literature is genetic in nature and contains less detailed biochemistry. In addition, the presence of multiple organelles presented the challenge of accounting for the trafficking of metabolites in and out of organelles. These challenges have been overcome and functional models have been formulated and used.

A few years after the publication of the yeast genome in 1997 [143], the first genome-scale reconstruction appeared in 2003 [129] followed by the formulation of a GEM [109]. This GEM had several predictive capabilities for growth and metabolic characteristics of yeast. The first large-scale gene deletion study that analyzed 599 knock-out strains was performed around the same time [130]. The results from this analysis were encouraging as the function of 526 (88%) of the KO strains analyzed were correctly predicted by the yeast GEM.

This sparked a series of expansions and alterations to this initial GEM (Figure 6.1). This reconstruction process happened in parallel in different laboratories and thus differed from the development of the *E. coli* reconstruction that represented a linear, step-by-step process of expansion in scope and coverage of the network. Each of these reconstructions and models have their pros and cons, as discussed in [301]. This discord in the yeast systems biology community eventually led to a reconstruction jamboree, as described below, to achieve the best possible network reconstruction based on community consensus. We note, though, that the rate of incorporation of new metabolic genes has been surprisingly slower than that for *E. coli*, and that the total number of genes accounted for in the latest reconstruction is still less than 1000. Based on this experience with *S. cerevisiae*, a network reconstruction tool, called RAVEN [7], that has advantages for yeast and fungi metabolism has been developed.

Uses   The applications of the *S. cerevisiae* GEM can be grouped into four different categories that are generally similar to those described for *E. coli* and prokaryotes in the previous two chapters: (1) developing strategies for metabolic engineering and strain improvement to achieve overproduction of a specific compound; (2) the use of metabolic models as tools for interpretation and integration of experimental data in order to improve understanding of physiology and biological processes; (3) the development and application of computational methods for genome-scale metabolic model reconstruction and analysis; and (4) the use of GEMs to study evolution both between different species, i.e., evolutionary relationships, but also within *S. cerevisiae*, including the identification and analysis of duplicate genes.

   This success with yeast metabolism represents a development of GEMs from bacteria to that of a single-cell eukaryotic organism. This achievement increased the confidence that GEMs could be successfully developed for even more complex multicellular organisms, as has proven to be the case as described below.

6.1.2   Community-based reconstruction

Due to different approaches utilized in the reconstruction as well as different interpretations of the literature, the various yeast genome-scale metabolic network reconstructions have many differences. In addition, the names chosen for metabolites and enzymes in these reconstructions were inconsistent. This lack of consistency between the reconstructions complicated comparison between them and the use of the corresponding GEMs for data analysis and integration. Thus, members of the yeast systems biology community recognized that a single 'consensus' reconstruction and annotation of the metabolic network was of great community interest.

A reconstruction jamboree   The goal of a network reconstruction jamboree is to reconcile and refine the currently available knowledge base for the target organism in a community consensus manner [428]. To date, reconstruction jamborees have been carried out for three target organisms, *S. cerevisiae* (see below), *Salmonella typhimurium* LT2 [422], and *Homo sapiens* [424].

   If available, multiple existing metabolic network reconstructions made by individual research groups provide a starting point for a jamboree. A jamboree should then, in a systematic manner, update, re-evaluate, refine, and expand the content of the reconstruction. A current workflow for reconstruction jamborees is illustrated in Figure 6.2. It consists of three phases.

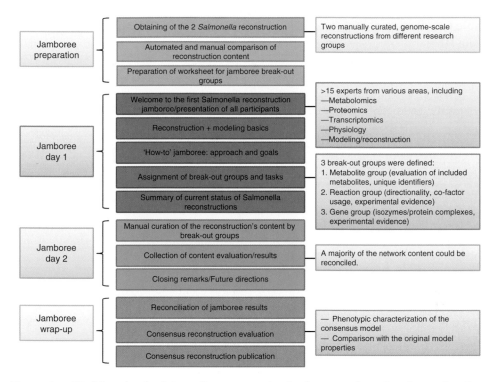

Figure 6.2 Workflow for the *Salmonella* reconstruction jamboree conducted on September 5 and 6, 2008 at the University of Iceland. Taken from [428].

Phase 1: This pre-jamboree phase should establish similarities and differences between existing metabolic reconstructions in terms of metabolites, reactions, and GPRs. This phase may require significant manual evaluation of the content but no reconciliation. The preparation phase should result in worksheets that state the problem that the jamboree team needs to address.

Phase 2: During the jamboree meeting, the participants are divided into at least three groups based on preference and expertise (metabolite, reaction, and GPR group). A fourth group may be established for evaluation of reactions that have no evidence but may be needed for mathematical modeling. Each group should evaluate the material based on evidence given by the reconstruction and available resources (literature, databases, and annotations).

Phase 3: A follow-up phase should include testing of the network functionality and comparison with the prediction capabilities of the initial reconstruction(s). Reconstruction dissemination will also take place in this post-jamboree phase.

Follow-up meetings should be organized to plan further refinement of the consensus reconstruction and form the basis for a new jamboree. Thus, this process leads to a community-based iterative reconstruction process that assembles every few years and curates the information that has been generated on the target organism in the intervening years.

Table 6.1 Summary of the consensus reconstruction by cellular compartment. From [165].

Compartment	Reactions	Metabolites
Cytoplasm	835	590
Extracellular	15	158
Golgi	2	13
Mitochondrion	188	235
Nucleus	30	42
Endoplasmic reticulum	32	28
Vacuole	2	22
Peroxisome	77	80
Mitochondrial membrane	142	0
Plasma membrane	311	0
Peroxisomal membrane	44	0
ER membrane	17	0
Vacuolar membrane	35	0
Golgi membrane	5	0
Nuclear membrane	26	0

Community-based reconstruction for yeast A consensus reconstruction for yeast was developed using the 'jamboree' approach. The overall goal of the jamboree was, by careful curation and comprehensive annotation of the network and its components, to make the consensus reconstruction useful for the broadest possible set of users. The general reconstruction could then be used directly in bioinformatics applications aimed at polyomic data integration or used as a starting point for building GEMs. Thus, a jamboree recognized explicitly the difference between a network reconstruction and its GEM. This first yeast jamboree successfully produced the Yeast 1.0 reconstruction [165].

Contents of the consensus reconstructions The consensus network resulting from the first yeast jamboree consisted of 1168 metabolites, 832 genes, 888 proteins, and 96 catalytic protein complexes, for a total of 2153 species. It had 1857 reactions, including 1761 metabolic reactions and 96 complex formation reactions. The reconstruction contained 15 cellular compartments (see Table 6.1) including membrane compartments to which reactions and metabolites are localized.

The reconstructed network had 664 distinct chemical entities. This is fewer than the total as some metabolites are present in more than one compartment. For example, ATP is present in the nucleus, cytoplasm, Golgi, mitochondrion, peroxisome, and vacuole, and as such represents six metabolites in the total tally, but represents just one chemical species. This reconstruction emphasized the use of unique chemical specifiers that should be included in well-curated models. Unique specifiers allow for direct mapping of metabolomic data onto the reconstructed networks and enables network comparison.

The reconstructed network included 1312 unique chemical transformations, of which 911 occur within a single compartment and the remaining 401 are transport reactions. The transport reactions are the most difficult to identify. As with metabolites, the same reaction can take place in more than one cellular compartment, thus the number of unique reactions and total number of reactions are different.

The overall distribution of metabolites and reactions between the various compartments is summarized in Table 6.1. EC number and PubMed reference annotations are provided for 738 and 478 unique transformations in the network, respectively. Each reaction includes all of its co-factors such as ATP, NADH, and CoA.

Subsequently, the jamboree was repeated, and an expansion in scope and content produced Yeast 5.0, which is the most current version of the yeast metabolic reconstruction [160].

6.2   Metabolism in *Chlamydomonas reinhardtii*

Green algae   Since the 1970s, algae have garnered significant interest for their potential industrial applications in biofuels and nutritional supplements. Among eukaryotic microalgae, *Chlamydomonas reinhardtii* has emerged as the model organism [156]. *C. reinhardtii* has been widely used to study photosynthesis, cell motility and phototaxis, cell wall biogenesis, the circadian clock, and other fundamental cellular processes. In addition to its utility as a model for studying biological processes, *C. reinhardtii* is currently being used as a platform for developing biofuels [137, 383], as an expression system for human protein therapeutics [354], and as a resource for developing optogenetic therapy [59].

Characterizing *C. reinhardtii* metabolism is important in order to study its photosynthesis, understand its biology, and is intimately tied to its industrial applications and engineering production strains. Extensive literature on *C. reinhardtii* metabolism [405] and multiple metabolic mutants [155] provides a solid foundation for the detailed characterization of its metabolic functions. The availability of the complete genome sequence for *C. reinhardtii* [265] and its functional annotation have enabled the identification of a metabolic gene portfolio [151,259]. These resources were employed to reconstruct and validate experimentally the genome-scale metabolic network of *C. reinhardtii* [65], the first network to account for detailed photon absorption permitting growth simulations under different light sources.

6.2.1   Metabolic network reconstruction

The *i*RC1080 genome-scale *C. reinhardtii* metabolic network reconstruction accounts for the activity of 1080 distinct proteins. Of the putative protein-coding genes in the *C. reinhardtii* genome, an estimated 20% function in metabolism. *i*RC1080 accounts for the activity of more than 32% of the estimated genes with metabolic functions. A major feature emerging from *i*RC1080 is the relative centrality of the chloroplast and its importance in light-driven metabolism (Figure 6.3). The chloroplast accounts for more than 30% of the total reactions in the network and 9 of the 10 photon-utilizing reactions. The thylakoid contains essential pathways for photoautotrophic

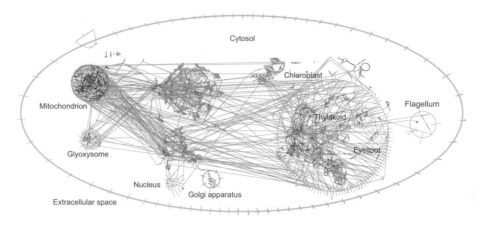

Figure 6.3   *C. reinhardtii* metabolic network reconstruction (*i*RC1080). From [65].

growth including photosynthesis, chlorophyll synthesis, and carotenoid synthesis. The eyespot accounts for retinol metabolism, the mechanistic basis for phototaxis. *i*RC1080 contains a detailed reconstruction of lipid metabolic pathways. It accounts explicitly for all metabolites in these pathways, providing sufficient detail to specify completely all individual molecular species.

6.2.2   Description of photon usage

The quality of light sources used in photobioreactors largely determines the efficiency of energy usage in industrial algal cultures. Light spectral quality also affects how photon absorption induces various metabolic processes: photosynthesis, pigment and vitamin synthesis, and the retinol pathway. *i*RC1080 integrates biological and optical data to produce a novel light-modeling approach that enables quantitative growth predictions for any light source of interest, resolving wavelength and photon flux. This unique detail for photosynthetic metabolism is summarized in Figure 6.4.

Metabolic reaction activity spectra are used to bin light source emission spectra into effective ranges that can drive photon-dependent reactions in the network. The proportion of light emitted within an effective range relative to the entire visible spectrum defines an effective bandwidth coefficient. Each range of photons effective in driving a specific photon-dependent reaction in the network is represented as a metabolite in the network (e.g., photon298). Prism reactions of the form described in Figure 6.4B are used to model a specific light source to be used in growth simulations, such that total photon flux emitted from the simulated light source is distributed among the effective ranges to supply the necessary wavelengths of photons to act as reactants in photon-utilizing reactions.

Using prism reactions to describe distinct light sources, growth can be simulated under different lighting conditions. Simulated oxygen photoevolution during simulated growth under sunlight of varying intensities showed close agreement with experimentally measured data (Figure 6.5A). Predicted biomass yields under incandescent white light and 674nm peak LED light were on the same order of magnitude as experimentally measured yields for those lighting conditions. This agreement is notable given that the network explicitly accounts for the spectral photon

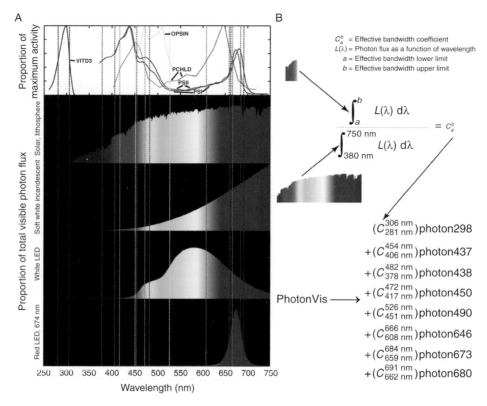

Figure 6.4   Analysis of light spectra. (A) Activity and irradiance spectra. (B) Prism reaction derivation. From [65].

flux of these light sources and the subsequent processing of this energy to generate all of the constituents of biomass without any parameter fitting to the experimental data.

The efficiency of light utilization was also evaluated using the *C. reinhardtii* GEM under several light sources for which no experimental growth measurements were available. The biomass yields on light were computed for each light source given the minimum incident photon flux required to achieve growth rate saturation (Figure 6.5B). This analysis demonstrates the prospective extensibility of the network and modeling approach to any possible lighting condition and also the potential for designing optimal light source parameters, for example, as depicted for the 'efficient LED design' in Figure 6.5B.

The *i*RC1080 network and light modeling framework represent the first attempt to explicitly account for photon usage in a GEM. This achievement is encouraging for further incorporation of physical environmental parameters into *in silico* metabolic models.

6.3   Metabolism in *Homo sapiens*

The Human Genome Project (HGP)   The HGP took 20 years and $3billion dollars to complete. The sequence was built in a step-wise fashion with each iteration issued as a Build *X*, where *X* represented the particular version. Build 35 appeared in 2004 [81],

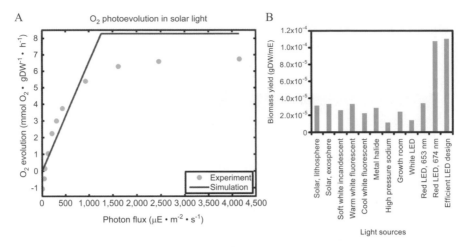

Figure 6.5 Photosynthetic model simulation results. (A) O_2 photoevolution under solar light. Simulated (blue line) and experimentally measured (green dots) O_2 evolution are compared. (B) Efficiency of light utilization. The minimum photon flux required for biomass yield is presented for 11 light sources derived from measured spectra and for the designed growth-efficient LED. From [65].

contained well over 99% of the euchromatin sequence, and was accompanied by a number of online databases containing its annotation. The metabolic genes could thus be identified and the human metabolic network reconstructed on a genome-scale [100]. Since then, an updated reconstruction has appeared [424].

Metabolism is foundational to human health and disease Metabolism plays an important role in all aspects of human physiology. Metabolic function is important for understanding disease states and progression, aging and nutrition, and for improving the performance of people operating at the limits of human capabilities such as athletes, astronauts, and soldiers. Metabolism is known to be involved in the major human diseases: diabetes, obesity, cancer, and cardiovascular disease. More recently, growing evidence shows the involvement of metabolism in physiological and patho-physiological brain functions, from schizophrenia to neurodegenerative disorders. Successful implementation of molecular systems biology of human metabolism is thus likely to have broad consequences.

6.3.1   Recon 1

The reconstruction process The reconstruction process started with an automated extraction of genes annotated to have metabolic functions (step 1 in Figure 6.6). This was followed by manual curation (step 2) where a team of researchers simultaneously curated network components by evaluating over 50 years of biological evidence from the extensive data available at the time. Quality control/quality assurance methods were used throughout the reconstruction. A manual literature-based reconstruction ensured that the network components and their interactions were based on direct physical evidence and reflected the current knowledge of human metabolism. An initial GEM was formulated (step 3) and used to validate the basic functionality of the human metabolic network by simulating 288 known metabolic functions in humans

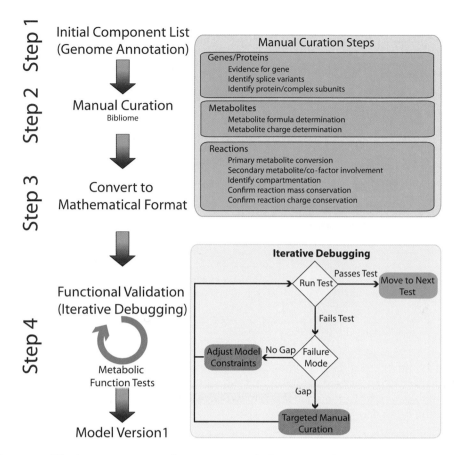

Figure 6.6   The four-step process of reconstructing the human metabolic map deploying 288 functional tests. Prepared by Neema Jamshidi.

(step 4). It took five iterative rounds of reconstruction and validation to produce Recon 1, the first genome-scale reconstruction of *H. sapiens* metabolism. Recon 1 was almost entirely constructed from human-specific data.

Recon 1 is a multi-scale knowledge base   Recon 1 has extensive molecular component information. It accounts for 1496 open reading frames, 2004 proteins, 2712 metabolites, and 3311 metabolic reactions (Figure 6.7). Detailed information about the metabolites and their chemical structures is included. The reconstructed network contains proper chemical reactions that are mass- and charge-balanced. The genetic basis for metabolism is also included. Gene–protein–reaction (GPR) annotations are provided, linking genes to the reactions. GPRs are a Boolean representation of the gene, transcript, protein, and reaction relationship accounting for splice variants, isozymes, and protein complexes (Figure 6.7A).

Basic cell biology is represented. Recon 1 is compartmentalized, and accounts for seven cellular compartments, including the cytoplasm, nucleus, mitochondria, lysosome, peroxisome, Golgi apparatus, and endoplasmic reticulum. Reactions are assigned a cellular localization. Some reactions take place in many compartments, and many metabolites are found in multiple compartments. Knowledge of intracompartmental transport of metabolites needs to be improved.

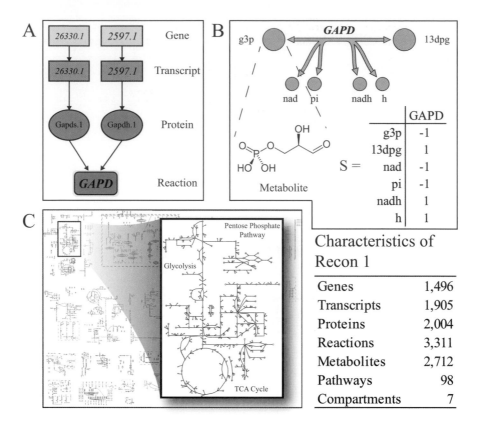

Figure 6.7   Recon 1 is a global human metabolic network reconstruction containing the known biochemical and physiologic data. (A) Gene–protein–reaction (GPR) associations can be represented in Boolean logic. GPRs are essential for determining phenotypes of genetic perturbations as well as understanding the underlying mechanisms of a particular phenotype. (B) Recon 1 accounts for 3,311 metabolic reactions and their associated metabolites. The reactions in the network can be represented in a mathematical format called the stoichiometric matrix. (C) Recon 1 is a thorough and very complex assessment of human metabolism accounting for 1,496 genes and 7 cellular compartments. From [53].

The reconstruction can be represented by a large map. This map introduces the hierarchical concepts in network biology, namely those of a reaction, pathway, module, subsystem, and genome-scale network. Once inputs and outputs are defined, this network can be converted into a computational model enabling many systems biology studies of metabolic functions.

6.3.2   Uses of Recon 1

As for the reconstructions discussed in this and previous chapters, Recon 1 has been used for many applications. These are summarized in Figure 6.8 and described in a bit more detail below [53].

Integration of high-throughput data for model construction   Recon 1 has been used for constructing cell- and tissue-specific models to study their unique physiology (Figure 6.8I). To do so, omics data from the target tissues are mapped onto Recon 1 to

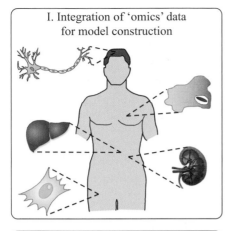

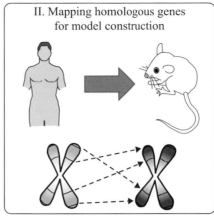

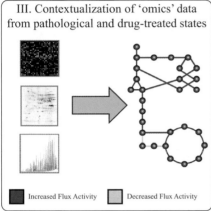

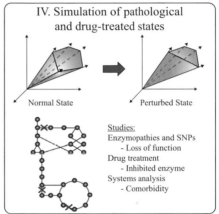

Figure 6.8 Applications of Recon 1. (I) Utilizing high-throughput data, Recon 1 can be tailored to cell- and tissue-specific networks. (II) Recon 1 has been transformed into other mammalian metabolic network reconstructions, particularly *Mus musculus*. The high overlap of homologous genes in Recon 1 with similar mammals enables the reconstruction of accurate mammalian models quickly. (III) Omics data can be interpreted by mapping the data onto Recon 1's metabolic network backbone. (IV) Recon 1 can be used to simulate and predict phenotypes, providing biological clues to physiology and pathology as well as guiding experimental design. From [53].

determine the cell-specific reactions needed to build a cell-specific model. Algorithmic approaches have been developed to quickly build cell- and tissue-specific models. An atlas of such tailored models has been developed [6], and models of cancer cell lines can be developed [124]. Such models have been used for myriad applications ranging from host–pathogen interactions [48], to brain metabolism [233], to the discovery of drug targets in cancer [133].

Mapping homologous genes for model construction   As detailed in the next section, the homology of metabolic genes between mammalian species allows one to map the gene content of Recon 1 onto other mammalian genomes to generate a list of metabolic genes in that mammal. Specific legacy data from the target mammal are then used to curate the reconstruction.

Contextualization of omics data from pathological and drug-treated states As stated above, metabolism plays a prominent role in many human disorders and diseases, including cancer, diabetes, obesity, infection, neural disorders, and inherited gene and enzyme deficiencies. Many studies have used Recon 1 and its derivatives for (i) interpreting context omics data and experiments for biological discovery, or (ii) simulating phenotypes mathematically to guide experiments. In particular, this approach has been used to study pathology and drug effects. Some of the algorithms developed are summarized in Table 6.2.

Simulation of pathological and drug-treated states Recon 1 has been used to simulate many medical situations of interest. We will describe two such applications, and many more are described in [53].

The systemic effect of inherited inborn errors of metabolism has been studied. The Online Mendelian Inheritance of Man (OMIM) database [17] catalogs all known hereditary morbid SNPs. The effects of causal SNPs can be simulated by reducing the flux of the affected reactions [395] identifying a systemic effect on the network, particularly in changes to the variability of the substrate uptake and secretions. Thus, the network-based predictions can be used to determine changes in biofluid metabolite levels revealing potential biomarkers.

Recon 1 has also been used to study cancer metabolism and identification of potential pharmaceuticals. Synergies between drug targets were computed using a generic cancer metabolic network and Recon 1 [124]. It was found that selective targets could be predicted that affected the cancer model preferentially, based on metabolic auxotrophies. This approach was used to predict selective drug targets in renal cell cancer that is deficient in fumarate hydratase. A pathway for heme biosynthesis and degradation was implicated as essential in renal cell cancer, but not normal cells. The computational prediction was followed up with experimental validation in both cell line and mouse models [133].

6.3.3   Building multi-cell and multi-tissue reconstructions

Multiple cell types The tissue microenvironment is characterized by many cell types that can be tightly coupled metabolically. A workflow for the construction of multi-cell type models of metabolism has been developed [233]. This workflow for generating multi-cell type models (Figure 6.9) consists of the following four steps.

Step 1: Reconstruct a metabolic network for the target organism; Recon 1 was used in this case.

Step 2: Develop cell-type specific models based on context-specific data; as discussed above.

Step 3: The different cell types are linked with transport reactions, as supported by literature and experimental data. Initial context-specific reconstructions are incomplete and may contain false positives owing to contamination from proximal tissue.

Step 4: Simulation and analysis. A validated model is then ready for use. For example, a multi-cell type model of cell types in the human brain has been used to: (i) predict disease-associated genes, such as glutamate decarboxylase;

Table 6.2 Building submodels using omics data.

Type	Context-specific model		Global model		Differential models
Algorithm	GIMME	iMat	GimmeP	MBA	MADE
Omic data type	Transcriptomic	Transcriptomic	Trans/Proteomics	Multi-omics	Transcriptomic
Citation	[36]	[396]	[50]	[191]	[187]
Description	Reconciles expression data with a user-defined objective function	Reconciles expression data with number of active/inactive reactions, ignoring functionality	Reconciles expression data with proteomic data-defined reactions	Constructs a minimal, functional model from user-defined 'core reactions'	Builds two models based on differential expression, removing need to define expression thresholds of activity

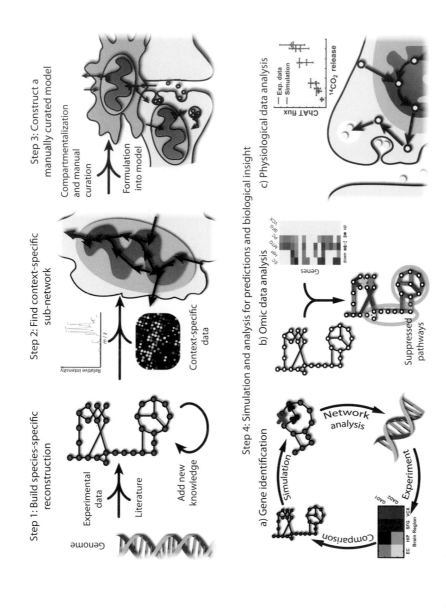

Figure 6.9 Schematic of a workflow for reconstruction of a multi-cell metabolic model and its uses. From [233].

(ii) analyze omics data to identify sets of genes that change together and affect specific pathways (such as the brain-region-specific suppression of central metabolism in Alzheimer's disease patients); (iii) analyze physiological data in the context of the model, thereby enabling, for example, the identification of tissue properties relevant to disease treatment, such as the calculation of the percentage of the brain that is cholinergic (see [233]).

The detailed multi-cell type model of the human brain was the first of its kind. Such models of metabolic interactions and subtasking in the tissue microenvironment is now possible using this workflow provided that there are sufficient polyomic data available and extensive bibliomic data on physiological tissue functions.

Multiple tissue types Similar procedures have been developed to study metabolic interactions between multiple organs. The steps taken are similar to those taken for studying the multiple cell type interaction in the tissue microenvironment. As an example, we describe below the formulation of a multi-tissue type reconstruction that contains the major tissues relevant to the study of diabetes.

Setting up individual tissue type models. First, three cell-specific metabolic networks of the dominant cell type in the three major tissues (hepatocyte from liver, myocyte from skeletal muscle, and adipocyte of adipose tissue) are reconstructed (Figure 6.10A). These three tissue-specific reconstructions have properties that can be broken up into three main categories as summarized (Figure 6.10B). The first category

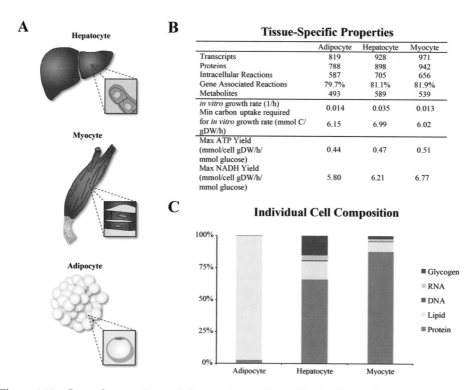

A

Hepatocyte

Myocyte

Adipocyte

B

Tissue-Specific Properties

	Adipocyte	Hepatocyte	Myocyte
Transcripts	819	928	971
Proteins	788	898	942
Intracellular Reactions	587	705	656
Gene Associated Reactions	79.7%	81.1%	81.9%
Metabolites	493	589	539
in vitro growth rate (1/h)	0.014	0.035	0.013
Min carbon uptake required for *in vitro* growth rate (mmol C/gDW/h)	6.15	6.99	6.02
Max ATP Yield (mmol/cell gDW/h/ mmol glucose)	0.44	0.47	0.51
Max NADH Yield (mmol/cell gDW/h/ mmol glucose)	5.80	6.21	6.77

C

Individual Cell Composition

- Glycogen
- RNA
- DNA
- Lipid
- Protein

Figure 6.10 General properties and characteristics of individual cell-specific metabolic reconstructions. From [49].

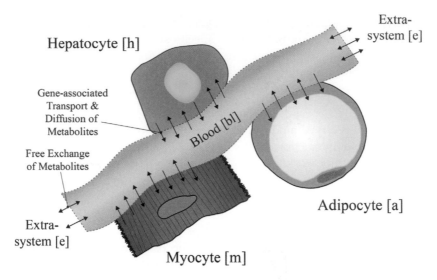

Figure 6.11 Schematic of the multi-tissue modeling approach. From [49].

gives the content of the metabolic networks reconstructed. The second category has the *in vitro* growth rate and the required non-glucose carbon amount to maintain that growth rate. This growth rate was used as a maintenance function to model biological turnover *in vivo*. The third category details the energy and oxidative capacities of the networks. Note that the growth rate was fixed as a constraint for these simulations as a metabolic demand. The biomass maintenance functions for the three metabolic networks were built based on the individual dry cell weight compositions given in Figure 6.10C. Note that the adipocyte is primarily composed of lipids while the myocyte is mostly protein, but the hepatocyte is a more balanced composition of protein, glycogen, and lipids. Thus, the biomass composition of the three cell types places very different metabolic demands on the corresponding tissue.

Integration of the tissue type models. The three cell-specific reconstructions representing the three major tissues are then combined into a multi-tissue model by connecting them all to a common blood compartment. Metabolites enter the multi-tissue metabolic network through the extra-system indicated, through exchange reactions. Metabolites are then imported into the three different cell types through gene-associated intercellular transporters and/or free diffusion. Note that for differentiating the cell-specific models, all reactions in the reconstruction were annotated with [a], [h], [m], and [bl] for the adipocyte, hepatocyte, myocyte, and the blood compartment, respectively (Figure 6.11). A unique metabolite molecule can exist in the different cell types and can be separately accounted for. The same reaction can take place in different cell types. This gives a multi-tissue type reconstruction a similar characteristic as described for the multi-organelle reconstruction for yeast earlier in the chapter. We note that the multi-tissue model is not simply a trivial sum of the cell-specific reconstructions. For example, a bicarbonate buffering system was required in the blood compartment to obtain a functioning simulator of this multi-tissue reconstruction, thus introducing a key physiological function of blood.

Studying physiological states. The integrated reconstruction of the three tissue types can be used to describe certain systemic features of human metabolism as shown in

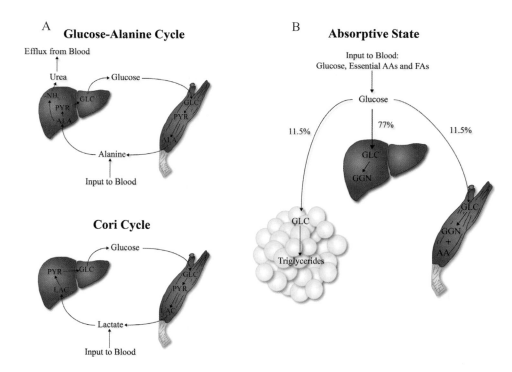

Figure 6.12   Physiological states of systemic human metabolism. (A) The Alanine and Cori cycles of human metabolism. (B) The absorptive state of human metabolism. From [49].

Figure 6.12. The alanine and Cori cycles are cyclic metabolic processes that enable peripheral tissues to receive glucose under nutrient-limited situations. Gluconeogenic substrates (e.g., alanine and lactate) are released from peripheral tissues and absorbed by the liver to produce glucose. The glucose is then returned to the peripheral tissues for their metabolic requirements (Figure 6.12A).

The adsorptive state of human metabolism can also be described with a multi-tissue metabolic reconstruction (Figure 6.12B). In the absorptive state, food is digested and absorbed into blood primarily as glucose and amino acids. The figure shows the influx of glucose into the blood and the computed fractions of uptake by each of the three tissue-specific compartments. Essential amino acids and fatty acids are also provided. Physiological information is included in the reconstruction of the multi-tissue model: the adipose tissue stores triacylglycerol, the muscle stores protein and glycogen, and the liver stores glycogen. Some of the glucose delivered to the liver was converted to fatty acids that are transported to the adipose for triacylglycerol production.

Applications. A multi-tissue-type reconstruction can be converted into a mathematical model that can simulate various systemic metabolic processes. Genetic, nutritional, and other perturbations can be introduced to this model and the physiological consequences assessed.

The multi-tissue metabolic reconstruction can be used to assess the metabolic differences in obese and diabetic obese individuals [49]. Gene expression data from cohorts of obese and diabetic obese individuals were used as a constraint to generate specific versions of the reconstruction for each group. These cohort-specific reconstructions were consistent with physiological observations and they provided suggestions of potential underlying mechanisms for known macroscopic physiological changes seen in diabetic patients.

This multi-tissue model was the first of its kind, and it is likely that increasingly more detailed and validated reconstructions will appear that can describe systemic metabolic functions in humans in more and more detail.

6.3.4   Mapping Recon 1 onto other mammals

The contents of Recon 1 have been mapped to other mammals [397]. Such mapping is performed by finding the corresponding metabolic genes in the genome sequence of the target mammal by a homology search. This mapping process enabled the creation of initial draft metabolic reconstructions of five mammals, including the mouse, the rat, the cow, the chimpanzee, and the dog.

Extensive legacy information for the mouse enabled a curation process of its draft metabolic reconstruction. A functional mouse metabolic model was built by iterative testing until it passed a predefined set of 260 validation tests through a process similar to that used for the formulation of Recon 1 (Figure 6.6). This effort resulted in the largest and most comprehensive metabolic reconstruction available for the mouse, accounting for 1415 genes coding for 2212 gene-associated reactions and 1514 non-gene-associated reactions. This reconstruction was named *i*MM1415, for *Mus musculus*.

*i*MM1415 can be used for a range of applications as those that have appeared for human metabolism discussed above. Of particular interest is the availability of genetically altered strains for the mouse. Thus, the *i*MM1415 mouse metabolic model offers a better opportunity than the human model to examine the detailed genetic basis for some metabolic functions in mammals. A computational model of *i*MM1415 was examined for its phenotype prediction capabilities from known metabolic gene deletion strains. The majority of computed essential genes were also essential *in vivo*; thus, the initial true positive rate was high. However, non-tissue-specific models were unable to predict gene essentiality for many of the metabolic genes found to be essential *in vivo*. 'Softer' phenotypes can also be simulated, such as those associated with the deletion of the lipoprotein lipase gene [397].

Thus, initial progress is being made to use the detailed and validated human reconstruction for other mammalian species.

6.3.5   Recon 2

Recon 1 represents the initial reconstruction of the global human metabolic map, published in the fall of 2004. Since then, much more information for human metabolism has become available. Given the successful uses of Recon 1, an updated version was generated through a jamboree process [424]. Recon 2 accounts for 1789 enzyme-encoding genes, 7440 reactions, and 2626 unique metabolites distributed over 8 cellular compartments and thus represents a notable increase in scope and coverage relative to Recon 1. A detailed comparison of the content of Recon 1 and 2 is given in Table 6.3. Recon 2 is thus an updated and improved version of the global human metabolic network. It has several new features and capabilities, three of which we describe here.

Table 6.3   Comparison of the contents of Recon 1 and Recon 2. From [424].

Property	Recon 1	Recon 2
Total number of reactions	3744	7440
Total number of metabolites	2766	5063
Number of unique metabolites	1509	2626
Number of metabolites in extracellular space	404	642
Number of metabolites in cytoplasm	995	1878
Number of metabolites in mitochondrion	393	754
Number of metabolites in nucleus	95	165
Number of metabolites in endoplasmic reticulum	235	570
Number of metabolites in peroxisome	143	435
Number of metabolites in lysosome	217	302
Number of metabolites in Golgi apparatus	284	317
Number of transcripts	1905	2194
Number of unique genes	1496	1789
Number of subsystems	90a	99
Number of blocked reactions (% of all reactions)	1270 (34%)	1603 (22%)
Number of dead-end metabolites	339b	1176b
Size of **S** (the stoichiometric matrix) (m, n)	(2766, 3744)	(5063, 7440)
Number of linearly independent mass balances	1070	2774
Sparsity (% of non-zero entries in **S**)	0.1385	0.0837
Number of accomplished metabolic tasks	294	354
Mapped IEMs (% of all IEMs)	233 (71%)	248 (76%)
Number of unique genes causing IEMs	255	272
Number of IEMs affecting metabolic tasks (% of effective IEMs)	71 (36%)	98 (44%)

Recon 2 covers the majority of known exometabolites   Recon 2 accounts for 642 extracellular metabolites. They should be detected experimentally in cell culture media or in biofluids, such as plasma and urine. Recon 2 extracellular metabolites can be compared with a reported 140 metabolites found in cancer exometabolomes [180]. The majority of these 140 metabolites are present in the extracellular compartment of Recon 2. A further comparison can be made between the Recon 2 exometabolome and those reported in the Human Metabolome Database (HMDB) [458] as being detectable in biofluids. Biofluid information could be found for about half of the metabolites in Recon 2 identified in the HMDB database.

Recon 2 includes mapping of drug actions to enzymes   Many of the 2657 metabolic enzymes and 1052 enzymatic complexes in Recon 2 are known drug targets. DrugBank

[457], a comprehensive resource that includes drug-to-enzyme mappings for over 6000 small molecule and peptide/protein drugs, can be compared to the content of Recon 2. Such comparison reveals that 1290 drug targets map to 308 enzyme and enzymatic complexes. As described in the following chapter, such associations allow the incorporation of 'ligand to protein to reaction' associations (LPRs) in metabolic reconstructions that, in turn, allow for the assessment of drug intervention in metabolic functions.

Recon 2 and inborn errors of metabolism (IEMs)   A gold standard compendium for IEMs [368] can be used to assess Recon 2's ability to analyze IEMs and how it exceeds that of Recon 1. This compendium accounts for 235 IEMs, such as phenylketonuria and orotic aciduria, along with their known metabolite biomarkers. As Recon 2 captured more metabolic genes than Recon 1, more IEMs could be mapped and studied using Recon 2 (Table 6.3), and the vast majority of the mapped IEMs affect reaction activity in Recon 2. A comparison between the predictive potential of Recon 2 and Recon 1 for IEM-associated biomarkers has been performed [424] using a process that is analogous to gene-deletion studies used in microbial modeling and as described in Chapter 4. Remarkably, Recon 2 predicted 54 reported biomarkers for 49 different IEMs, with an accuracy of 77%.

Recon X   As for other model organisms discussed in this text, the reconstruction of the global human metabolic map is expected to go through a series of updates. Given the growing recognition of the importance of human metabolism in health and disease, it is likely that we will see a series of steadily improved reconstructions of human metabolism appear in the years to come.

6.4   Summary

- Genome-scale metabolic reconstructions are now available for several eukaryotic organisms, including yeast, green algae, and mammals.
- For prokaryotes, GEMs built from these reconstructions can be used for many purposes and have been demonstrated to have both basic and applied uses.
- For multicellular organisms, a genome-scale reconstruction can be tailored to a particular cell or tissue type based on expression profiling or proteomic data. Several tissue-specific models have appeared for humans.
- Multiple cell-specific models can be linked together to study the tissue-microenvironment.
- Multiple tissue-specific models can be linked together to study systemic metabolism.
- A variety of objective functions can be used for computation, other than the biomass growth objective function that is so frequently used to study the behavior of prokaryotes

7 Biochemical Reaction Networks

Studies at higher system levels are likely to inform those at the simpler level of the cell and vice versa – Sir Paul Nurse

The procedures for the bottom-up reconstruction of metabolic networks are well developed. While it is a major and universal cellular process, metabolism is only one example of a cellular function for which we would like to have a genome-scale reconstruction. The established reconstruction procedures for metabolism are now being expanded and adapted for other biochemical reaction networks. Although these efforts are in their early stages, we include in this chapter a description of their status because they are indicative of what is to come. We first describe proteins and some of their properties and how they can be included in a metabolic reconstruction. Then we describe how the process of transcription and translation can be detailed in chemical, and thus stoichiometric, terms. All reconstructed networks in a target organism can be integrated seamlessly within a stoichiometric framework.

7.1 Protein Properties

Metabolic network reconstructions describe the chemical transformations that take place among the metabolites in a metabolic network. These reactions occur on the surface of a protein (Figure 7.1). Thus, if properties of proteins can be included in the way a reaction is carried out, then such properties can be included in a reconstruction. If these properties can be stated mathematically then they can be included in a computational model of the network reconstruction.

The chemistry that takes place on the surface of a protein is influenced by many characteristics of the protein molecule. There are many properties of enzymes (Table 7.1) that would be desirable to include in a network reconstruction. Some properties are already built into the GPRs. These would include the specificity versus promiscuity of an enzyme.

Enzyme specificity and efficiency   The need for enzyme efficiency and activity can thus be assessed using a network model. During the early stages of evolution, enzymes are expected to have exhibited broad substrate specificity and low catalytic efficiency.

Table 7.1 Enzyme properties of interest. Modified from [242].

Activity	Stability	Specificity	Efficiency
Turnover frequency (k_{cat})	Temperature stability	Substrate range	Space–time yield
Specific activity (kat/kg, U/mg)	pH stability	Substrate specificity (K_m, k_{cat}/K_m)	Product inhibition
Temperature profile	Ingredient/ by-product stability	Substrate regioselectivity and enantioselectivity	By-product/ ingredient inhibition
pH profile	Solvent stability	Substrate, conversion (%), yield	Producibility/ expression yield

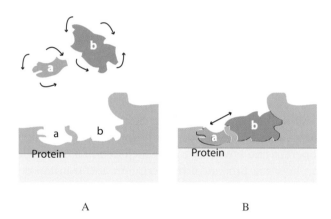

A   B

Figure 7.1 A schematic showing how the binding sites of two molecules on an enzyme bring them together to collide at an optimal angle to produce a reaction. (A) Two molecules can collide at random and various angles in free solution. Only a fraction of the collisions lead to a chemical reaction. (B) Two molecules bound to the surface of a reaction can only collide at a highly restricted angle, substantially enhancing the probability of a chemical reaction between the two compounds. Redrawn based on [240].

Through genetic changes and selection, promiscuous enzymes are thought to have evolved to exhibit specific and more efficient catalytic abilities. Thus, today's metabolic enzymes are commonly assumed to be 'specialists,' having evolved to catalyze one reaction on a unique primary substrate in an organism.

However, many of today's enzymes are 'generalists' that promiscuously catalyze reactions on a variety of substrates *in vivo* or exhibit multifunctionality by catalyzing multiple classes of reactions. In fact, by analyzing the GPRs in a genome-scale *E. coli* metabolic reconstruction, one finds that 37% of the enzymes act on a variety of substrates and catalyze 65% of the metabolic reactions [277]. Thus, a fundamental question arises: why do some enzymes evolve to become specialists, whereas others retain generalist characteristics?

By analyzing enzyme functions and properties from experimental data and by using GEMs, it has been shown that the *in vivo* biochemical network context in which an enzyme operates may influence the evolution of enzyme specificity [277]. Such a systems biology-based assessment of the need for specialization and efficiency shows that there are marked differences between generalist enzymes and specialist enzymes [277]. Specialist enzymes (i) are frequently essential, (ii) maintain higher metabolic flux, and (iii) require more regulation of enzyme activity to control metabolic flux in dynamic environments than do generalist enzymes. Specialist enzymes thus are found at locations in the network where selection pressure on performance may be higher than elsewhere in the network.

It is interesting to note that mutations can come with gain-of-function (GOF) or loss-of-function (LOF) of catalytic capabilities of enzymes. Both LOF and GOF mutations would alter the reactome and would be reflected in an altered association between a protein complex and the reactions that it can carry out. A well-known recent example is a GOF mutation that leads to a new reaction catalyzed by isocitrate dehydrogenase (ICDH) that produces a novel metabolite that in turn inhibits DNA methyl-transferases and thus the epigenetic status of a cell. Such changes are implicated in tumorigenesis [250, 461].

Enzyme specificity and efficiency can thus be described and studied within a metabolic reconstruction. This ability provides an impetus for analyzing other enzyme properties in a network context. Including other protein properties requires us to have structural information about the protein molecule.

7.2   Structural Biology

Textbooks on biochemistry and molecular biology are dominated by two types of images: pathway maps and structures of molecules, such as those shown in Figure 7.2. The reconstruction process described in previous chapters leads to a genome-scale representation of the former. Recently, the vast amount of structural data on macromolecules in a target organism is finding its way into reconstructions enabling expansion in the scope and use of genome-scale network reconstructions. This combination of network and structural biology is opening up new vistas in systems biology. We discuss three new capabilities that such integrated reconstructions offer.

Evolutionary distribution of folds   *Thermotoga maritima* is a thermophile that has been the subject of large-scale protein crystallization and structural determination efforts. The inclusion of a large number of protein structures enables the study of the distribution of protein folds throughout a network and the associated evolutionary implications [473]. The conceptual basis for such integration is shown in Figure 7.2, and it forms the basis for what is called *structural systems biology*. Achieving a description on these two familiar views of molecular biology enables large-scale analyses, such as the network-scale comparison of correlations between protein fold conservation and biochemical function.

Protein folds describe similar spatial arrangements of secondary structures found in the proteins. They are used for structural classification of proteins. Protein evolution

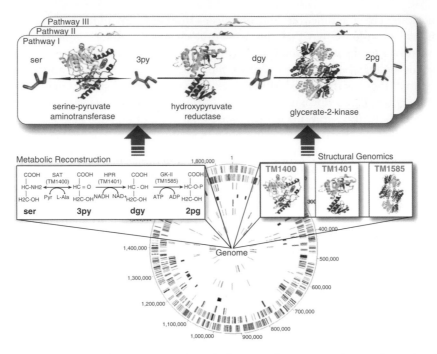

Figure 7.2 Genome-scale integration of metabolic network reconstructions and associated protein structures. Underlying genomics information (bottom) enabled both a metabolic reconstruction (left) and an atomic-level structure determination/modeling of *T. maritima* proteins (right). Integration of these two approaches enabled detailed information to be acquired for every reaction in the network (top); an example from the *T. maritima* serine degradation pathway is illustrated. Taken from [473].

is thought of as having occurred not just through point mutations, but systematic rearrangement of such folds or structural motifs to form new protein functions.

A study [473], enabled by the integrated structural–metabolic network reconstruction of *T. maritima*, provided a quantitative estimate of the dominance of the patchwork model [188] versus the retrograde model [172] of metabolic enzyme evolution [473]. The conceptual basis for pathway emergence for these two models is shown in Figure 7.3. Furthermore, this study showed that the set of proteins responsible for the central metabolism in *T. maritima* is highly non-random and dominated by a small number of protein folds. This finding, in turn, suggests that the central metabolism network has evolved mainly from a set of the most ancient proteins that have had sufficient time to develop divergent functionalities.

Thermal stability If protein structures are available, their temperature sensitivity can be predicted with a reasonable degree of accuracy [96, 337]. For the *E. coli* reconstruction, there are experimental structures available for about one-third of the enzymes, and 59% have computed structures (Figure 7.4). Based on this information, one can compute the temperature-dependence of all these enzymes, and overlay them as temperature-dependent functions of all the reactions in *i*JO1366. The temperature-dependency of the growth rate can then be computed (Figure 7.5A),

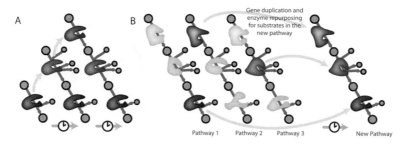

Figure 7.3 Various theories exist to describe the emergence of metabolic pathways. (A) The retrograde model suggests that through evolution, duplication events of neighboring genes extended pathways, starting from a key metabolite. (B) In the patchwork model, it is believed that novel pathways arise from the recruitment and duplication of broad-specificity enzymes, thus forming a new pathway. Gray arrows represent duplication events and mutation. Adapted from [276]. Prepared by Nathan Lewis.

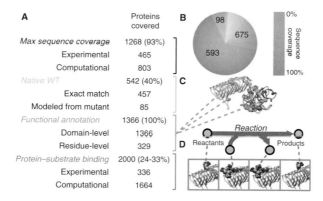

Figure 7.4 Properties of the *E. coli* metabolic model integrated with protein structures. (A) The *E. coli* GEM-PRO model provides maximal amino acid sequence coverage, native WT structures, functional annotation, and protein–substrate binding for proteins included in *i*JO1366. Percentages are out of 1366 total proteins, except for the percentage of protein–substrate binding pairs, which is out of an estimated total between 6144 and 8448 such pairs. (B) The distribution of maximum amino acid sequence coverage of proteins by structures included in the GEM-PRO. Numbered wedges indicate the number of proteins with 0%, 100%, or partial sequence coverage. (C) Example of a native WT structure included in the GEM-PRO. Green highlighted residues denote annotated functional sites. (D) Protein–substrate binding is structurally represented as the pairwise interactions between each protein and the reactants or products of the catalyzed metabolic reaction. Taken from [66].

and the computations compare well with the measured growth rate of the optimal temperature range where the thermal stability predictions apply.

The network properties that constrain optimal growth can be identified by COBRA methods covered in Part III. These methods can be used to identify the reactions in the network that constrain growth as the temperature is raised from 37°C to 42°C. These reactions are called *hot spots* and are shown on a network map (Figure 7.5B). These hot spots occur predominantly in biosynthetic pathways for cofactors and prosthetic groups. These predictions have been addressed and supported

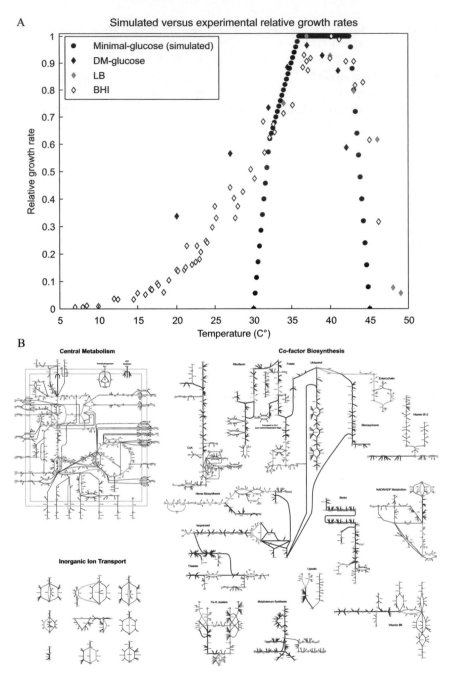

Figure 7.5 Computed growth rates as a function of temperature and the identification of the growth-constraining metabolic pathways ('hot spots'). Panel A: growth rates as a function of temperature are depicted relative to maximum growth rates under each condition. Circles are growth on minimal media with glucose simulated in this study, and diamonds are experimentally measured growth on Davis minimal medium (DM) with glucose, lysogeny broth (LB), and brain heart infusion (BHI) broth. Panel B: network hot spots at 42°C in *i*JO1366 subsystems. Predicted network hot spots are shown as red reactions. These subsystems include central carbon metabolism, inorganic ion transport, but mostly co-factor synthesis pathways. Taken from [66].

using medium supplementation experiments and mutational analysis of resequenced strains that have been adaptively evolved to grow at 42°C.

Off-target binding effects of drug molecules The ability to include protein structures in metabolic network reconstructions allows one to predict binding sites of drug molecules computationally and thus enable a network-based assessment of drug intervention. Surprisingly, pharmaceutical science is in the early stages of determining the exact mechanisms of drug action, both intended and side effects, and a predictive structural systems biology approach can address these issues.

An integrated network and protein structure reconstruction approach to assess how a drug molecule may affect a particular system has been performed [64]. This study adapted Recon 1 to metabolism in the human kidney, and its capacity for filtration of the contents of blood. The Protein Data Bank (PDB) was used to get structures of human protein and computational methods were used to predict potential drug binding sites. The combined model was used to retrospectively investigate potential causal drug targets leading to increased blood pressure in participants of clinical trials for a particular pharmaceutical, Torcetrapib, that had been withdrawn from the market due to side effects.

As with the two cases discussed above, the inclusion of protein structure data in a network reconstruction environment produced novel results. Using the structural systems biology approach, causal drug off-targets were predicted that had previously been observed to impact renal function in gene-deficient patients and may play a role in the adverse side effects observed in clinical trials. Genetic risk factors for drug treatment were also predicted that correspond to both characterized and unknown renal metabolic disorders as well as cryptic genetic deficiencies that are not expected to exhibit a renal disorder phenotype except under drug treatment [64].

Thus, an integration of structural and systems biology is a step towards computational systems medicine. It also has important implications for drug development and personalized medicine.

7.3 Transcription and Translation

The inclusion of protein structure and properties in network reconstructions increases their scope, the number of biological phenomena that can be analyzed, and introduces an orthogonal data source. It is now clear that the entire process of *protein synthesis* can be included in a network reconstruction and thus explicitly included in a genome-scale view of a cell.

Scope The scope of a protein synthesis network would include transcription and translation (Tr/Tr), any post-translational modification, and can provide the entire metabolic requirement for synthesizing the entire proteome of the target cell. Such expansion in network scope and coverage will increase the scope of reconstructed cellular functions and enable direct integration with other reconstructed networks in the target cell.

The scope of a Tr/Tr reconstruction is the synthesis of all proteins, mRNA, tRNA, and rRNA involved in the functions shown in Figure 7.6A. The metabolites that are consumed by the network to produce the functional proteins, e.g., amino acids, and

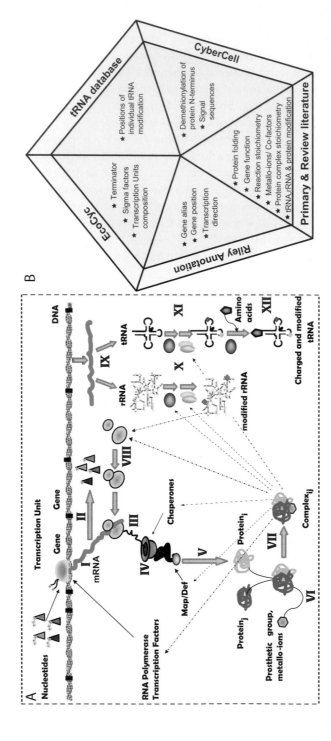

Figure 7.6 The scope of a genome-scale protein synthesis network in bacteria that has mechanistic chemical reactions and a full genome-sequence basis. (A) Schematic representation of the network components and reactions. (B) The five main data sources used for the reconstruction. From [420].

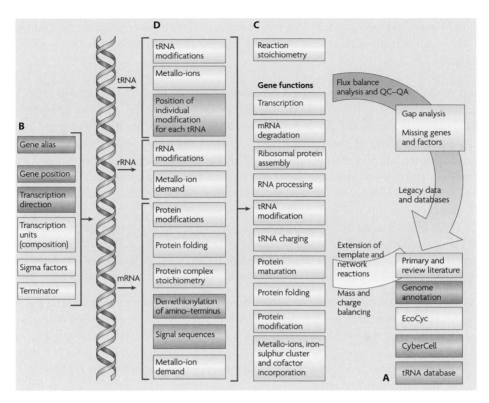

Figure 7.7   Schematic of a workflow for the reconstruction of transcriptional and translational networks (Tr/Tr). The reconstruction of a Tr/Tr network can now be performed in an algorithmic manner as illustrated in this figure for *E. coli* and described in detail in the text. From [117].

functional ribonucleic acids (rRNA, mRNA, and tRNAs), and nucleotides, are inputs into the network. Such a reconstructed Tr/Tr network does not contain transcriptional regulators and their functions. Regulatory functions need to be reconstructed separately and they would represent inputs to a Tr/Tr network as they would specify which proteins need to be synthesized.

This mechanistic and genetically based approach to the reconstruction of Tr/Tr networks is fundamental and on a genome-scale. It has been developed for *E. coli* [420] and can be applied to other bacteria. We note that secreted proteins are highly valuable products and systems analysis of the protein secretion pathway in yeast is advancing.

Workflow   Reconstructions of Tr/Tr networks at a genome-scale follow a similar four-step procedure as has been established for metabolism and described in Chapter 3. Tr/Tr network reconstructions can be generated using a genome annotation and the genome sequence as a scaffold. A Tr/Tr network reconstruction will thus contain sequence-specific synthesis reactions for every included gene and gene product participating in Tr/Tr functions. Thus, a full genetic basis for these functions is included and it can immediately account for sequence variations in the target organism. A schematic of the iterative multi-data source workflow needed for Tr/Tr reconstruction is shown in Figure 7.7.

Step 1: Automated genome-based reconstruction   Information about the components of the Tr/Tr network can be directly extracted from the genome annotation. This automated step should provide details for (i) gene function, (ii) gene type (e.g., protein coding or functional RNA), (iii) start and stop codons, (iv) direction of transcription, and (v) transcription unit association. Genome annotations and databases provide information about the type of transcription terminator (e.g., rho-dependent, attenuation, etc.) and sigma factors for transcriptional initiation (e.g., σ^{70}, σ^H, etc.).

Tr/Tr reactions can be formulated in an automated fashion using this information, the genome sequence, and *template reactions*. The manually formulated template reactions make use of the fact that the Tr/Tr reactions are very similar for most genes and thus make automation easier. Two examples are as follows.

- The binding of the RNA polymerase follows the same reaction schema for every promoter but with detailed differences. As will be discussed in the following chapter, in *E. coli*, initiation of transcription requires the binding of the holo-enzmye RNA polymerase ($\alpha_2\beta\beta'$) to a sigma factor (*e.g.*, σ^{70}). Then, this complex binds to a promoter site that has a recognition site for this particular sigma factor.
- Gene-specified quantities such as amounts of different amino acids for polypeptide or nucleotide triphosphates (NTPs) for mRNAs replace placeholders in a template reaction based on gene information. Subsequently, corresponding reactions can thus be formulated accurately and in a gene-specific manner.

Thus, the automation of the initial set of reactions underlying the Tr/Tr network is easier than for metabolism, which is fortunate as there are literally tens of thousands of reactions that are involved in this network.

Step 2: Curation and formulation based on bibliomic data   The template reactions are formulated manually and curated using detailed bibliomic data. Manual curation is also required for protein complex stoichiometry, the presence and stoichiometry of metallo-ions, and any prosthetic groups and co-enzymes involved. Most databases do not contain this information. Challenges unique to the reconstruction of this network are reaction mechanisms involved in macromolecule modifications, such as rRNA and tRNA modifications. Ambiguities in the bibliome need to be tracked by using notes or a confidence score. They help with the diagnosis of network functions that come under question and facilitate updates as new information becomes available.

Step 3: Converting a genome-scale reconstruction to a computational model   The reaction list generated through steps 1 and 2 can be readily converted into a mathematical format using bioinformatically based programming that extracts the stoichiometric coefficients from each network reaction and transfers them into a matrix format. The stoichiometric matrix that describes the Tr/Tr network has been called the *Expression* or E matrix.

The network boundary for Tr/Tr networks includes metabolism, i.e., metabolic components are imported or exported across this systems boundary. The uptake rates into the Tr/Tr network for these metabolites can be derived from experimental data. The definition of a systems boundary is important in formulating models of reconstructed networks as discussed in Part III of the book. Because protein synthesis creates demand reactions for the metabolic network (that now are detailed as boundary fluxes into the Tr/Tr network) the integration of metabolic and Tr/Tr networks can be achieved readily, and is discussed later in the chapter.

Step 4: Reconstruction uses and integration of omics data types   As with metabolic networks, once a reconstructed Tr/Tr network has been converted into a mathematical format a computational model can be formulated and various applications are enabled. The E matrix is the basis for such a model and we illustrate three applications of a computational model based on it.

A. Topological properties   Correlated reaction sets (co-sets), discussed in Chapter 13, are sets of reactions that are used together in all physiological states of a network. They thus give an unbiased definition of a network *module* [325]. The occurrence of co-sets can be calculated in the model of the Tr/Tr network to identify functional coupling between proteins. A total of 14 multi-protein modules were identified in this way that contained 91 of 153 proteins or protein complexes in the network (Figure 7.8A). Interestingly, many of these modules contained proteins from different subsystems based on classical pathway designation.

B. Omics data mapping   Reconstructed Tr/Tr networks enable quantitative integration of omics data to both expand and refine the knowledge about the networks and its components. While the integration of transcriptomic and proteomic data may be more straightforward, the integration procedure of ChIP-Chip data quantifying binding affinities of the RNA polymerase or other transcription factors needs to be established. Figure 7.8B schematically shows the mapping of omics data types enabled by the E matrix, as well as the M (metabolic) and O (operon) matrices, discussed elsewhere in the book. These three matrices can be integrated into a larger network.

C. Growth and cellular composition   Cell growth rate is directly correlated with its protein synthesis capacity and thus with the number of active ribosomes. The E matrix-based model can be used to relate ribosome content to growth rate. The *in silico* computed ribosome production capabilities showed very good agreement with the reported *in vivo* ribosome production capabilities for all investigated doubling times (Figure 7.8C). This consistency between computations that use molecular composition as constraints on the model and the observed growth rate show the overall consistency in disparate data types that can be achieved by relating them through a reconstruction and a computational model.

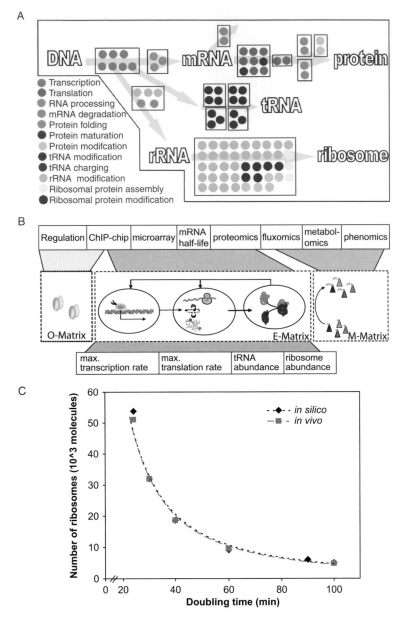

Figure 7.8 The Tr/Tr networks can be converted into computational models and used. (A) Schematic representation is shown of the calculated functional modules, the associated proteins, and their canonical assignments. (B) Integration of omics data into 'E-matrix' as reaction constraints. (C) Comparison of *in vivo* and *in silico* maximal number of ribosomes as a function of doubling time. From [420].

7.4 Integrating Network Reconstructions

Concept With increasingly detailed molecular and genetic data becoming available for various cellular functions, procedures are emerging for their reconstruction akin to what is in place for metabolism. These networks share compounds and will thus form

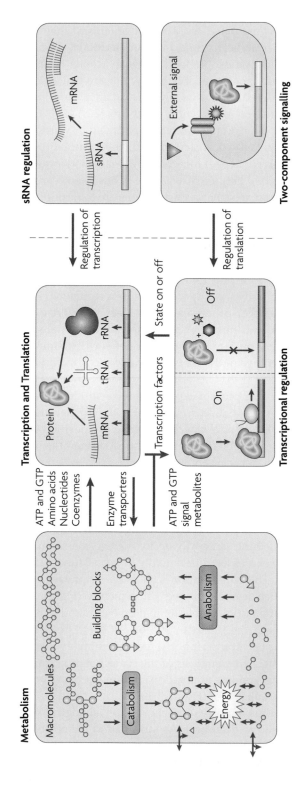

Figure 7.9 The networks that underlie cellular function can be reconstructed and integrated into a coherent whole. Such networks include the transcription–translation machinery, the transcriptional. regulatory network, two-component signaling, and sRNA regulatory networks. Today, metabolism and transcription and transcription/translation are available in mechanistic detail, genome-based, and genome-scale. From [117].

a unified network. Once two or more of the five different types of networks described in Figure 7.9 have been reconstructed for a target organism, they can be merged to form integrated network reconstructions and computational models that include a large number of cellular activities. Network reconstructions for metabolism and protein synthesis are the only ones available today that are genomically based and contain chemical reaction detail. Integrated reconstructions have been formed by merging an M and an E matrix for a target organism [229, 423]. E matrices only exist today for prokaryotes. The protein secretion pathway in yeast has been elucidated [120], and this represents the first step towards the formulation of an E matrix for eukaryotes. Although this concept of integration of many networks into a coherent whole was articulated in 2003 [358], only a part of it has been realized, which is a reflection of how difficult genome-scale reconstructions are to formulate and validate.

Integrating metabolism and protein synthesis in a thermophile Integration of metabolism and transcription/translation processes is, in principle, quite straight-forward. Transcription and translation requires energy and building blocks such as nucleotides and amino acids as inputs and hence these processes are constrained by the ability of the metabolic network to produce these precursors [12]. In addition, the transcription and translation processes can be seen to exert demands on the metabolic network function and thus limit other metabolic functions. The transcription/translation network feeds back to the metabolic network by controlling the levels of the enzymes and their activity in the metabolic network. This functional coupling is shown in Figure 7.9.

An integrated metabolic and protein expression reconstruction has been built for the thermophile *T. maritima* that we discussed earlier in the chapter, and an ME-model has been constructed [229]. This model can compute the composition of the proteome needed to execute a cellular function and thus the needed rate of gene expression to generate a functioning cell under a given condition. *Thermotoga* can grow on various sugars and the expression rate can be computed under two growth conditions and compared.

The expression rate for all genes in *Thermotoga* were computed for growth on arabinose and cellobiose and compared (Figure 7.10A). The results show that most of the genes need to be expressed the same way under the two conditions, but a handful of genes are computed to be expressed differentially to form a functional proteome for growth on these two sugars. This computation is the *in silico* equivalent of a differential expression experiment, which was performed subsequently (Figure 7.10B), and the measured and computed differential expression patterns are similar. By analyzing the sequences of the promoters in the upstream regions of the differentially expressed genes, binding motifs for the two transcriptional regulators involved (CelR and AraR) can be identified (Figure 7.10C). The analysis of the corresponding transcription units then leads to the correction of genome annotation and the discovery of putative genes involved in growth on these substrates (Figure 7.10D).

The ME matrix model thus enables prediction of physiological states, that once examined experimentally lead to discovery. As indicated in the center of Figure 7.10, this corresponds to an implementation of an iterative workflow that can systematically unravel the need for transcriptional regulation in this thermophilic organism.

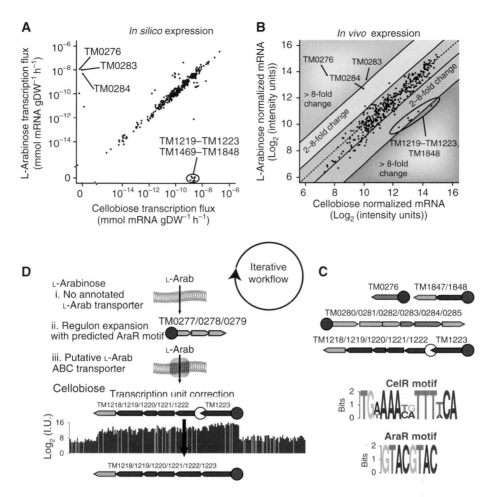

Figure 7.10 ME matrix computation can drive biological discovery. (A) Computation: *in silico* comparative transcriptomics identifies sets of genes that are differentially regulated for growth on L-arabinose (L-Arab) versus growth in cellobiose minimal media. (B) Measurement: *in vivo* transcriptome measurements corresponding to panel (A). (C) Analysis: two distinct putative TF-binding motifs are present upstream of the TUs containing the genes differentially expressed for the two regulators, AraR and CelR. (D) Discovery: searching *T. maritima* genome for additional AraR and CelR motifs results in new biological knowledge. From [229].

7.5   Signaling Networks

There has been great interest in reconstructing and modeling signaling networks in cells. However, as not all the components of a signaling pathway can be read from a genome annotation, the reconstruction procedures for signaling networks is not as mature as those devised for metabolic or Tr/Tr network reconstructions. Therefore, signaling networks are not yet available on a genome-scale. There is a great need and interest in reconstructing signaling networks and integrating them with other cellular functions as shown in Figure 7.9. We will briefly discuss the challenges associated with large- or genome-scale reconstruction of signaling networks.

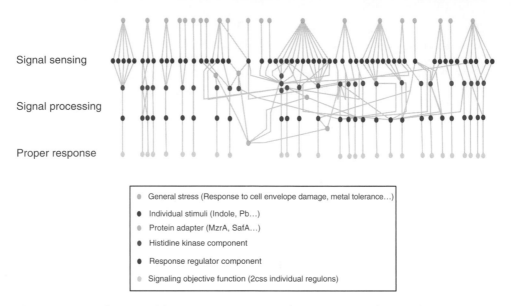

Figure 7.11   A schematic of the two-component signaling systems (2css) in *E. coli*. Prepared by Juan Nogales Enrique.

Regulation and signaling in bacteria   The metabolic, transcriptional regulation, translation, and transcription processes together represent a sizable fraction of the genes in a microbial genome. However, there are other networks that are currently the subject of intensive study that will likely be the subject of future network reconstruction efforts. Such efforts are likely to develop four-step reconstruction processes, paralleling those described above.

Two-component signaling systems are an example of this type of network (Figure 7.11). While the components of two-component signaling pathways (histidine kinases and response regulators) can be identified relatively easily by sequence homology, the connectivity of these pathways is not known completely, even in *E. coli*. Progress has recently been made towards systematically mapping the connectivity of two-component pathways in *E. coli* [462] as well as in other bacteria [399] using a variety of experimental methods. It is expected that in the future, comprehensive reconstructions of two-component systems can be achieved by combining literature-based information with these types of high-throughput data [390]. The early models of TRNs in *E. coli* did include some of the known two-component signaling pathways that respond to metabolic stimuli [83], and thus such integration is achievable.

Signaling in eukaryotes   Despite the lack of a full genetic basis, signaling network reconstructions have been built for eukaryotic cells. Boolean and stoichiometric representations are two common approaches for representing and analyzing a signaling network. The differences between the approaches are illustrated in Figure 7.12. An example signaling pathway or network is shown in panel A. This figure depicts how a signal (lightning bolt) is transduced through a series of phosphorylation,

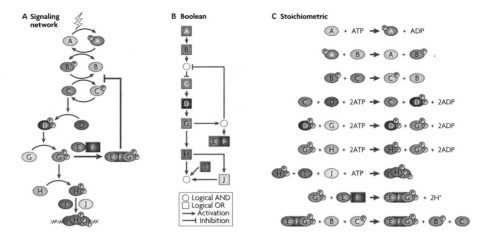

Figure 7.12   An example signaling network (panel A) and a corresponding Boolean (panel B) and stoichiometric (panel C) representation. From [176].

dephosphorylation, and binding events, that results in a phosphorylated protein complex (IHJ) binding to DNA. Phosphate groups are represented by a circled 'P'.

The Boolean description is found in Figure 7.12B in the form of series of logical statements, as GPRs are represented in metabolic reconstructions. The corresponding stoichiometric description is shown in panel C as a set of mass-balanced chemical equations. Although the stoichiometric approach provides a finer description of signaling network processes than the Boolean approach, there is a sizable increase in computational costs. However, the advantage of a stoichiometric representation of a signaling network is the ability to integrate it directly with a metabolic network, that, for instance, provides the ATP molecules needed to carry out the protein phosphorylation reactions.

7.6   Summary

- The procedures that have been developed for metabolic network reconstruction can be applied to other cellular networks provided that the appropriate data and information are available.
- Such workflows are developing for the structural proteome, translation and transcription in bacteria, and cell signaling.
- If such reconstructions can be represented stoichiometrically, then they can be integrated. Data can be mapped on such reconstructions and analyzed.

8 Metastructures of Genomes

The universe may be a 4-dimensional soap bubble in an 11-dimensional space. Who knows? – Christian Klixbull Jørgensen

With the publication of the first full genome sequence in the mid 1990s, it became possible, in principle, to identify all the gene products involved in complex biological processes in a single organism. In practice, almost 17 years later, this has proven difficult to accomplish using just sequence information. A toolbox of molecular biology methods that are implemented on a genome-scale are now available and it allows us to measure many properties of a genome with unprecedented resolution. A workflow that systematically integrates these data types can lead to the definition of the transcription unit (TU) architecture of a genome. This information, in turn, enables the reconstruction of transcriptional regulatory networks (TRNs) and other features. The full application and integrative analysis of genome-wide measurements led to the concept of a *metastructure* of a prokaryotic genome as there are many more attributes to a genome than its sequence and 3D arrangement.

8.1 The Concept of a Metastructure

With the plethora of genome-scale measurement methods, genomes can be characterized at multiple different organizational levels. Genomes used to be thought of in terms of their base sequence and three-dimensional structure. It is now clear that there are many more dimensions to genome organization, use, and information content (Figure 8.1). This realization has led to the concept of a metastructure of a bacterial genome [74].

Detailed examination of genome-wide data sets results in the definition of structural, operational, and functional annotations [74, 349]. Structural genome annotation provides the foundation for further operational and functional annotation and consists of coding (open reading frames, or ORFs) and non-coding genes, as well as intergenic regions. Elucidating the precise structural genome annotation subsequently allows the decoding of the operational genome annotation, which consists of operons and TUs. As a higher level of genome organization, the operon structure is a key to deciphering the flow of information encoded in the genome. A functional genome annotation

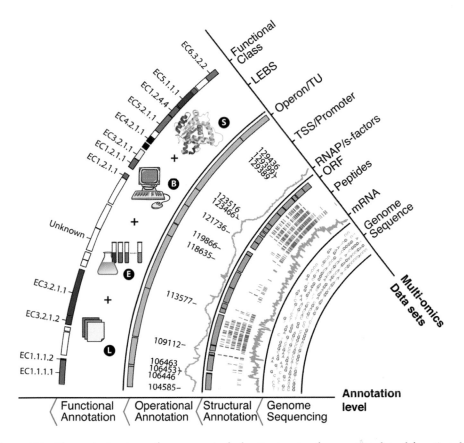

Figure 8.1   The metastructure of a genome includes its structural, operational, and functional annotation. Data sets include genome sequence, transcription profiles, peptide reads, RNA polymerase (RNAP), sigma factor binding profiles, transcription start site (TSS) reads, as well as literature data (L), experimental data (E), bioinformatic data (B), and structural information (S). From [349].

assigns a function to an ORF and describes the biochemical properties of the gene products.

Measurements and computation of properties as a function of genome location
To establish the metastructure we need to deploy a variety of experimental and bioinformatic analysis methods. This series of methods is illustrated in Figure 8.2. The integration through manual curation and evaluation is very labor-intensive. To perform this integration, one needs to master many facets of genomics and bioinformatics, including:

- omics approaches that include genome resequencing, TSS determination, and RNA-seq and peptide mapping, generated with an LC-MS/MS (LC, liquid chromatography; MS, mass spectrometer); and
- bioinformatics approaches that include genome re-annotation, functional RNA prediction, ribosome binding site energy calculations, and determination of intrinsic terminators.

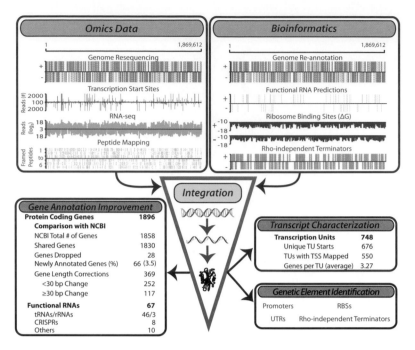

Figure 8.2   Generation of multiple genome-scale data sets integrated with bioinformatics predictions reveals the genome organization of *T. maritima*. From [220].

This multifaceted approach results in an improved gene annotation, transcriptome and proteome characterization, and genetic element identification. The summary numbers given in the figure correspond to the genome of the thermophile *T. maritima*. Similar data sets are now available for other prokaryotes [74, 349, 391]. As discussed below, such polyomic data integration represents a grand challenge in the field.

Some details of metastructures   The details of a metastructure are quite intricate and can be hard to decipher. A schematic of the genetic elements upstream and downstream of a transcription start site is shown in Figure 8.3A. The promoter is characterized by the -35 and -10 boxes. The 5′ untranslated region (5′UTR) spans the difference between the start sites for transcription and translation. The 5′UTR often encodes small RNA (sRNA) molecules and has the ribosome binding site (RBS).

One can graph some of the details of a metastructure for a target genome against the sequence as template. Figure 8.3 shows many of the properties of the *T. maritima* genome as a function of genome location [220] that we now describe.

Panel (A)   shows a prototypic layout of functional elements on DNA.

Panel (B)   shows the results of a motif search for a bipartite promoter with a hexamer -35 box and nonamer -10 box. The distribution of each promoter element is shown relative to the TSS. The motif determined for each promoter element is displayed as a sequence logo in the two inserts.

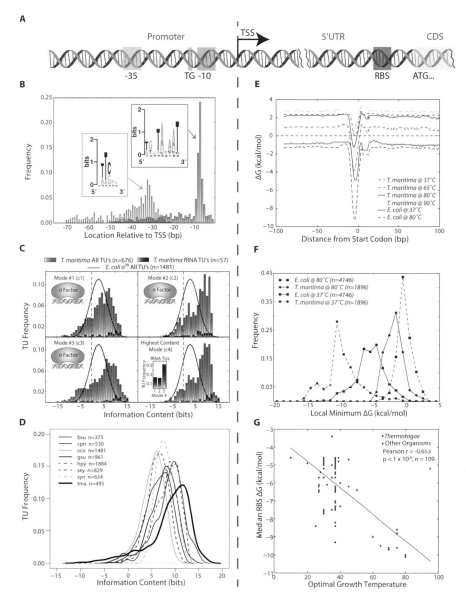

Figure 8.3 Identification and quantitative comparison of genetic elements for transcription and translation initiation. From [220].

Panel (C) shows three information content-based modes of σ factor binding as a way to classify promoters. These are histograms containing genome-wide sets of promoters. The first binding mode (Mode #1) contains the $\sigma 70$ contacts with the -35 and -10 hexamer promoter elements. Mode #2 represents binding to the extended -10 promoter (5'TGn and -10 hexamer). Mode #3 represents $\sigma 70$-binding to both the -35 and the extended -10 promoter elements. The content mode for each TU is presented with an inset showing the distribution of functional RNAs across the different modes.

Panel (D)   shows the highest scoring mode for σ70-like promoter elements calculated for seven additional bacterial species. The legend shows the KEGG Genome ID for each organism. Thus, this is a comparative metastructure characterization. Note that the curve for *T. maritima*, a thermophile, is different from the others.

Panel (E)   shows the calculated median RBS ΔG for all genes based on the position relative to the start codon. This binding energy is temperature-sensitive. Temperature profiles are shown for *T. maritima* at 37°C (for comparison only), 65°C (lower growth limit), 80°C (growth optimum) and 90°C (upper growth limit). Similar profiles are shown for *E. coli* at 37°C (optimal) and 80°C (for comparison only).

Panel (F)   shows the local minimum RBS ΔG for all genes in a 30-nucleotide window upstream of the annotated start codon generated for *T. maritima* and *E. coli* at 37°C and 80°C.

Panel (G)   shows the median local minimum RBS ΔG calculation for 109 bacteria plotted against the optimal growth temperature. Species in the Thermotogae phylum ($n = 15$) are shown in red.

This detailed example illustrates the complexity of a genome and its metastructure. Many genome properties can be measured or computed.

8.2   Transcriptional Regulatory Networks

The metastructure will contain information about the location of the transcription units (TUs). Their transcription is condition-dependent and is determined by transcription factors (TFs) and the binding specificity of the σ factors. We can now measure how these proteins are bound to the genome on a genome-wide basis. By varying environmental conditions and through knock-out strain construction, we can now determine the regulatory logic on each promoter. When such data are mapped onto the metastructure and the gene products are put in contact with the network where they function, one can reconstruct portions of the transcriptional regulatory network (TRN). No TRN in a target organism has been reconstructed in a bottom-up fashion, but progress is being made as described below. A fairly comprehensive view of TRNs can be obtained by a combination of bottom-up and top-down methods [45].

Reconstructing operons   One result from this challenging polyomic data integration process is the TU architecture of the genome. A specific example relating to the *thr* operon in *E. coli*, the first operon clockwise downstream from the origin, is shown in Figure 8.4. Some of the details of this process are as follows.

Modular units (MUs):   co-transcribed segments (modular units) consist of RNA polymerase binding regions (RBRs), RNAP-guided transcript segments (RTSs), potential ORFs (pORFs), and TSSs and are defined based on polyomic data integration. Each module can be represented by various regulatory elements initiating transcription at different TSSs, representing different TUs.

Definition of TUs: under log-phase growth conditions, modular units FWD-1 (containing thrA) and FWD-2 (containing thrBC) are transcribed together forming contiguous TUs (TU-1,2,3, based on TSS data) (Figure 8.4A). However, in the

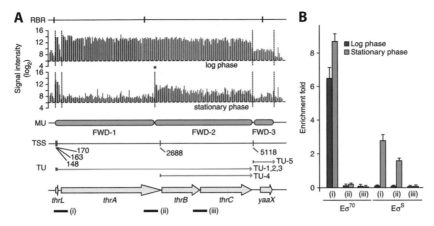

Figure 8.4 Determination of transcription units and use of alternative TSSs. See text for details. From [74].

stationary phase, transcription of modules FWD-1 and FWD-2 is triggered separately, defining an additional transcription unit (TU-4). Module FWD-3, resulting in TU-5, is used similarly under log and stationary phases. Dotted lines in the transcription profiles indicate change points of transcription. The change point under stationary phase led to the determination of one additional transcription unit (TU-4).

Regulatory elements: the differential usage of modular units result from regulation (Figure 8.4B). In this case, the measured binding of σ^{70} and σ^{S} holoenzyme ($E\sigma^{70}$ and $E\sigma^{S}$) occupancy within the promoter regions (i, ii) and control region (iii) in log and stationary phases, respectively. Significant occupation preferences of σ^{70} and σ^{S} holoenzymes confirmed the TU architecture.

What used to be thought of as an operon is now known to be composed of at least five different TUs, each representing an *independently addressable element* on the genome. Thus, the definition of operons is fundamentally changing based on integration of the multiple genome-wide data sets that are now available. Defining the TUs and how they are regulated is a complex undertaking requiring multiple data types, but one that leads to the reconstruction of the TRN.

Reconstructing regulons The definition of the TUs emanates from the metastructure, and we see that the classically defined operons have multiple TUs that are regulated differentially. To look at the coordinated regulation of all the TUs in the operons that are in a regulon, one needs to obtain a series of genome-wide data sets and integrate them systematically to build a model that describes its regulatory logic.

Broad-acting TFs in bacteria form regulons, including the leucine-responsive protein (Lrp) that is involved in nitrogen and amino acid metabolism. A four-step method to fully reconstruct the Lrp regulon in *E. coli* is shown in Figure 8.5A.

Step 1: involves obtaining high-resolution ChIP-chip (or ChIP-exo) data for Lrp, the RNA polymerase, and expression profiles under multiple environmental conditions. These data sets lead to the identification 138 unique and reproducible

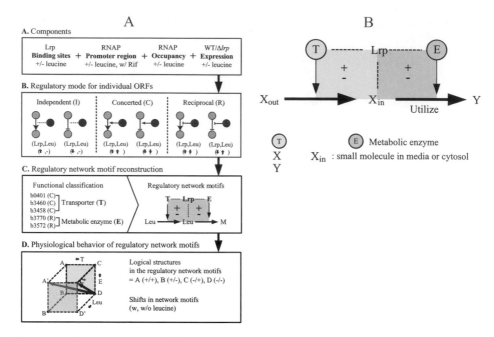

Figure 8.5   Reconstructing and characterizing a regulon. Panel A, overview of a workflow used for reconstruction. Panel B, a regulatory model for the regulon. From [73].

Lrp-binding regions and the classification of their binding state under different conditions.

Step 2:   the analysis of these condition-specific polyomic data sets revealed six distinct regulatory modes for individual TUs.

Step 3:   the functional assignment of the regulated ORFs is used to reconstruct four types of regulatory network motifs around the metabolites that are affected by the corresponding gene products. The regulation can be feedforward or feedback, and the mode of regulation can be positive or negative.

Step 4:   the determination of how a regulatory molecule (leucine in this case), as a signaling molecule, shifts the regulatory motifs for particular metabolites. A regulatory structure for the regulon emerges that shows the regulatory motifs for different amino acids fall into the traditional classification of amino acid families, thus elucidating the structure and physiological functions of the Lrp-regulon.

The same procedure can be applied to other broad-acting TFs, opening the way to a full bottom-up reconstruction of the transcriptional regulatory network in bacterial cells.

Reconstructing stimulons   The reconstruction procedure for a regulon can be scaled up to reconstruct a stimulon consisting of many regulons. Similar data as for Lrp can be generated for other transcription factors involved in nitrogen and amino acid metabolism in *E. coli*. Such data obtained for ArgR, Lrp, and TrpR show that 19 of 20 amino acid biosynthetic pathways are controlled either directly or indirectly by these transcriptional regulators.

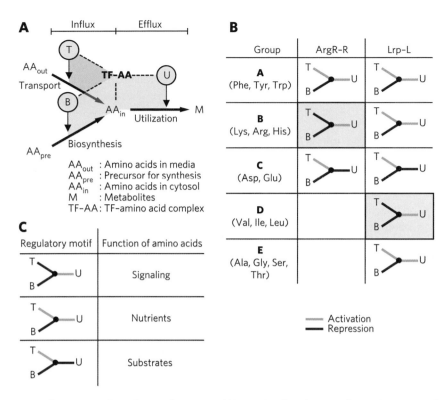

Figure 8.6 Reconstruction of a regulatory motif in a stimulon that regulates nitrogen and amino acid metabolism. (A) Schematic diagram for the generic regulatory motif reconstruction in feedback circuit. (B) Logical structures of the regulatory motif in response to the exogenous amino acids for specific cases. (C) The classification of function of amino acids derived from the logical structures of the connected feedback circuit motif. From [75].

Classifying the regulated genes into three functional categories of transport, biosynthesis, and metabolism leads to the elucidation of regulatory network motifs, akin to that described for Lrp in Figure 8.5, that constitute the building blocks of the regulatory structure of a stimulon. The regulatory logic of these motifs was determined based on the relationships between transcription factor binding and changes in the amount of transcript in response to exogenous amino acids (see Figure 8.6). Remarkably, the resulting logic shows how amino acids are differentiated as signaling and nutrient molecules, revealing the overarching regulatory principles of the amino acid stimulon.

The sigma factor network A sigma factor is a variable member of the bacterial RNA polymerase complex, the so-called holoenzyme, that confers binding specificity, highlighted in Figure 8.4. There are different numbers of sigma factors in different bacterial species. In *E. coli*, for example, there are seven sigma factors.

The determination of the genome-wide binding patterns of the holoenzyme can be determined using a combination of omics data types through a complex workflow (Figure 8.7B). This workflow basically overlays the genome-binding data (ChIP-chip) for the RNA polymerase and the σ factors. Then TSS data are related to the binding map of the various forms of the holoenzyme. These TSSs are then related to the TU architecture. The result is the basic data that is needed to reconstruct the entire sigma factor network.

Such a network is very complicated (Figure 8.7C). Only one form of the holoenzyme binds to some promoters, while more than one form can bind to others. The binding to a promoter by a particular form of the holoenzyme leads to a particular TU being synthesized. The TU0001 and onward correspond to those shown in Figure 8.4.

Recapitulation New genome-wide data types allow us to reconstruct the TRN for a target organism. The metastructure gives the TU architecture. The binding of regulators to the promoters and TSS data obtained in a condition-dependent manner give the regulatory logic of the promoter that leads to the formation of the transcript. The TRN reconstruction is now possible on a genome-scale, but no formal descriptions have yet been developed and no chemically accurate and mechanistic models have yet been built for a genome-scale TRN. Thus, we do not have O matrices available yet. R matrices based on Boolean representation of regulatory rules have appeared as a precursor for building an O matrix [141].

8.3   Refactoring DNA for Synthetic Biology

The organization of information and functional elements on genomes is much more complicated than previously thought. Although we do know a fair amount about metastructures of simple genomes, there are still unknown interactions among genetic elements that lead to unpredictable results. Synthetic biologists have recognized the difficulty of working with native metastructures and developed a *refactoring* approach to synthetic DNA and synthetic operons (Figure 8.8). This figure represents a workflow that relies on many experimental and computational methods and is taken from [264]. The procedure involves finding the needed biological parts and then reassembling them with genetic parts into novel synthetic constructs designed to have a certain functionality.

Finding parts The top of Figure 8.8 shows three strategies for the identification of libraries of variants of pathway-specific genetic parts using genetic and biochemical knowledge. The left panel shows how operons or gene clusters can be found using neighborhood orthology analysis. The middle panel shows the identification of orthologs of a characterized gene by homology search, multiple sequence alignment, and phylogenetic tree construction. The right panel shows a method for identifying enzymes with identical substrate specificity to a model enzyme. These are thus parts that are needed as elements of a synthetic operon. Coding sequences (CDSs) are then optimized and a library is synthesized.

Building synthetic operons Known promoters and candidate RBSs are then obtained from genetic libraries of these parts. Then, after extensive *in silico* and *in vivo* testing and debugging, oligonucleotides are designed to synthesize the final design, which is then introduced and tested for biological function in the host organism.

The advantage of this approach is that the origin of all parts is known and the design should maximize known interactions among the parts. As all possible interactions or the general interactions with the host are not fully known, a library of

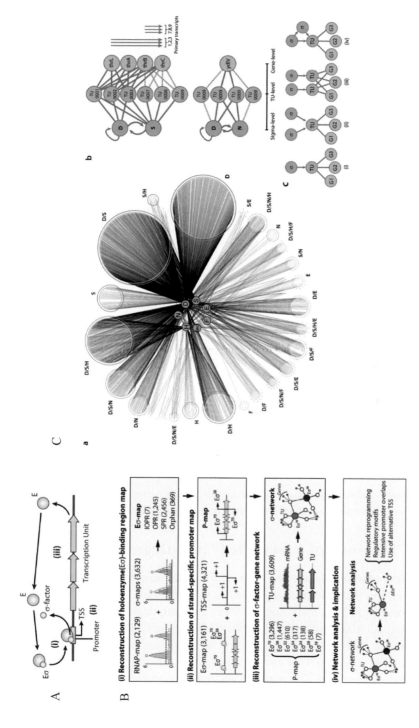

Figure 8.7 Reconstruction of the sigma factor network. (A) The template reaction that shows the function of a sigma factor. (B) A schematic of a workflow used to integrate multiple data types to reconstruct the network. (C) The sigma factor network resulting from the reconstruction process, both an overall genome-scale view as well as some detailed examples. Prepared by Dong Hyuk Kim and Byung Kwan Cho.

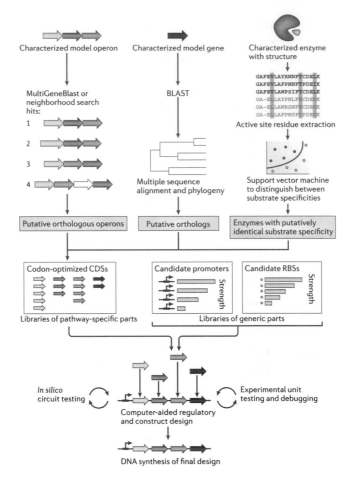

Figure 8.8   A workflow showing the steps involved in the identification of various biological components, their refactoring and their integration into transcriptional units. From [264].

constructs is made and screened for function in the host. An example of this refactoring approach is seen in the nitrogen fixation gene cluster from *Klebsiella oxytoca* [415].

8.4   The Challenge of Polyomic Data Integration

Growing number of omics data types   Over the past 20 years there have been incredible advances in the technology that generates genome-scale data sets. In addition to these component profiling technologies, there are component interaction measurements emerging, such as ChIP-chip (or now ChIP-seq or ChIP-exo) measurements. This 'embarrassment of riches' of omics data is satisfying, but without integrated analysis of multiple data types we cannot fully bring out the complete information that they individually contain. The previous section gave an idea of how challenging it is to integrate multiple genome-wide data types. Well-defined and quality-controlled workflows are needed.

Pairwise omic data integration   Omics data types can be integrated in a pairwise fashion. Perhaps the most familiar example is the correlation of transcriptomic data and proteomic data, frequently found to be dissatisfying given the lack of expected concordance between the two. This expectation breaks down when one considers all the processes that the E matrix contains, the post-transcriptional processing of transcripts, and the differential half-lives of a transcript and the protein it encodes. Many other pairwise integration methods of two different omics types have been described (Figure 8.9). Clearly, two collated data types are more informative if the two are treated separately.

As satisfying as such pairwise analyses might be, one will have to be able to integrate many more omics data types, such as is shown in Figure 8.1. This process proves to scale geometrically with omics data types.

Polyomic data integration The complexity of data sets at the genome-scale is compounded by the availability of multiple omics data types and the concomitant increase in time and effort that data integration requires. These challenges have been met, as the examples in the previous section show, but the experience of researchers is that this effort is extremely time- and resource-consuming. These concerns raise the question of resource allocation: how can we determine the relative resource allocation for data processing and integration versus data generation? To help answer this question, a formal analysis has been performed for relative resource allocation for data generation versus data integration [319].

This analysis of the resource requirements of a workflow show that the resource requirement for data integration grows surprisingly rapidly with the number of data sets and the number of different data types (Figure 8.10). For instance, if there are three data types to be integrated ($m = 3$) then after six or more data sets ($n = 6$), the time and effort it takes to integrate the data exceed those needed for data generation. The results of such formal analyses of resourcing a workflow was found to be consistent with empirical experience.

Similar analyses has been performed for a single data type and how resource allocation changes over time as experimental technologies improve [374]. The relative efforts of the experimental design and data sample collection vs. sequencing vs. data reduction and management vs. downstream data analysis has changed dramatically over time. The quote by Sydney Brenner at the beginning of Chapter 28 ("We are drowning in a sea of data and thirsting for knowledge. Most biology today is low input, high-throughput, and no output biology") was certainly true in the year 2000, but will not necessarily be so in 2020, as illustrated in Figure 8.11.

8.5   Building Mathematical Descriptions

In Chapter 15 we will describe three generations of genome-scale models: those that describe the phenotypic potential of a cell, those that describe which phenotypic states are chosen, and those that describe the dynamic states of a cell in molecular detail. Furthermore, one can seek to simulate the function of a single cell or a population of cells. These applications call for the use of different forms of mathematics, depending on what cellular functions are to be simulated (Figure 1.6).

The phenotypic capacities of cells can be obtained using constraint-based models. These are detailed in Part III of this book. Such analysis relies on the use of linear algebra and constraint-based optimization. Dynamic states are typically simulated with

	Genomics	Transcriptomics	Proteomics	Metabolomics	Protein–DNA interactions	Protein–protein interactions	Fluxomics	Phenomics
Genomics	Genomics (sequence annotation)	• ORF validation • Regulatory element identification	• SNP effect on protein activity or abundance	• Enzyme annotation	• Binding-site identification	• Functional annotation	• Functional annotation	• Functional annotation • Biomarkers
Transcriptomics		Transcriptomics (microarray, SAGE)	• Protein: transcript correlation	• Enzyme annotation	• Gene-regulatory networks	• Functional annotation • Protein complex identification		• Functional annotation
Proteomics			Proteomics (abundance, posttranslational modification)	• Enzyme annotation	• Regulatory complex identification	• Differential complex formation	• Enzyme capacity	• Functional annotation
Metabolomics				Metabolomics (metabolite abundance)	• Metabolic-transcriptional response		• Metabolic pathway bottlenecks	• Metabolic flexibility • Metabolic engineering
Protein–DNA interactions					Protein–DNA interactions (ChIP–chip)	• Signaling cascades		• Dynamic network responses
Protein–protein interactions						Protein–protein interactions (yeast 2H, coAP–MS)		• Pathway identification activity
Fluxomics							Fluxomics (isotopic tracing)	• Metabolic engineering
Phenomics								Phenomics (phenotype arrays, RNAi screens, synthetic lethals)

Figure 8.9 Pairwise omic data integration helps to address biological questions at the systems level. This table summarizes some of the potential biological insights that can be gleaned from the pairwise omics data integration. Empty elements represent more challenging integrations where an application is less obvious. Modified from [195].

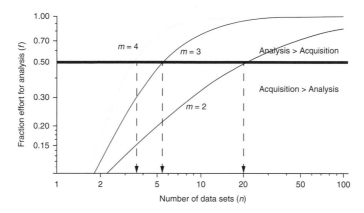

Figure 8.10 Estimation of the relative resource allocation for experiments and for analysis (x axes are shown in log scale). The fraction (f) of resource allocation for analysis as a function of the number of data sets (n); and the number of disparate data types being integrated (m). Modified from [319].

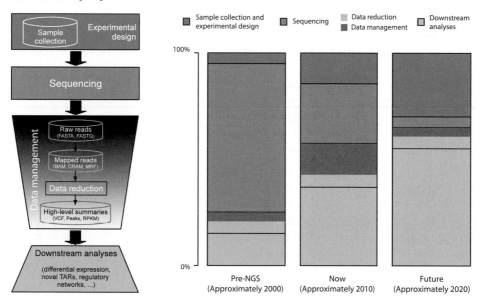

Figure 8.11 The estimated relative resource allocation for sample collection/experimental design, sequencing, data reduction/management, and downstream analysis in 2000, 2010, and 2020. Modified from [374].

ordinary differential equation models, and these are described in an accompanying book [317]. One should note that when it comes to multi-scale analysis, some cellular processes, or modules, can be considered to be in a steady state relative to others that are changing dynamically. For instance, metabolism tends to be much faster than the process of protein synthesis. Such multi-scale analysis calls for the use of the classical quasi-steady state assumption and formulation of hybrid models.

If the functions of a single cell are to be considered, one is often faced with the challenge of representing a finite number of molecules, and thus the continuum

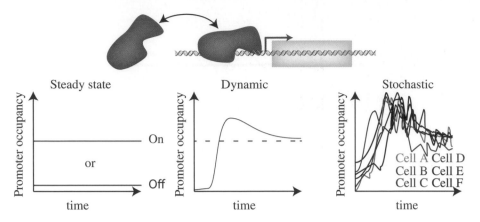

Figure 8.12   Illustration of the properties of different types of mathematical descriptions: steady state, dynamic, and stochastic. Prepared by Nathan Lewis.

assumption breaks down. In such situations, one has to consider stochastic simulation methods. Single-cell models may have to use all three forms of mathematics (stochastic, steady, and dynamic states; see Figure 8.12). Finally, the 3D arrangement of the cell and its compartments may have to be considered, calling for mathematical methods that yield spatial resolution.

Clearly, much work is still ahead of us to determine how to integrate many such mathematical methods and their numerical characteristics. Regardless of which method or methods are deployed, reliable genome-scale network reconstructions for the target organism are needed as well as their mathematical representations, a topic that we describe in the next part of this book.

8.6   Summary

- Multiple genome-wide data types can now be obtained and used to characterize genomes.
- The integration of such data sets shows that there are many dimensions to the function of a genome, leading to the concept of the *metastructure* of a genome.
- The metastructure includes the transcription unit architecture. The condition-dependent expression and binding state of regulators can be determined. When such data are analyzed in the context of the metastructure, one can reconstruct the transcriptional regulatory network.
- Systematic integration of polyomic data sets against a structured background represents a grand challenge in the field.
- If such reconstructions can be represented stoichiometrically, then they can be integrated with other reconstructed networks.
- Integrated network reconstructions can be converted into a mathematical format and used to formulate *in silico* cells. Such models can be simulated in various ways depending on the intended purpose of the simulator.

PART II
Mathematical Properties of Reconstructed Networks

During the past 30 years biology has become a discipline for people who want to do science without learning mathematics – Marvin Cassman and colleagues.

The set of chemical reactions that comprise a network can be represented as a set of chemical equations. Embedded in these chemical equations is information about reaction stoichiometry. All this stoichiometric information can be represented in a matrix form; the stoichiometric matrix, denoted by **S**. The stoichiometric matrix is thus a mathematical representation of a reconstructed network that, in turn, represents curated and organized biological knowledge about a target organism. It is the starting point for various mathematical analyses used to determine network properties.

Part II of this text will summarize the basic properties of the stoichiometric matrix. As **S** is a mathematical object, the treatment is necessarily mathematical. However, **S** represents biochemistry. We will thus relate the mathematical properties of **S** to the biochemical and biological properties that it fundamentally represents. The two last chapters of this part of the book will illustrate two approaches that can be used to characterize the contents of the spaces that are associated with **S**.

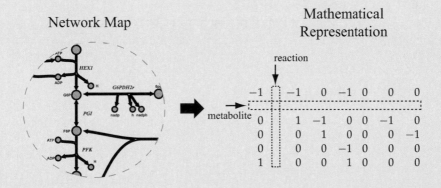

9 The Stoichiometric Matrix

Mathematics is the door and key to the sciences – Roger Bacon

The reactions that comprise a biological network can be represented by chemical equations. The stoichiometric matrix is formed from these chemical equations. It has several important attributes. In this chapter we focus on four principal views of the stoichiometric matrix and its content: (i) it is a data matrix, (ii) it is a connectivity matrix, (iii) it is a mathematical mapping operation, and (iv) it is a central part of *in silico* models used to compute steady and dynamic network states. These features are summarized in Figure 9.1.

9.1 The Many Attributes of S

The stoichiometric matrix is formed by the stoichiometric coefficients of the reactions that constitute a reaction network. It is organized such that every column corresponds to a reaction and every row corresponds to a compound. The entries in the matrix are stoichiometric coefficients that are integers. Each column that describes a reaction is constrained by the rules of chemistry, such as elemental balancing. Every row thus describes all the reactions in which the corresponding compound participates, and therefore how the reactions are interconnected. This deceptively simple matrix has many noteworthy attributes that are summarized in Table 9.1.

Informatic attributes. The stoichiometric matrix is a data matrix. The data that go into building a genome-scale stoichiometric matrix come primarily from the annotated genomic sequence and detailed assessment of the literature (bibliomic data) that is available about the target organism. Often, inferences from phylogenetics are used as well. All this information is the basis for the reconstruction process described in Part I.

Physical/chemical attributes. The stoichiometric coefficients represent counts of molecules that are involved in a chemical reaction. Chemical reactions come with conservation relationships of elements, charge, and other properties. These properties must be represented accurately. The cellular location of a reaction is included through the assignment of a metabolite to a cellular compartment.

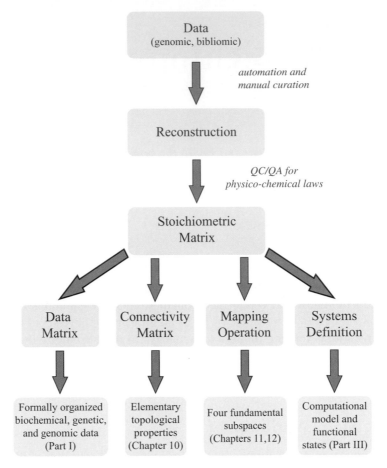

Figure 9.1   Four main features of the stoichiometric matrix and the corresponding portions of this textbook.

Genetic/genomic attributes.   A genome-scale network reconstruction effectively represents a two-dimensional annotation of a genome [313]. It contains information not only about the components of a cell, but also about how they are connected. **S** is a species property; all members of a species share this matrix. In principle, there is only one network encoded by the genome of a particular organism.

Biological attributes.   Because the stoichiometric matrix is specific to a species, it can be used to study the phenotypic differences and capabilities of various species. As evolution can occur by changing the number of reactions encoded in a genome (e.g., with horizontal gene transfer or gene deletion), the stoichiometric matrix can be used to study distal causation.

Mathematical attributes.   The stoichiometric matrix consists of integers. If the reactions represented are elementary, their numerical values are (–2, –1, 0, 1, 2). It therefore has no error associated with it and is thus a *knowable* matrix. In a similar way, a genomic sequence is knowable because each position can only be occupied by one of four bases.

Table 9.1 The many attributes of the stoichiometric matrix.

Attribute	Represents
Informatic	Annotated genome Bibliomic data Comparative genomics
Physico-chemical	Chemistry, cellular location conservations
Genetic	Genomic characteristics Represents a species
Biological	Species differences Distal causation
Mathematical	Integer entries 'Knowable' matrix
Systemic	Pool formation network structure
Numerical	Integers, sparse well-conditioned

Systemic attributes. The stoichiometric matrix is a connectivity matrix and gives the structure (or the topology) of a network. This structural information results in the definition of pools and pathways that are associated with the null spaces of the stoichiometric matrix.

Numerical attributes. For large networks, the stoichiometric matrix has mostly zero elements. It is a sparse matrix and may require sparse matrix representation and computational procedures as it reaches the genome-scale. All the elements of the matrix are of the same order of magnitude, making it a numerically well-conditioned matrix.

Thus, there are many different attributes to the stoichiometric matrix. All are addressed and studied in this book.

9.2 Chemistry: S as a Data Matrix

Every column in **S** represents a chemical reaction. It must be consistent with the chemistry that it represents. For instance, every reaction has to be elementally and charge-balanced. In addition, from a data-mapping standpoint, it is useful to have unique chemical specifiers associated with the compounds that the rows represent, and it can be useful to have EC numbers associated with the columns of the matrix to classify the type of chemical transformation that a column represents. This section illustrates the incorporation and representation of chemical properties in the formulation of **S**.

9.2.1   Elementary biochemical reactions

There is a limited number of elementary types of biochemical reactions that take place in cells. These fall into the three categories below. In the examples of each, derived for metabolic transformations, we used C to denote a primary metabolite, P as a phosphate group, and A as a co-factor such as the adenosine moiety in AMP, ADP, and ATP.

Reversible conversion   Transformation between two compounds consisting of the same two chemical moieties C and P can be written as

$$CP \rightleftharpoons PC \qquad (9.1)$$

representing two elementary reactions (forward and reverse). Although such reversible conversions are often used to generically describe reactions, they can only represent a simple chemical rearrangement of the molecule without any change in its elemental composition. For instance, isomerases catalyze such reactions. The stoichiometric matrix that describes this reaction is

$$\mathbf{S} = \begin{pmatrix} -1 & 1 \\ 1 & -1 \end{pmatrix} \begin{array}{l} : CP \\ : PC \end{array} \qquad (9.2)$$

where the first column of the matrix represents the forward reaction and the second column the reverse reaction. These are elementary reactions that have non-negative flux values. The first row represents CP and the second row PC. Concentrations are also non-negative quantities. Under certain circumstances, one may wish to combine the two elementary reactions into a net reaction that can take on positive or negative values.

Bi-molecular association   Many biochemical reactions involve the combination of two moieties, C and P, to form a new compound:

$$C + P \rightleftharpoons CP \qquad (9.3)$$

Sometimes, such reactions may not involve forming and breaking of covalent bonds, but a series of hydrogen bonds to form a complex, such as the dimerization of two protein molecules, or the initial binding of a substrate to an active site on an enzyme molecule. The stoichiometric matrix that describes a bi-molecular association is

$$\mathbf{S} = \begin{pmatrix} -1 \\ -1 \\ 1 \end{pmatrix} \begin{array}{l} : C \\ : P \\ : CP \end{array} \qquad (9.4)$$

where the rows represent C, P, and CP, respectively, and the single column represents the net reaction.

A co-factor-coupled reaction   A frequent reaction in biochemical reaction networks is one in which one compound (AP) donates a moiety (P) to another compound (C):

$$C + AP \rightleftharpoons CP + A \tag{9.5}$$

In reality, such reactions have an intermediate, and can be decomposed into two bi-molecular association reactions. Strictly speaking, this is not an elementary reaction. The stoichiometric matrix that describes the co-factor-coupled (or moiety exchange) reaction is

$$\mathbf{S} = \begin{pmatrix} -1 \\ -1 \\ 1 \\ 1 \end{pmatrix} \begin{matrix} : C \\ : AP \\ : CP \\ : A \end{matrix} \tag{9.6}$$

where the rows represent C, AP, CP, and A, respectively, and the column represents the net reaction. The word 'co-factor' is used synonymously with 'carrier.'

Combining stoichiometric matrices   A stoichiometric description of multiple reactions is formed easily by combining the individual matrices. The three types of transformations can be combined into one matrix (Equation (9.7)). Note that the net reaction rate has been used in forming **S** in this equation.

$$\begin{matrix} v_1 & v_2 & v_3 \\ \begin{pmatrix} 0 & 0 & 1 \\ 0 & -1 & -1 \\ 0 & -1 & 0 \\ -1 & 1 & 1 \\ 0 & 0 & -1 \\ 1 & 0 & 0 \end{pmatrix} & \begin{matrix} A \\ C \\ P \\ CP \\ AP \\ PC \end{matrix} \end{matrix} \tag{9.7}$$

It should be clear to the reader that stoichiometric matrices for two separately reconstructed networks that share compounds can be integrated easily. This easy integration makes the stoichiometric representation of networks highly scalable.

9.2.2   Basic chemistry

The chemical reactions that form the columns in **S** have the basic rules of chemical transformation associated with them. There are conservation quantities (such as elements and charge) and there are non-conserved quantities (such as osmotic pressure and free energy) associated with chemical transformations. These properties must be accounted for in the construction of a biochemically meaningful stoichiometric matrix.

The elemental matrix   Metabolites consist of six chemical elements. The elemental matrix, **E**, gives the composition of all the compounds considered in a network. A column of **E** corresponds to a compound, x_i, and the rows correspond to the elements, typically only six of them found in organic compounds: carbon, oxygen, nitrogen, hydrogen, phosphorous, and sulfur. Compounds can be represented as points in a space formed by the elements as the axes (Figure 9.2).

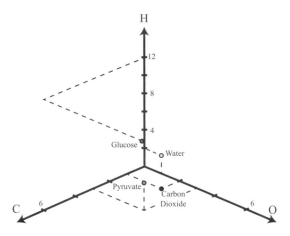

Figure 9.2   The elemental space. Representation of compounds containing carbon, hydrogen, and oxygen in a three-dimensional space. The coordinates are: glucose (6,12,6), pyruvate (3,3,3), water (0,2,1), and carbon dioxide (1,0,2).

It is important to note that the elemental composition of a molecule does not uniquely specify its chemical structure. For instance, glucose and fructose have the same elemental composition. Thus, associating unique chemical identifiers with the columns of $\mathbf{E}$ is desirable. The elemental composition of common metabolites is shown in Table 9.2.

Example: Consider the simple chemical reaction

$$2H_2 + O_2 \rightarrow 2H_2O \tag{9.8}$$

that involves only two elements, oxygen and hydrogen. The elemental matrix for this chemical reaction is

$$\mathbf{E} = \begin{pmatrix} 0 & 2 & 1 \\ 2 & 0 & 2 \end{pmatrix} \tag{9.9}$$

where the first row corresponds to oxygen, the second row to hydrogen, and the columns correspond to the compounds, in this case ordered as H_2, O_2, and H_2O.

Conserved quantities   A chemical reaction cannot create or destroy elements. Thus, the inner product of the rows, $\mathbf{e}_i$, in the elemental matrix and the reaction vectors, and the columns in $\mathbf{S}$ (called $\mathbf{s}_j^v$) must be zero, or

$$\langle \mathbf{e}_i, \mathbf{s}_j^v \rangle = 0 \tag{9.10}$$

for all the elements found in the compounds that participate in the reaction. This inner product simply adds up the number of elements in the compounds on each side of the reaction. Because the stoichiometric coefficients are negative for the reactants (the

Table 9.2 The elemental composition of some common metabolites. Adapted from [285].

Compound	Elemental composition	Compound	Elemental composition
Glucose	$C_6H_{12}O_6$	Alanine	$C_3H_7NO_2$
Glucose-6-phosphate	$C_6H_{11}O_9P$	Arginine	$C_6H_{14}N_4O_2$
Fructose-6-phosphate	$C_6H_{11}O_9P$	Asparagine	$C_4H_8N_2O_3$
Fructose-1, 6-phosphate	$C_6H_{10}O_{12}P_2$	Cysteine	$C_3H_7O_2NS$
Dihydroxyacetone phosphate	$C_3H_5O_6P$	Glutamic acid	$C_5H_9NO_4$
Glyceraldehyde-3-phosphate	$C_3H_5O_6P$	Glycine	$C_2H_5NO_2$
1,3-Diphosphoglycerate	$C_3H_4O_{10}P_2$	Leucine	$C_6H_{13}NO_2$
2,3-Diphosphoglycerate	$C_3H_3O_{10}P_2$	Isoleucine	$C_6H_{13}NO_2$
3-Phosphoglycerate	$C_3H_4O_7P$	Lysine	$C_6H_{14}N_2O_2$
2-Phosphoglycerate	$C_3H_4O_7P$	Histidine	$C_6H_9N_3O_2$
Phosphoenolpyruvate	$C_3H_2O_6P$	Phenylalanine	$C_9H_{11}NO_2$
Pyruvate	$C_3H_3O_3$	Proline	$C_5H_9NO_2$
Lactate	$C_3H_5O_3$	Serine	$C_3H_7NO_3$
6-Phosphogluco-lactone	$C_6H_9O_9P$	Threonine	$C_4H_9NO_3$
6-Phosphogluconate	$C_6H_{10}O_{10}P$	Tryptophane	$C_{11}H_{12}N_2O_2$
Ribulose-5-phosphate	$C_5H_9O_8P$	Tyrosine	$C_9H_{11}NO_3$
Ribulose-5-phosphate	$C_5H_9O_8P$	Valine	$C_5H_{11}NO_2$
Xylulose-5-phosphate	$C_5H_9O_8P$	Methionine	$C_5H_{11}O_2NS$
Ribose-5-phosphate	$C_5H_9O_8P$	Sedoheptulose-7-phosphate	$C_7H_{13}O_{10}P$
Erythrose-4-phosphate	$C_4H_7O_7P$	5-Phosphoribosyl-1-pyrophosphate	$C_5H_8O_{14}P_3$
Inosine monophosphate	$C_{10}N_4H_{12}O_8P$	Ribose-1-phosphate	$C_5H_9O_8P$
Hypoxanthine	$C_5N_4H_4O$	Inosine	$C_{10}H_{12}N_4O_5$

compounds that disappear in the reaction) and positive for the products (the compounds that appear in the reaction), this sum is zero. The number of atoms of an element on each side of the reaction is the same. For the elemental matrix in Equation (9.9) and the reaction vector $\mathbf{s}_i^v = (-2, -1, 2)^T$, we see that

$$\langle (0, 2, 1), (-2, -1, 2)^T \rangle = 0 \quad \text{and} \quad \langle (2, 0, 2), (-2, -1, 2)^T \rangle = 0 \qquad (9.11)$$

All elemental balancing equations taken together lead to the simple matrix equation:

$$\mathbf{ES} = \mathbf{0} \qquad (9.12)$$

Table 9.3   The elemental composition of the glycolytic intermediates. This table represents the matrix $\mathbf{E}$. Note that NAD is treated as one chemical moiety as it never changes in this system.

	Glu	G6P	F6P	FBP	DHAP	GAP	PG13	PG3	PG2	PE	PYR	LAC	NAD	NADH	AMP	ADP	ATP	P_i	H^+	H_2O
C	6	6	6	6	3	3	3	3	3	3	3	3	0	0	10	10	10	0	0	0
H	12	11	11	10	5	5	4	4	4	2	3	5	0	1	13	13	13	1	1	2
O	6	9	9	12	6	6	10	7	7	6	3	3	0	0	7	10	13	4	0	1
P	0	1	1	2	1	1	2	1	1	1	0	0	0	0	1	2	3	1	0	0
N	0	0	0	0	0	0	0	0	0	0	0	0	0	0	5	5	5	0	0	0
S	0	0	0	0	0	0	0	0	0	0	0	0	0	0	0	0	0	0	0	0
NAD	0	0	0	0	0	0	0	0	0	0	0	0	1	1	0	0	0	0	0	0

Although not shown here, the same must be true of compound electrical charge, as it is balanced during a chemical reaction. The rows of $\mathbf{E}$ are row vectors that are in the left null space of $\mathbf{S}$.

9.2.3   Example: glycolysis

We now give an example of how to formulate a stoichiometric matrix. For this purpose, we pick one of the most familiar pathways of all: glycolysis. This example originates from [317].

Defining the system   The glycolytic pathway degrades glucose (a six-carbon compound) to form pyruvate or lactate (three-carbon compounds) as end products. During this degradation process, the glycolytic pathway builds redox potential in the form of NADH and high-energy phosphate bonds in the form of ATP via substrate-level phosphorylation. Glycolysis also assimilates an inorganic phosphate group that is converted into a high-energy bond and then hydrolyzed in the ATP use reaction (or the 'load' reaction). Glycolysis as a system is shown in Figure 9.3.

The compounds or the nodes in the network   In its simplest form, glycolysis has 12 primary metabolites, 5 co-factor molecules (ATP, ADP, AMP, NAD, NADH), inorganic phosphate (P_i), protons (H^+) and water (H_2O). The system thus has 20 compounds (see Table 2.1). The elemental composition of these compounds is found in Table 9.3.

The reaction or the links in the network   The links formed between these compounds are the glycolytic reactions. There are 21 reactions, including all the transport reactions into and out of the system. The reactions are summarized in Table 2.2.

The stoichiometric matrix   The stoichiometric matrix, $\mathbf{S}$, can be formulated for the glycolytic system (see Table 9.4). Its dimensions are 20×21, representing the 20 metabolites and the 21 fluxes given in Tables 2.1 and 2.2, respectively.

Table 9.4 An annotated stoichiometric matrix for the glycolytic system in Figure 9.3. The matrix is partitioned to show the co-factors separate from the glycolytic intermediates and to separate the exchange reactions and co-factor loads. The last column has the connectivities, ρ_i, for a compound, and the last row has the participation number, π_i, for a reaction. These two quantities are described in Chapter 10. The second block in the table is the product **ES** to evaluate the elemental balancing status of the reactions. All exchange reactions have only one participating compound (i.e., a participation number of unity) and are not balanced elementally.

	Glycolytic reactions											AMP metabolism		Primary export		Co-factors		Primary inputs		Inorganic		
	υ_{HK}	υ_{PGI}	υ_{PFK}	υ_{TPI}	υ_{ALD}	υ_{GAPDH}	υ_{PGK}	υ_{PGLM}	υ_{ENO}	υ_{PK}	υ_{LDH}	υ_{AMP}	υ_{APK}	υ_{PYR}	υ_{LAC}	υ_{ATP}	υ_{NADH}	υ_{GLUin}	υ_{AMPin}	υ_{H^+}	υ_{H_2O}	ρ_i
Glu	-1	0	0	0	0	0	0	0	0	0	0	0	0	0	0	0	0	1	0	0	0	2
G6P	1	-1	0	0	0	0	0	0	0	0	0	0	0	0	0	0	0	0	0	0	0	2
F6P	0	1	-1	0	0	0	0	0	0	0	0	0	0	0	0	0	0	0	0	0	0	2
FBP	0	0	1	0	-1	0	0	0	0	0	0	0	0	0	0	0	0	0	0	0	0	2
DHAP	0	0	0	-1	1	0	0	0	0	0	0	0	0	0	0	0	0	0	0	0	0	2
GAP	0	0	0	1	1	-1	0	0	0	0	0	0	0	0	0	0	0	0	0	0	0	3
PG13	0	0	0	0	0	1	-1	0	0	0	0	0	0	0	0	0	0	0	0	0	0	2
PG3	0	0	0	0	0	0	1	-1	0	0	0	0	0	0	0	0	0	0	0	0	0	2
PG2	0	0	0	0	0	0	0	1	-1	0	0	0	0	0	0	0	0	0	0	0	0	2
PEP	0	0	0	0	0	0	0	0	1	-1	0	0	0	0	0	0	0	0	0	0	0	2
PYR	0	0	0	0	0	0	0	0	0	1	-1	0	0	-1	0	0	0	0	0	0	0	3
LAC	0	0	0	0	0	0	0	0	0	0	1	0	0	0	-1	0	0	0	0	0	0	2
NAD	0	0	0	0	0	-1	0	0	0	0	1	0	0	0	0	0	1	0	0	0	0	3
NADH	0	0	0	0	0	1	0	0	0	0	-1	0	0	0	0	0	-1	0	0	0	0	3
AMP	0	0	0	0	0	0	0	0	0	0	0	-1	1	0	0	0	0	0	1	0	0	3
ADP	1	0	1	0	0	0	-1	0	0	-1	0	0	-2	0	0	1	0	0	0	0	0	6
ATP	-1	0	-1	0	0	0	1	0	0	1	0	0	1	0	0	-1	0	0	0	0	0	6
P_i	0	0	0	0	0	-1	0	0	0	0	0	0	0	0	0	1	0	0	0	0	0	2
H^+	1	0	1	0	0	1	0	0	0	-1	-1	0	0	0	0	1	1	0	0	-1	0	8
H_2O	0	0	0	0	0	0	0	0	1	0	0	0	0	0	0	-1	0	0	0	0	-1	3
π_j	5	2	5	2	3	6	4	2	3	5	5	1	3	1	1	5	3	1	1	1	1	
C	0	0	0	0	0	0	0	0	0	0	0	-10	0	-3	-3	0	0	6	10	0	0	
H	0	0	0	0	0	0	0	0	0	0	0	-13	0	-3	-5	0	0	12	13	-1	-2	
O	0	0	0	0	0	0	0	0	0	0	0	-7	0	-3	-3	0	0	6	7	0	-1	
P	0	0	0	0	0	0	0	0	0	0	0	-1	0	0	0	0	0	0	1	0	0	
N	0	0	0	0	0	0	0	0	0	0	0	-5	0	0	0	0	0	0	5	0	0	
S	0	0	0	0	0	0	0	0	0	0	0	0	0	0	0	0	0	0	0	0	0	
NAD	0	0	0	0	0	0	0	0	0	0	0	0	0	0	0	0	0	0	0	0	0	

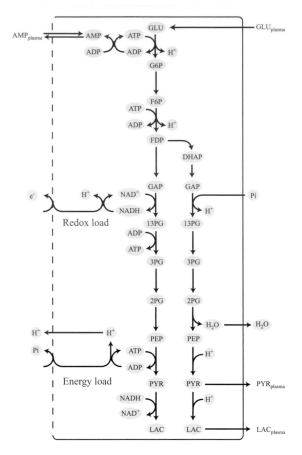

Figure 9.3   Glycolysis as a system: the reaction schema, co-factor interactions (across dashed line), and environmental exchanges (across solid line).

Elemental balancing   The stoichiometric matrix needs to be quality-controlled to make sure that the chemical equations are mass-balanced. The elemental compositions of the compounds in the glycolytic system are given in Table 9.3. This table is the elemental matrix, $\mathbf{E}$, for this system. We can multiply $\mathbf{ES}$ to quality control the reconstructed network for elemental balancing properties of the reactions (i.e., verify that $\mathbf{ES} = \mathbf{0}$). The results are shown in Table 9.4. All the internal reactions are balanced elementally. The exchange reactions are not balanced elementally as they represent net addition to or removal from the system as defined.

Charge balancing   In this example, we treat the compounds as being uncharged. This assumption is not physiologically accurate, but it will not affect many types of computations. If significant changes in pH are to be considered, then the charged state of the molecules needs to be established. A row can be added into $\mathbf{E}$ representing the charges of the molecules. Then, charge balance is ensured by making sure that

$\mathbf{ES} = \mathbf{0}$. Charge balancing the whole system for transporters can be difficult, because some of the transport systems, co-transport ions, and individual ions can cross the membrane by themselves. Overall, the system has to be charge-neutral. Accounting for full charge balances and the volume of a system can be quite involved mathematically, see [192,193].

9.3   Network Structure: S as a Connectivity Matrix

A network can be visually represented as a *map*. Each *node* in the map corresponds to a row in a *connectivity matrix*, and each column corresponds to a *link* in the map.

9.3.1   The maps of S

The reaction map   S represents a map where a compound is a node and the reactions connect (link) the compounds (Figure 9.4A). This map is the *reaction map* (also called reaction-centered map) and is the standard way of viewing metabolic reactions and pathways in biochemistry textbooks.

The compound map   The negative of the transpose of the stoichiometric matrix, $-\mathbf{S}^T$, also represents a map (Figure 9.4B), which we will call the *compound map* (also referred to as the metabolite-centered map). The map that $-\mathbf{S}^T$ represents has the reactions (now the rows in $-\mathbf{S}^T$) as the nodes in the network and the compounds (now the columns of $-\mathbf{S}^T$) as the connections, or the links. This representation of a biochemical reaction network is unconventional, but useful in many circumstances.

Examples: Simple examples of reaction and compound maps are shown in Figure 9.5. The compound map for glycolysis is shown in Figure 9.6. The compound map can be complicated notably by highly connected co-factor molecules.

9.3.2   Biological quantities displayed on maps

It is worth examining the columns $(\mathbf{s}_i^v)$ and rows of $\mathbf{S}$ a bit more closely. Let's examine a reaction:

$$x_1 + x_2 \overset{v_i}{\rightarrow} x_3 + x_4 \tag{9.13}$$

with the corresponding column of $\mathbf{S}$, $\mathbf{s}_i^v = (-1, -1, 1, 1)^T$. This vector is in the column space of $\mathbf{S}$. Moving along this vector is like carrying out this reaction. Note that motion along this vector will conserve the sum $x_1 + x_2 + x_3 + x_4$. Thus, a column in $\mathbf{S}$ represents a 'tie' between the compounds participating in a particular reaction. If these compounds participate in other reactions, there will be interactions between the motions along the columns of $\mathbf{S}$. These vectors, $\mathbf{s}_i^v$, span the column space of $\mathbf{S}$ and thus give a conceptually useful basis for the column space of $\mathbf{S}$. As we will see, certain combinations of the column vectors form pathways through the reaction map.

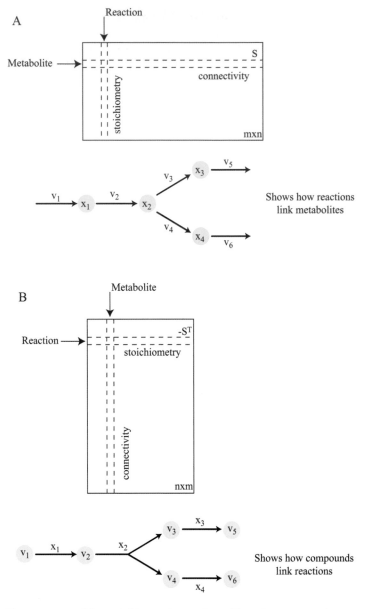

Figure 9.4   The structure of the stoichiometric matrix and how it corresponds to a map. Both the regular (reaction) and transpose (compound) maps are shown in panels A and B, respectively.

Similar observations apply to the rows of **S** (or the columns of $-\mathbf{S}^T$). A column in $-\mathbf{S}^T$ will 'tie' together, or connect, all the reactions in which a metabolite participates. These connections, however, do not imply any particular relationship among reactions, and therefore are not considered 'hard' connections. As will be further discussed in Chapter 11, metabolite pools form, which are the linear combinations of metabolite concentrations. These pools represent a conservation among metabolites that is

Reaction Maps

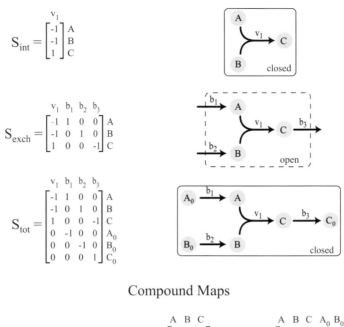

Compound Maps

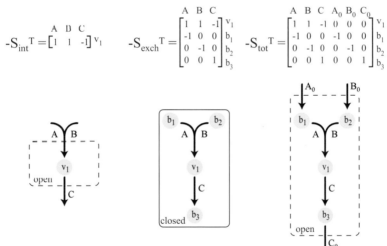

Figure 9.5 Simple examples of reaction maps versus compound maps. Reaction maps (top) show metabolites as nodes and reactions as directed edges. The reaction map includes both the internal and exchange fluxes, if present. In contrast, compound maps of the same systems (bottom) show the reactions as nodes and metabolites as directed edges. A systems boundary that allows for the exchange of the internal nodes is open on a reaction map. The compound map of an open reaction map is closed, and vice versa, as is shown by changing the network from (A) to (B) and to (C).

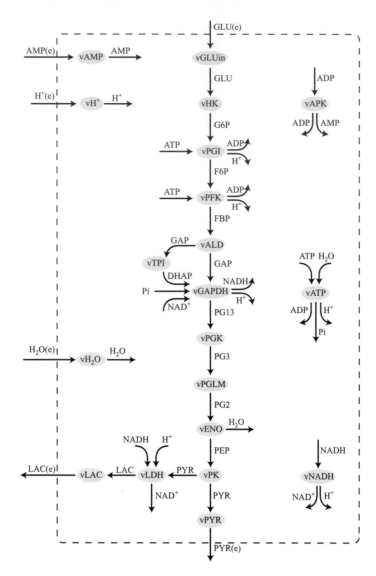

Figure 9.6   The compound map for glycolysis, shown in Figure 9.3. The connections have been broken up to simplify the appearance of the map. Prepared by Addiel De Alba Solis.

mediated by specific reactions, and therefore represents a more meaningful relationship among reactions in which the metabolites participate.

Note that the columns of **S**, in contrast, create a 'hard' connection between the metabolites, as a reaction will simultaneously use and produce the participating compounds. Conversely, the connectivities created between the reactions are 'soft,' as the reactions in which a compound participates can have varying flux levels that may not have fixed ratios. These ratios are determined by the kinetic properties of the reactions.

9.3.3   Linearity of maps

The topological structure of the maps formed by connectivity matrices are very important in determining the properties of the network. The topological properties of maps can be linear and non-linear.

Linear maps are made up of links that have only one input and one output. Thus, the columns of **S** will only have two entries, corresponding to the two nodes (metabolites) that the link (reaction) connects. Similarly for $\mathbf{S}^T$, if only one compound links two reactions, the map is linear. Although frequently used for illustrative purposes, the occurrence of such links in biological reaction networks is rare.

Non-linear maps are made up of links with more than one input or more than one output. The number of compounds that participate in a reaction can be found by adding up the non-zero elements in the corresponding column of **S**. In genome-scale metabolic models, the most common number of metabolites participating in a reaction is four, as in reaction (9.5). Thus, metabolic co-factors create non-linearity in the map of **S**.

Metabolites that participate in more than two reactions create a non-linearity in the map of $\mathbf{S}^T$. The participation number of a metabolite in genome-scale models can be as high as 150 (for ATP), but is 2 for most compounds found in the cell [105]. The metabolites that participate in many reactions thus create strong non-linearities in the compound map. The co-factors lead to strong non-linear topological features of metabolic networks.

9.4   Mathematics: S as a Linear Transformation

9.4.1   Mapping fluxes onto concentration time derivatives

Mathematically, the stoichiometric matrix **S** is a *linear transformation* (Figure 9.7) of the flux vector,

$$\mathbf{v} = (v_1, v_2, \ldots, v_n) \tag{9.14}$$

to a vector of time derivatives of the concentration vector

$$\mathbf{x} = (x_1, x_2, \ldots, x_m) \tag{9.15}$$

as:

$$\frac{\mathrm{d}\mathbf{x}}{\mathrm{d}t} = \mathbf{S}\mathbf{v} \tag{9.16}$$

The reader may also be familiar with other notations of time derivatives

$$\frac{\mathrm{d}\mathbf{x}}{\mathrm{d}t} = \mathbf{x}' = \dot{\mathbf{x}} \tag{9.17}$$

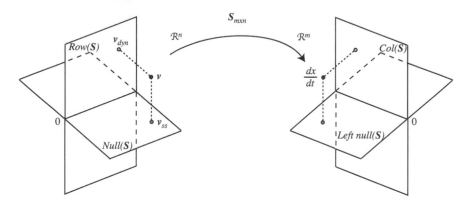

Figure 9.7   The stoichiometric matrix as a linear transformation. The four fundamental subspaces of **S** are shown. Prepared by Iman Famili.

which perhaps makes it clearer that the $d\mathbf{x}/dt$ is a vector, and that

$$\dot{\mathbf{x}} = \mathbf{S}\mathbf{v} \tag{9.18}$$

is a linear transformation.

Dimensions   There are m metabolites (x_i) found in the network and n reactions (v_i), thus:

$$\dim(\mathbf{x}) = m, \quad \dim(\mathbf{v}) = n, \quad \dim(\mathbf{S}) = m \times n \tag{9.19}$$

For a typical biological network there are more reactions than compounds, or $n > m$. The matrix **S** may not be full rank, and therefore $\mathrm{rank}(\mathbf{S}) = r < m$.

9.4.2   The four fundamental subspaces

There are four fundamental subspaces associated with a matrix. Figure 9.7 shows the four fundamental subspaces of **S**, which have important roles in the analysis of biochemical reaction networks, as detailed in the following chapters. The vector produced by a linear transformation is in two orthogonal spaces (the column and left null spaces), and the vector being mapped is also in two orthogonal spaces (the row and null spaces).

Dimensions of the fundamental subspaces   The mapping that the stoichiometric matrix represents is illustrated in Figure 9.7. The stoichiometric matrix is typically rank-deficient. The rank r of a matrix denotes the number of *linearly independent* rows and columns that the matrix contains. Rows are linearly dependent if any one row can be computed as a linear combination of the other rows. Linear dependency between the compounds and reactions determines the dimensionality of each of the four fundamental subspaces.

The dimensions of both the column and row space is r:

$$\dim(\mathrm{Col}(\mathbf{S})) = \dim(\mathrm{Row}(\mathbf{S})) = r.$$

Because the dimension of the concentration vector is m, we have

$$\dim(\mathrm{Left\ Null}(\mathbf{S})) = m - r.$$

Similarly, the flux vector is n-dimensional, thus

$$\dim(\mathrm{Null}(\mathbf{S})) = n - r.$$

Contents of the fundamental subspaces The four fundamental subspaces contain important information about a reaction network. Their contents are as follows.

Null space: the null space of **S** contains all the steady-state flux distributions allowable in the network. The steady state is of much interest as most homeostatic states are close to being steady states.

Row space: the row space of **S** contains all the dynamic flux distributions of a network, and thus the thermodynamic driving forces that change the rate of reaction activity.

Left null space: the left null space of **S** contains all the conservation relationships, or *time-invariants*, that a network contains. The sum of conserved metabolites or conserved metabolic pools do not change with time and are combinations of concentration variables.

Column space: the column space of **S** contains all the possible time derivatives of the concentration vector, and thus how the thermodynamic driving forces move the concentration state of the network.

Basis for vector spaces A *basis* for a space can be used to *span* the space. Thus, a basis describes all of the contents of a space. Different bases can be used for this purpose, including a linear basis like the commonly used orthonormal basis, and a convex basis for finite linear spaces. The choice of basis for the four fundamental subspaces becomes important because it influences the interpretation of the contents of a space. Singular value decomposition gives simultaneous orthonormal bases for all the four fundamental subspaces (Chapter 11). Chapter 12 gives alternative and more biologically meaningful sets of basis vectors.

9.4.3   Looking into the four fundamental subspaces

The column and left null spaces The time derivative is in the column space of **S** (denoted by $\mathrm{Col}(\mathbf{S})$) as can be seen from the expansion of $\mathbf{Sv}$:

$$\frac{d\mathbf{x}}{dt} = \mathbf{s}_1^v v_1 + \mathbf{s}_2^v v_2 + \cdots + \mathbf{s}_n^v v_n \qquad (9.20)$$

where the s_i^v are the *reaction vectors* that form the columns of **S**. The Col(**S**) is therefore spanned by the reaction vectors, s_i^v. The reaction vectors are structural features of the network, and are fixed. However, the fluxes v_i are scalar quantities and represent the flux through reaction i. The fluxes are variables. We do note that each flux has a maximal value, $v_i \leq v_{i,\max}$, and this limits the size of the time derivatives. Thus, only a portion of the column space is explored, i.e., we can cap the size of the column space of **S**. The vectors in the left null space (l_i) of **S** are orthogonal to the column space, i.e., $\langle l_j \cdot s_i^v \rangle = 0$. The vectors l_i represent a mass conservation (see Chapter 11).

The row and null spaces The flux vector can be decomposed into a dynamic component and a steady-state component,

$$\mathbf{v} = \mathbf{v}_{\text{dyn}} + \mathbf{v}_{\text{ss}} \tag{9.21}$$

The steady-state component satisfies

$$\mathbf{S}\mathbf{v}_{\text{ss}} = 0 \tag{9.22}$$

and $\mathbf{v}_{\text{ss}}$ is thus in the null space of **S** (see Chapter 12). The dynamic component of the flux vector, $\mathbf{v}_{\text{dyn}}$, is orthogonal to the null space and consequently is in the row space of **S**.

Recapitulation Each pair of subspaces, where **v** and $\dot{x}$ reside, form orthogonal sets to each other, and their dimensions sum up to the dimension of their corresponding vectors, i.e., dim(Null(**S**)) + dim(Row(**S**)) = n and dim(Left null(**S**)) + dim(Col(**S**))= m. These are introductory observations about **S** and its fundamental subspaces. In Chapter 11, we will study the individual fundamental subspaces in more detail.

9.5   Systems Science: S and Network Models

Dynamic mass balances A dynamic mass balance on a compound is formed by summing up the fluxes through all the reactions that form the compound and subtracting those that degrade it (see Figure 9.8). In general, such a dynamic mass balance is described by the ordinary differential equation:

$$\frac{dx_i}{dt} = \sum_k s_{ik} v_k = \langle s_i^x \cdot \mathbf{v} \rangle \tag{9.23}$$

representing a summation of all fluxes v_k that form compound x_i, and those that degrade it. s_{ik} is the corresponding stoichiometric coefficient. The set of all such differential equations that describe the dynamic mass balance of every compound in a network is represented by the matrix equation

$$\frac{dx}{dt} = \mathbf{S}\mathbf{v} \tag{9.24}$$

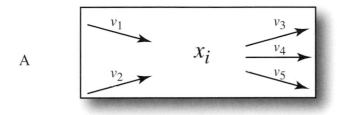

A

B

$$\frac{dx_i}{dt} = \underbrace{v_1 + v_2}_{formation} - \underbrace{v_3 - v_4 - v_5}_{degradation}$$

$$= \left\langle (1,1,-1,-1,-1) , (v_1,v_2,v_3,v_4,v_5)^T \right\rangle$$

$$= \left\langle \mathbf{s}_i^x \cdot \mathbf{v} \right\rangle$$

Figure 9.8 The dynamic mass balance on a single compound. Panel A shows all the rates of formation and degradation of a compound x_i (a graphical representation called a node map). Panel B shows the corresponding dynamic mass balance equation that simply states that the rate of change of the concentrations x_i is equal to the sum of the rates of formation minus the sum of the rates of degradation. This summation can be represented as an inner product between a row vector, $\mathbf{s}_i^x$ and the flux vector, $\mathbf{v}$. This row vector becomes a row in the stoichiometric matrix in Equation (9.16).

and thus $\mathbf{s}_i^x$ is a row in **S**. Equation (9.24) represents the fundamental equation of the *dynamic mass balances* that characterizes all functional states of a reconstructed biochemical reaction network. The stoichiometric matrix is a key component of this relationship.

Systems boundary The boundaries around a network can be drawn in different ways (see Figure 9.9). When defining a network, a *systems boundary* is drawn. The reactions are then partitioned into internal and exchange reactions. Exchange, or *boundary*, fluxes are denoted with b_i and internal fluxes with v_i. Similarly, the concentration vector is partitioned into internal (x_i) and external (c_i) concentrations. There are several different versions of **S** depending on what is encompassed by a network. A specific example is provided in Figure 9.10.

Defining the systems boundary Note that in the above consideration we have drawn a systems boundary around the cell. Such definition is common because it is consistent with physical realities. However, because the definition of a systems boundary can be chosen, we can segment any network into subnetworks by drawing 'virtual' boundaries. This property is useful in defining subsystems that may be 'fast' (i.e.,

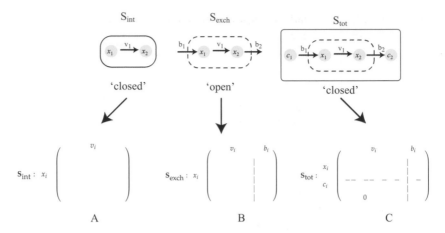

Figure 9.9   Schematic illustration of open and closed networks. (A) The internal stoichiometric matrix: considers the cell as a closed system; we focus just on the internal fluxes. This form is useful to define pools of compounds that are conserved (Chapter 11) and closed loop pathways (Chapter 12). (B) The exchange stoichiometric matrix: does not consider the external compounds, and contains the internal fluxes and the exchange fluxes with the environment. This form of the matrix is frequently used in pathway analysis of a network (see Chapter 12). (C) The total stoichiometric matrix. This form of **S** is the most general one. The dashed lines show the partitioning of the internal elements in the matrix. This form accounts for the internal reactions (v_i), the exchange reactions (b_i), the internal compounds (x_i), and the external compounds (c_i).

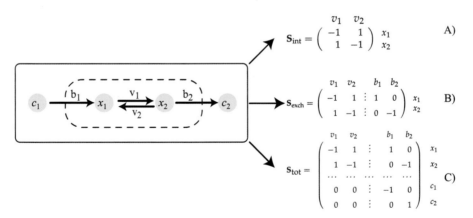

Figure 9.10   An example network. (A) The internal stoichiometric matrix has $m = 2$, $n = 2$, and $r = 1$. Thus, all the fundamental subspaces have a dimension of 1. The internal stoichiometric matrix can be further partitioned. The compounds that cannot be exchanged with the environment form one group, and those that can, form another. (B) The exchange stoichiometric matrix has $m = 2$, $n = 4$, and $r = 2$. It is full rank. The null space has a dimension of 2 (= 4 − 2), while the left null space has a dimension of 0 (= 2 − 2). (C) The total stoichiometric matrix has $m = 4$, $n = 4$, and $r = 3$. Thus, both of the null spaces are one-dimensional. In later chapters, we will learn how the dimensions of the null spaces relate to pools and pathways.

have rapid dynamics) and lead to temporal decomposition, and subsystems that have biochemical relevance (e.g., fatty acid biosynthesis).

9.6   Summary

- The stoichiometric matrix is a mathematical representation of a reconstructed network. It represents the biochemical, genomic, and genetic information on which the reconstruction is based.
- The stoichiometric matrix consists of stoichiometric coefficients that are integer numbers. The columns of the stoichiometric matrix represent chemical reactions while the rows represent compounds.
- The stoichiometric matrix includes informatic, chemical, physical, genetic, genomic, mathematical, numerical, and systemic attributes of the biological system that it represents.
- The stoichiometric matrix has many important features: (1) it is a data matrix, (2) it gives the structure of a network, (3) it is a mathematical mapping operation with four fundamental subspaces, and (4) it forms a key part of *in silico* models describing the functional states of networks.
- The stoichiometric matrix is a data matrix. The columns, $\mathbf{s}_i^v$, represent chemical transformations and thus come with chemical information. They are reaction vectors, that imply elemental and charge balance. The reaction vectors are thus orthogonal to the rows of the elemental matrix. These conservation quantities are in the left null space of the stoichiometric matrix. Some quantities, such as free energy, are non-conserved during a chemical reaction. These quantities will be in the row space of the stoichiometric matrix.
- The stoichiometric matrix is a connectivity matrix; it represents a reaction map. The transpose of the stoichiometric matrix represents a compound map. Both maps are topologically non-linear, as they contain joint edges between nodes.
- Mathematically, the stoichiometric matrix represents a transformation, or a mapping, of one vector (the flux vector) to another (the vector of time derivatives of the concentrations). Such a mapping operation comes with four fundamental subspaces (the row, the null, the column, and the left null spaces). Each one of these spaces contains chemically and physically meaningful quantities.
- The stoichiometric matrix is a key component of a systems model of network functions that is in the form of dynamic mass balances. The boundaries of a reaction network can be drawn in different ways and lead to three fundamental forms of $\mathbf{S}$.

10 Simple Topological Network Properties

Topology is the property of something that doesn't change when you bend it or stretch it as long as you don't break anything – Edward Witten

The stoichiometric matrix is a connectivity matrix. Elementary topological properties of the network it represents can be computed directly from the individual elements of **S**. Direct topological studies are interesting from a variety of standpoints. They focus on relatively easy to understand and intuitive properties of the structure of the network. Elementary topological properties relate to how connected a network is, and how its components participate in forming the connectivity properties of the network. There may be many *functional states* for a given network structure (see Chapter 16). Topological properties are thus global and less specific than functional states of networks. Some of the differences between functional states and network topology are covered in Part III.

10.1 The Binary Form of S

The elementary topological properties are determined based on the non-zero elements in the stoichiometric matrix. Thus, we define the elements of a new matrix $\hat{\mathbf{S}}$ as

$$
\begin{aligned}
\hat{s}_{ij} = 0 \quad &\text{if} \quad s_{ij} = 0 \\
\hat{s}_{ij} = 1 \quad &\text{if} \quad s_{ij} \neq 0
\end{aligned}
\tag{10.1}
$$

which is the *binary form* of **S**. This matrix is composed of only zeros and ones. If $\hat{s}_{ij}$ is unity, it means that compound i participates in reaction j. Note that in the rare case where a homodimer is formed, i.e., in a reaction of the type $2A \rightarrow A_2$, the stoichiometric coefficient of two becomes unity in the binary form of **S**.

S is a sparse matrix A number of genome-scale stoichiometric matrices have been reconstructed (see [2]). As there are typically only a handful of compounds that participate in a reaction out of hundreds of compounds participating in a network, the stoichiometric matrix is *sparse*. A sparse matrix is mostly composed of zero elements. For instance, if there are on average 3 compounds that participate in a reaction, but there are m compounds in the network, then the fraction of non-zero elements in the

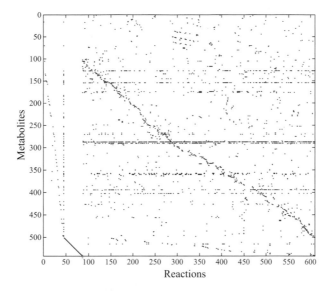

Figure 10.1 A stoichiometric matrix for *Geobacter sulfurreducens*. The dimensions of this matrix are $m = 541$ and $n = 609$, giving rise to 329,469 elements in the matrix. Of these, 2655, or 0.81%, of the elements are non-zero. Image provided by Radhakrishnan Mahadevan.

matrix is $3/m$. If m is 300, then only 1% of the elements are non-zero and the matrix is sparse.

A pictorial representation of a genome-scale stoichiometric matrix for *Geobacter sulfurreducens* is shown in Figure 10.1. The 2655 non-zero entries are indicated. They represent only 0.81% of the total of 329,469 elements in the matrix.

10.2 Participation and Connectivity

One can easily define and compute simple properties of the stoichiometric matrix that describe its topological features. The number of non-zero entries in a row and a column of **S** give two elementary topological properties that are easy to understand.

Reaction participation number The sum of the non-zero entries in a column:

$$\pi_j = \sum_{i=1}^{m} \hat{s}_{ij} \tag{10.2}$$

gives the number of compounds that participate in reaction j. This quantity, π_j, can be called the *participation number* for a reaction. For elementary reactions, this number is most likely three. Note that all compounds have to participate in a reaction for it to take place. It represents the number of nodes that form an edge in the reaction map.

Compound connectivity The sum of the number of non-zero entries in a row:

$$\rho_i = \sum_{j=1}^{m} \hat{s}_{ij} \tag{10.3}$$

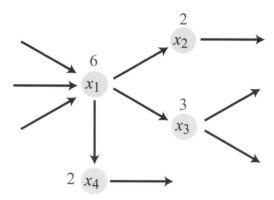

Figure 10.2 Simple reaction network illustrating connectivities of network nodes. The network shown has four nodes (compounds). The connectivities, ρ_i, for each node are given in the red numbers next to a node. x_1 represents the most connected node.

gives the number of reactions in which compound i participates. This number, ρ_i, is a measure of how connected, or linked, a compound is in the network. A compound that participates in a large number of reactions will form a highly connected node on the reaction map (see Figure 10.2). This number is called the *connectivity number*, for the node, or simply its *connectivity*.

10.2.1   Rearranging the stoichiometric matrix

The order of the rows and the columns in **S** is arbitrary. We can reorder them without altering the information content of the matrix. Clearly, the participation numbers and connectivities are not altered because they are computed based on columns or rows, respectively.

Mathematical operators to shift rows and columns   The rows and columns of a matrix can be moved around by simple matrix operations. If we multiply a matrix that is composed of rows $\mathbf{r}_i$ by the identity matrix, the matrix is intact:

$$\begin{pmatrix} 1 & 0 & 0 & 0 \\ 0 & 1 & 0 & 0 \\ 0 & 0 & 1 & 0 \\ 0 & 0 & 0 & 1 \end{pmatrix} \begin{pmatrix} \mathbf{r}_1 \\ \mathbf{r}_2 \\ \mathbf{r}_3 \\ \mathbf{r}_4 \end{pmatrix} = \begin{pmatrix} \mathbf{r}_1 \\ \mathbf{r}_2 \\ \mathbf{r}_3 \\ \mathbf{r}_4 \end{pmatrix} \tag{10.4}$$

if we switch rows in the identity matrix, then the multiplication will shift the rows in a corresponding way. For instance, if we want to switch rows 2 and 3 we compute

$$\begin{pmatrix} 1 & 0 & 0 & 0 \\ 0 & 0 & 1 & 0 \\ 0 & 1 & 0 & 0 \\ 0 & 0 & 0 & 1 \end{pmatrix} \begin{pmatrix} \mathbf{r}_1 \\ \mathbf{r}_2 \\ \mathbf{r}_3 \\ \mathbf{r}_4 \end{pmatrix} = \begin{pmatrix} \mathbf{r}_1 \\ \mathbf{r}_3 \\ \mathbf{r}_2 \\ \mathbf{r}_4 \end{pmatrix} \tag{10.5}$$

Columns, $\mathbf{c}_i$, can similarly be rearranged by post-multiplication

$$(c_1, c_2, c_3, c_4) \begin{pmatrix} 1 & 0 & 0 & 0 \\ 0 & 0 & 1 & 0 \\ 0 & 1 & 0 & 0 \\ 0 & 0 & 0 & 1 \end{pmatrix} = (c_1, c_3, c_2, c_4) \tag{10.6}$$

More complex sorting operations can be performed in spreadsheet programs if the matrix is available in such a format.

Example: reordering the S for the core *E. coli* model based on participation numbers and connectivities Reordering of the stoichiometric matrix for the core *E. coli* model (see Chapter 2) is shown in Figure 10.3. The figure shows the relationship between the connectivity distribution (on the right of the image) and the rows. Similarly, the ordering of the columns and the reaction participation numbers is shown. This arrangement of **S** gives a visualization of the data that the matrix represents. In particular, the inorganics and the co-factors are found in the top rows. The exchange reactions are found in the last set of columns as their participation number is one.

10.2.2 Connectivities in genome-scale matrices

As soon as the first genome-scale matrices had been reconstructed, the connectivities for all the metabolites were computed (see Figure 10.4A). Such computations show that there are relatively few metabolites (two dozen or so) that are highly connected, while most of the metabolites participate in only two reactions. This result is a reflection of the fact that few carrier molecules participate in a large numbers of reactions and a few metabolites are key to certain metabolic functions, such as nitrogen, one-carbon, and two-carbon metabolism (see Table 10.1).

A surprising finding was the approximate linear appearance of the curve of the connectivities when the metabolites were rank-ordered by decreasing connectivity, when plotted on a log–log scale [105] (see Figure 10.4A). This curve can be redrawn based on the probability that a metabolite has a certain connectivity. Thus, there is a high probability of low connectivity and a low probability of high connectivity. Plotting these probabilities as a function of connectivity gives an approximate straight line on a log–log plot (Figure 10.4B). Networks that show such a power-law distribution are said to be *scale-free* [27].

The most highly connected nodes are carrier molecules and due to their high connectivity they form the dominant features of **S** (see Chapter 12). Such biochemical insight has led to the analysis of the connectivity distributions of decomposed forms of **S** based on biochemical classification of reactions. Such analysis has concluded that genome-scale stoichiometric matrices are actually *scale-rich* and that the overall power-law property is the result of the amalgamation of the connectivity properties of biochemically classified modules [414].

Biological interpretation The biological significance of the power-law distribution is not clear. It has been suggested that the most highly connected nodes in a network may represent the compounds that were in the network 'first' in evolutionary time [448]. Such interpretations must of course be made in view of the constraining chemistry; for instance, there is a finite number of chemical transformations that a particular

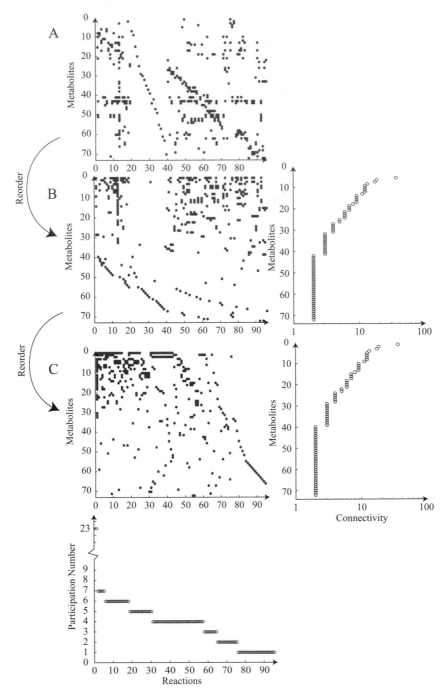

Figure 10.3 Reordering the rows and columns in the stoichiometric matrix for the core *E. coli* metabolic network. Panel A: an image of **S** for the core *E. coli* metabolic model. Panel B: the same matrix ordered by the metabolite connectivities. They are shown in the graph to the right of the row ordered matrix. Panel C: the matrix of panel B with the columns ordered based on compound participation numbers in the reactions. The compound participation numbers are shown in the graph at the bottom of the matrix and the metabolic connectivities to the right. Prepared by Harish Nagarajan.

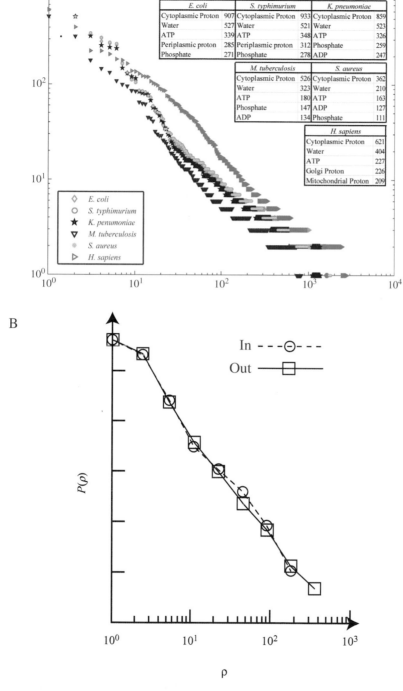

	E. coli		*S. typhimurium*		*K. pneumoniae*	
Cytoplasmic Proton	907	Cytoplasmic Proton	933	Cytoplasmic Proton	859
Water	527	Water	521	Water	523
ATP	339	ATP	348	ATP	326
Periplasmic proton	285	Periplasmic proton	312	Phosphate	259
Phosphate	271	Phosphate	278	ADP	247

M. tuberculosis		*S. aureus*	
Cytoplasmic Proton	526	Cytoplasmic Proton	362
Water	323	Water	210
ATP	180	ATP	163
Phosphate	147	ADP	127
ADP	134	Phosphate	111

H. sapiens	
Cytoplasmic Proton	621
Water	404
ATP	227
Golgi Proton	226
Mitochondrial Proton	209

◇	*E. coli*
○	*S. typhimurium*
★	*K. penumoniae*
▽	*M. tuberculosis*
●	*S. aureus*
▷	*H. sapiens*

Figure 10.4 (A) The distribution of node connectivities in the reconstructed first four genome-scale matrices. Data taken from [105, 106, 109, 130, 380]. Prepared by Harish Nagarajan. (B) Connectivity probability distributions for metabolites. ρ represents the connectivity of a node, and $P(\rho)$ represents the probability that a node has connectivity ρ. The connectivity distribution is averaged over 43 different organisms. The connectivity of reactions into a node (In) and out of a node (Out) are represented separately. From [190].

Table 10.1 Some activated carriers or co-enzymes in metabolism, modified from [219].

Co-enzyme/carrier	Examples of chemical groups transferred	Dietary precursor in mammals
Carbon		
Biocytin	CO_2	Biotin
Tetrahydrafolate	One-carbon groups	Folate (B9)
Coenzyme A (CoA)	Acyl groups	Pantothenic acid (B5) and other compounds
Thiamine pyrophosphate (TPP)	Aldehydes	Thiamine (B1)
S-adenosyl methionine	methyl groups	
cyano/5'-deoxyadenosyl cobalamin	rearrangement of vicinal -H and -R groups	Coenzyme B12
Energy		
ATP (NTPs)	Pi or PPi	
Redox		
Flavin adenin dinucleotide (FAD)	Electrons and protons	Riboflavin (B2)
Lipoate	Electrons and acyl groups	
Nicotinamide adenine dinucleotide (NAD or NADP)	Electrons and protons	
Coenzyme Q	Electrons and protons	Vitamin Q
Cytochromes / heme	Electrons	Iron
Nitrogen		
Pyridoxal phosphate	Amino groups	Pyridoxine (B6)
Glutamate (N_2-fixing plants)	Amino groups	

metabolite can undergo. Similarly, the 'attack tolerance' of a network is such that the removal of the most highly connected nodes has the broadest impact on network functions [189,190]. This consideration may apply to regulatory networks. Conversely, it is not possible to simply delete a metabolite from a network, but a link can be severed.

Node connectivity and network states It should be noted that highly connected nodes may represent effective targets for drug development. However, topological properties of networks must be interpreted in the context of the more biologically relevant functional network states and their properties. One such consideration, for instance, is that a metabolic network must make all the biomass components of the cell in order for it to grow. Therefore, even eliminating a step in a linear low-flux pathway leading to the synthesis of co-factors, vitamins, or amino acids, will prevent a genome-scale metabolic network from supporting growth. It has been found that the connectivity of a node does not correlate with the lethality of its links [255].

10.3   Linked Participation and Connectivities

The connectivities and participation numbers are very elementary topological properties. An expanded set of elementary topological network properties can be obtained from the two adjacency matrices of $\hat{\mathbf{S}}$. One relates to the columns of $\hat{\mathbf{S}}$, while the other relates to the rows.

10.3.1   The adjacency matrices of $\hat{\mathbf{S}}$

The reaction adjacency matrix A_v The pre-multiplication of a matrix by its transpose

$$\mathbf{A}_v = \hat{\mathbf{S}}^T \hat{\mathbf{S}} \tag{10.7}$$

leads to a symmetrical matrix whose elements are the inner product of its columns, $\hat{\mathbf{s}}_i$. The diagonal elements of $\mathbf{A}_v$ are:

$$(\mathbf{a}_v)_{ii} = <\hat{\mathbf{s}}_i^T \cdot \hat{\mathbf{s}}_i> = \sum_k \hat{s}_{ki}^2 \tag{10.8}$$

Thus, because the elements of $\hat{\mathbf{s}}_i$ are 0 or 1, this summation simply represents the number of non-zero elements in $\hat{\mathbf{s}}_i$, or the number of compounds that participate in the reaction. The diagonal elements of $\mathbf{A}_v$ are thus the same quantity as given in Equation (10.2).

The off-diagonal elements are given by:

$$(\mathbf{a}_v)_{ji} = <\hat{\mathbf{s}}_j^T \cdot \hat{\mathbf{s}}_i> = \sum_k \hat{s}_{jk}\hat{s}_{ki} \tag{10.9}$$

These elements can count how many compounds two reactions (reactions i and j) have in common.

The compound adjacency matrix, A_x The post-multiplication of a matrix by its transpose

$$\mathbf{A}_x = \hat{\mathbf{S}}\hat{\mathbf{S}}^T \tag{10.10}$$

leads to a symmetric matrix whose elements are the inner products of its rows. The diagonal elements are:

$$(\mathbf{a}_x)_{ii} = \sum_k \hat{s}_{ik}^2 \tag{10.11}$$

This summation gives the number of reactions in which compound x_i participates. This is the same quantity as computed in Equation (10.3). The off-diagonal elements are:

$$(\mathbf{a}_x)_{ij} = \sum_k \hat{s}_{ik}\hat{s}_{kj} \tag{10.12}$$

This is the number of reactions in which compounds x_i and x_j both participate, and thus how extensively the two compounds are connected topologically in the network.

10.3.2   Computation of the adjacency matrices

The reversible reaction   The stoichiometric matrix for a simple reversible reaction

$$\hat{\mathbf{S}} = \begin{pmatrix} 1 & 1 \\ 1 & 1 \end{pmatrix} \tag{10.13}$$

has two identical adjacency matrices:

$$\mathbf{A}_v = \mathbf{A}_x = \begin{pmatrix} 2 & 2 \\ & 2 \end{pmatrix} \tag{10.14}$$

As the matrices $\mathbf{A}_v$ and $\mathbf{A}_x$ are symmetric, the elements below the diagonal are left blank. Thus, each compound participates in two reactions (the forward and backward reactions are treated separately), and there are two compounds participating in each reaction.

The reversible bi-molecular reaction   The stoichiometric matrix for a reversible bi-molecular association

$$\hat{\mathbf{S}} = \begin{pmatrix} 1 & 1 \\ 1 & 1 \\ 1 & 1 \end{pmatrix} \tag{10.15}$$

has adjacency matrices:

$$\mathbf{A}_v = \begin{pmatrix} 3 & 3 \\ & 3 \end{pmatrix} \quad \text{and} \quad \mathbf{A}_x = \begin{pmatrix} 2 & 2 & 2 \\ & 2 & 2 \\ & & 2 \end{pmatrix} \tag{10.16}$$

Thus, $\mathbf{A}_v$ states that there are three compounds participating in each reaction (the forward and backward), and the two reactions have three compounds in common. Similarly, $\mathbf{A}_x$ states that each compound participates in two reactions (the diagonal) and that the first and second, first and third, and second and third compounds participate jointly in two reactions (forward and backward).

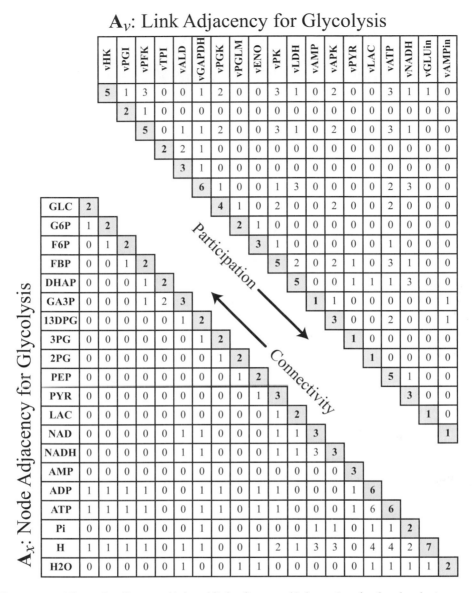

Figure 10.5 The node adjacency ($\mathbf{A}_x$) and link adjacency ($\mathbf{A}_v$) matrices for the glycolysis network. The adjacency matrices are symmetric allowing their simultaneous presentation in a single figure. Prepared by Harish Nagarajan.

Glycolysis As can be seen in Figure 10.5, the diagonal elements of $\mathbf{A}_x$ show the connectivity of the metabolites in the glycolysation pathway. The metabolites that dominate the network with high connectivities include the proton (7) and ATP, ADP (6). The diagonal elements of $\mathbf{A}_v$ show the participation number for the reactions involved in glycolysis. The reaction that has the highest number of metabolites participating in the glycolysis network is v_{GAPDH} with six metabolites. There are five reactions with the second-highest participation number of five in this network.

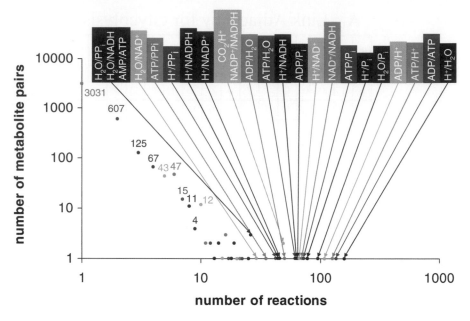

Figure 10.6   Metabolite coupling in *E. coli*. The number of metabolite pairs that share a given number of reactions are plotted and identified from the *E. coli* metabolic network. From [33].

Genome-scale matrices   The off-diagonal elements of $\mathbf{A}_x$ for the genome-scale metabolic networks for *E. coli*, *S. cerevisiae*, *H. pylori*, *S. aureus*, and the human cardiac mitochondrion have been studied [37]. When rank-ordered, they approximate a line on a log–log plot for all networks considered. This suggests the notion of metabolite coupling, the concept that pairs of metabolites influence network behavior on different scales. The results for *E. coli* are shown in Figure 10.6.

10.4   Summary

- The binary form of **S** is $\hat{\mathbf{S}}$, which has zeros everywhere except unity, where a non-zero element appears in **S**.
- The summation of the elements in column j of $\hat{\mathbf{S}}$ give the number of compounds, π_j, that participate in reaction j.
- The summation of the elements in row i of $\hat{\mathbf{S}}$ give the number of reactions, ρ_i, in which compound i participates, or how connected it is in the network.
- The binary stoichiometric matrix, $\hat{\mathbf{S}}$, has two adjacency matrices, $\mathbf{A}_v$ and $\mathbf{A}_x$, that are reaction- and compound-associated, respectively.
- A diagonal element of $\mathbf{A}_v$ gives the number of compounds (i.e., π_j) that participate in that reaction, and an off-diagonal element gives the number of compounds that the two corresponding reactions have in common.
- A diagonal element of $\mathbf{A}_x$ gives the number of reactions in which the corresponding compound participates (i.e., ρ_i), and an off-diagonal

element gives the number of reactions in which the two corresponding compounds both participate.

- The number of reactions in which compounds participate follow an approximate power-law distribution in genome-scale matrices of metabolism. The number of reactions in which pairs of metabolites participate also follows a power-law distribution.
- The rows and columns in a matrix can be ordered without loss of information in the matrix.

11 Fundamental Network Properties

The stoichiometric matrix is so informative about physiological states that we must study its fundamental properties – John Doyle

In the last chapter, we discussed the simple topological properties of the network that the stoichiometric matrix represents. In this chapter, we look deeper into the properties of the stoichiometric matrix, and how fundamental network properties can be used to obtain a more thorough understanding of the reaction network that it represents. This material is perhaps the most mathematical part of this book. It should be readily accessible to readers with formal education in the physical and engineering sciences, while readers with a life science background may find it challenging. The stoichiometric matrix is a mathematical mapping operation (recall Figure 9.7). Matrices have certain fundamental properties that describe this mapping operation. These properties are contained in the four *fundamental subspaces* associated with a matrix. This chapter discusses these subspaces and how we can mathematically define them and begin the process of interpreting their contents in biochemical and biological terms.

11.1 Singular Value Decomposition

Singular value decomposition (SVD) of a matrix is a well-established method used in a wide variety of applications, including signal processing, noise reduction, image processing, kinematics, and for the analysis of high-throughput biological data [15, 169]. Unlike matrices composed of experimentally determined numbers, the stoichiometric matrix is a 'perfect' matrix that is commonly composed of integers describing the structure of a reaction network. SVD of **S** can be used to analyze network properties and it is a particularly useful way to obtain the basic information about the four fundamental subspaces of **S**.

11.1.1   Decomposition into three matrices

Mathematical format   SVD states that for a matrix **S** of dimension $m \times n$ and of rank r, there are orthonormal matrices **U** (of dimension $m \times m$) and **V** (of dimension $n \times n$), and a matrix (of dimension $m \times n$) with diagonal elements $\Sigma = \mathrm{diag}(\sigma_1, \sigma_2, \ldots, \sigma_r)$ with

$\sigma_1 \geq \sigma_2 \geq ... \geq \sigma_r > 0$ such that:

$$\mathbf{S} = \mathbf{U\Sigma V}^T \tag{11.1}$$

where the superscript T denotes the transpose.

SVD of $\mathbf{S}$ is shown schematically in Figure 11.1. The columns of $\mathbf{U}$ and $\mathbf{V}$ are the left and right singular vectors of $\mathbf{S}$, respectively, and represent its *modes*, while the σ_i represent the singular values. The values in $\mathbf{\Sigma}$ give us the weight with which the modes contribute to the reconstruction of the matrix. These are rank-ordered by decreasing magnitude in $\mathbf{\Sigma}$ with the largest singular value being first.

SVD as a series of transformations The basic mathematical nature of these three transformations is shown in Figure 11.2. $\mathbf{V}^T$ represents orthonormalization of the flux space and these basis vectors are stretched by the singular values and mapped onto an orthonormal basis for the concentration space. The transformation $\mathbf{U}$ then converts the orthonormal combination of the concentrations back to the original coordinate system. This set of transformations is conceptually useful. SVD has certain properties that make it convenient for numerical and mathematical analysis. The requirement for orthonormality makes chemical and biological interpretation difficult, and, as we will see in the subsequent chapters, an alternate set of basis vectors can be used to further such interpretations.

A note on nomenclature The naming conventions of the right singular vectors $\mathbf{v}_k$ and the flux vector $\mathbf{v}$ may cause confusion. Unfortunately, the literature uses the symbol $\mathbf{v}$ for both quantities, a convention that we will not change here. Both are vectors, denoted with a bold face font, but one has a subscript and the other does not. Equation (11.15) should help illustrate the difference between the two.

11.1.2 The content of U, Σ, and V

The singular value spectrum Singular value spectrum are formed by plotting σ_i as a function of i. This spectra is often presented by fractional and cumulative fractional singular value spectra. The fractional singular values are calculated by

$$f_i = \frac{\sigma_i}{\sum_{k=1}^{r} \sigma_k} \tag{11.2}$$

thereby normalizing the spectra to its total length. The cumulative fractional singular values, F_i, are defined as the sum of the first i fractional singular values

$$F_i = \sum_{k=1}^{i} f_k \tag{11.3}$$

where i varies from 1 to r. Note that $F_r = 1$.

Two types of spectra are illustrated in Figure 11.3. In the spectrum represented by the squares, the singular values are of similar magnitude, and thus the cumulative fractional singular values form a linear curve. Conversely, in the spectrum shown with

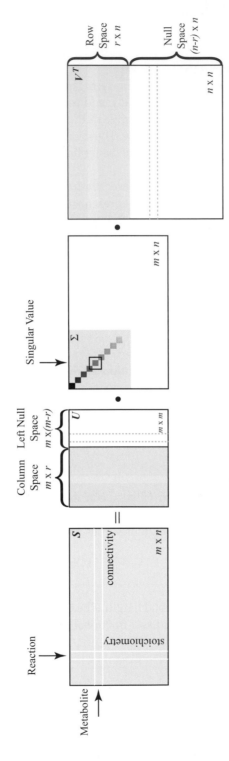

Figure 11.1 A schematic showing the singular value decomposition (SVD) of the stoichiometric matrix. The location of the orthonormal basis vectors for the four fundamental subspaces are indicated. Prepared by Iman Famili.

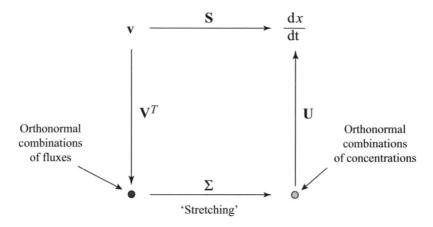

Figure 11.2   A schematic illustration of the singular value decomposition of **S**.

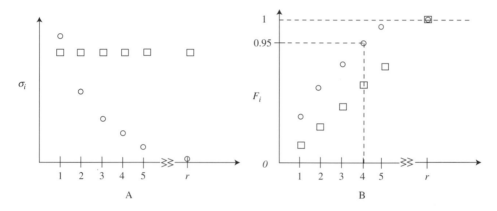

Figure 11.3   Singular value spectra. (A) The σ_i are rank-ordered and plotted as a function of i. (B) Cumulative fractional singular values, F_i. One can use a certain fraction, i.e., 0.95, to determine the 'effective dimensionality' representing 95% of the variance of the matrix being decomposed.

the open circles, the relative magnitude of the singular values drops quickly, and the cumulative fractional spectrum rises quickly, leading to a low effective dimensionality of the mapping that the matrix represents.

In data analysis, one often uses a numerical criterion (i.e., 0.95) to terminate the cumulative spectrum and define the number of modes that generate 95% of the reconstruction of the matrix, as illustrated in Figure 11.3B. Because the stoichiometric matrix is of perfect precision, meaning no 'measurement' noise in its elements, such cut-off may not be appropriate, depending on the information sought from the SVD.

Orthonormal bases for the four fundamental subspaces   The columns of **U** are called the *left singular vectors* and the columns of **V** are the *right singular vectors*. The columns of **U** and **V** give orthonormal bases for all the four fundamental subspaces of **S** (see Figure 11.1). The first r columns of **U** and **V** give orthonormal bases for the column and row spaces, respectively. The last $m - r$ columns of **U** give an orthonormal basis for the left null space, and the last $n - r$ columns or $\mathbf{V}$ give an orthonormal basis for the null space.

The inner product of orthonormal vectors is zero. The inner product of an orthonormal vector with itself is unity. Thus,

$$\mathbf{U}^T\mathbf{U} = \mathbf{I}_{(m \times m)} \quad \text{and} \quad \mathbf{V}^T\mathbf{V} = \mathbf{I}_{(n \times n)} \tag{11.4}$$

where we put the dimensions of the identity matrix as a subscript to emphasize their different dimensions. The transposes of $\mathbf{U}$ and $\mathbf{V}$ are thus their inverses as well.

11.1.3   Key properties of the SVD

Mapping between the singular vectors   The equation $\mathbf{S} = \mathbf{U}\mathbf{\Sigma}\mathbf{V}^T$ can be rewritten as:

$$\mathbf{S}\mathbf{V} = \mathbf{U}\mathbf{\Sigma} \tag{11.5}$$

which can be expanded in terms of a series of independent equations as:

$$\mathbf{S}\mathbf{v}_k = \sigma_k \mathbf{u}_k \tag{11.6}$$

In other words, $\mathbf{S}$ maps a right singular vector onto the corresponding left singular vector scaled by the corresponding singular value. A right singular vector (a column in $\mathbf{V}$) gives the weightings on the reaction vectors $\mathbf{s}_i^v$ needed to reconstruct each of the left singular vectors (a column in $\mathbf{U}$) as scaled by their respective singular values. Below we show how this relationship between the singular vectors can be interpreted in terms of network-level chemical transformation.

Mode-by-mode reconstruction of S   The stoichiometric matrix can be expanded into a sum as

$$\mathbf{S} = \sum_{i=0}^{r} \sigma_i (\mathbf{u}_i \otimes \mathbf{v}_i^T) \tag{11.7}$$

where $\otimes$ designates the outer product, and where each term in the summation successively adds the contribution of each mode (or singular vector) to the reconstruction of $\mathbf{S}$. As $\|\mathbf{u}_i\| = \|\mathbf{v}_i\| = 1$, the outer product is:

$$\|(\mathbf{u}_i \otimes \mathbf{v}_i^T)\| \approx o(1) \tag{11.8}$$

thus the σ_i values give the relative magnitude of the terms in the summation that reconstructs $\mathbf{S}$. Therefore, if the first three singular values are $\sigma_1 = 100$, $\sigma_2 = 10$, and $\sigma_3 = 1$, then the summation is

$$\mathbf{S} = 100 \cdot (\text{matrix } o(1)) + 10 \cdot (\text{matrix } o(1)) + 1 \cdot (\text{matrix } o(1)) + \cdots \tag{11.9}$$

Thus the singular values directly give the relative contribution of each outer product (i.e., each mode) to the reconstruction of the matrix. This property is the basis for truncating the singular value spectrum at a given point, as illustrated in Figure 11.3.

11.2   SVD and Properties of Reaction Networks

As discussed in Chapter 9, the stoichiometric matrix represents a transformation of the flux vector, $\mathbf{v}$, that resides in the row and the null space, to the vector of time derivatives, $\mathbf{x}'$, that resides in the column and left null space. These fundamental spaces of $\mathbf{S}$ all contain chemically and physically meaningful quantities. Here we show how the mapping from the row space to the column space fundamentally represents drivers and motions that correspond to an orthonormal set of *systems reactions*.

Orthonormal linear combinations of concentrations and fluxes The orthonormal basis vectors that are obtained from SVD do give useful information about the properties of the overall chemical transformations that characterize a network. The basic dynamic mass balance equation:

$$\frac{d\mathbf{x}}{dt} = \mathbf{Sv} \tag{11.10}$$

can be rearranged as

$$\mathbf{U}^T \frac{d\mathbf{x}}{dt} = \mathbf{U}^T \mathbf{S}\mathbf{V}\mathbf{V}^T \mathbf{v} \tag{11.11}$$

or

$$\frac{d(\mathbf{U}^T \mathbf{x})}{dt} = \mathbf{\Sigma}(\mathbf{V}^T \mathbf{v}) \tag{11.12}$$

Thus, the left singular vectors ($\mathbf{u}_i$) form linear combinations of the concentration variables and the right singular vectors ($\mathbf{v}_i$) form linear combinations of the fluxes.

The dynamic relationship between the groupings of fluxes and concentrations that correspond to the non-zero singular values can be written as:

$$\frac{d(\mathbf{u}_k^T \mathbf{x})}{dt} = \sigma_k(\mathbf{v}_k^T \mathbf{v}) \tag{11.13}$$

This simple derivation shows that a linear combination of concentrations:

$$\mathbf{u}_k^T \mathbf{x} = u_{k1}x_1 + u_{k2}x_2 + \cdots + u_{kr}x_m \tag{11.14}$$

is being uniquely moved by a linear combination of fluxes as:

$$\mathbf{v}_k^T \mathbf{v} = v_{k1}v_1 + v_{k2}v_2 + \cdots + v_{kn}v_n \tag{11.15}$$

and the extent of this motion is given by σ_k. An important feature of SVD is that the singular vectors are orthonormal to each other, and consequently each of the kth motions in Equations (11.14) and (11.15) are dynamically decoupled.

Forming independent motions or 'systems reactions:' Equation (11.13), therefore, defines a *systems reaction* as:

$$\sum u_{ki} x_i \quad \underset{\substack{\sum v_{kj} v_j \\ \text{for } v_{kj} < 0}}{\overset{\substack{\sum v_{kj} v_j \\ \text{for } v_{kj} > 0}}{\rightleftharpoons}} \quad \sum u_{ki} x_i \tag{11.16}$$

for $u_{ki} < 0$ for $u_{ki} > 0$

where the elements of $\mathbf{u}_k$ are equivalent to *systems stoichiometric coefficients* and the elements of $\mathbf{v}_k$ are *systems participation numbers*, showing how the kth reaction participates in moving the systems reaction (see Figure 11.4).

Analogy to chemical reactions   Note that the $\mathbf{u}_k$ vectors correspond to systemic reaction vectors that are analogous to the columns of the stoichiometric matrix, $\mathbf{s}_i^v$

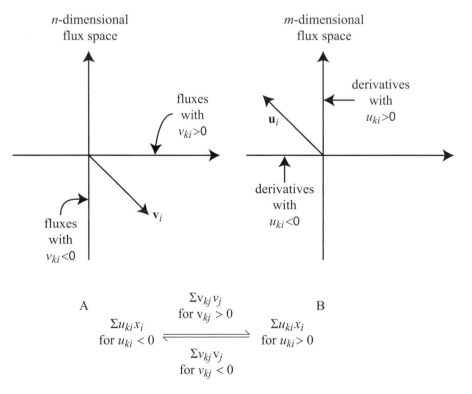

Figure 11.4   Systems reactions. The relationship between corresponding left ($\mathbf{u}_i$) and right ($\mathbf{v}_i$) singular vectors as a systems reaction. The right singular vector can be broken up into two parts, containing positive elements (on x-axis) and negative elements (on y-axis), panel A. Reactions with positive elements correspond to reactions driving the systems reaction forward, while those with negative elements drive it in the reverse direction. Analogously, the left singular vector, $\mathbf{u}_i$, can be broken into a part with positive elements (y-axis) and negative elements (x-axis), panel B. The former corresponds to compounds formed by the systems reaction, while the latter represents those disappearing. As $\mathbf{S}$ maps $\mathbf{v}_i$ onto $\mathbf{u}_i$, all points on $\mathbf{v}_i$ correspond to a point on $\mathbf{u}_i$. Further, because $\mathbf{v}_i$ and $\mathbf{v}_j$ are orthonormal, the systemic reactions are dynamically independent.

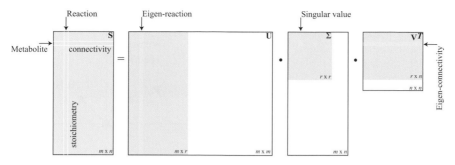

Figure 11.5 A schematic of the singular value decomposition of the stoichiometric matrix. Prepared by Iman Famili.

(see Figure 11.5). Thus, as we move a point along this vector, compounds with negative u_{ki} values decrease, while those with positive u_{ki} increase, and vice versa. Similarly, the reactions with positive v_{kj} values will drive a point in the increasing direction of $\mathbf{u}_k$, while those with negative values will act in the opposite direction. This relationship is graphically illustrated in Figure 11.4. Thus, Equation (11.16) describes a systemic reaction. These systems metabolic reactions can be used to describe the characteristics of the network as a whole. Note they are analogous to chemical reactions.

11.3 Studying Elementary Reactions using SVD

To develop an understanding of the information that SVD of $\mathbf{S}$ provides, we will apply it to elementary reactions. This application will illustrate the contents of the fundamental subspaces, and that, for some purposes, these subspaces may be better represented by a non-orthonormal set of basis vectors.

11.3.1 The linear reversible reaction

We now consider the reaction:

$$x_1 \underset{v_2}{\overset{v_1}{\rightleftharpoons}} x_2 \tag{11.17}$$

where we can think of $x_1 = CP$ and $x_2 = PC$ (e.g., Equation (9.1)), as this reaction simply represents the rearrangement of a molecule.

The SVD The corresponding stoichiometric matrix can be decomposed as:

$$\mathbf{S} = \mathbf{U}\boldsymbol{\Sigma}\mathbf{V}^T \tag{11.18}$$

that in this case takes the numerical form

$$\begin{pmatrix} -1 & 1 \\ 1 & -1 \end{pmatrix} = \frac{1}{\sqrt{2}} \begin{pmatrix} -1 & 1 \\ 1 & 1 \end{pmatrix} \begin{pmatrix} 2 & 0 \\ 0 & 0 \end{pmatrix} \begin{pmatrix} 1 & -1 \\ 1 & 1 \end{pmatrix} \frac{1}{\sqrt{2}} \tag{11.19}$$

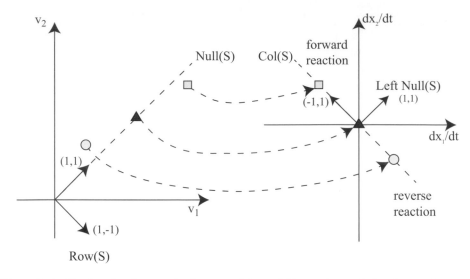

Figure 11.6 The four fundamental subspaces of the stoichiometric matrix for the reaction $x_1 \rightleftharpoons x_2$. Here, $n = m = 2$ and $r = 1$. Thus all four fundamental subspaces are one-dimensional. Note that we have multiplied all the basis vectors by $\sqrt{2}$ to make it easier to visualize the direction of the basis vectors of the four subspaces.

The subspaces   The four fundamental subspaces are shown in Figure 11.6. They are all one-dimensional.

The second column of $\mathbf{U}$ spans the left null space, and corresponds to a conservation relationship; namely

$$\langle \mathbf{u}_2^T, \mathbf{x} \rangle = x_1 + x_2 \tag{11.20}$$

being a constant. Similarly, the second row of $\mathbf{V}^T$ spans the null space of the stoichiometric matrix, and, representing

$$\langle \mathbf{v}_2^T, \mathbf{v} \rangle = v_1 + v_2 \tag{11.21}$$

that, as we will see in Chapter 12, corresponds to a type III pathway representing the equilibrium state.

The column and row spaces are both one-dimensional and related by:

$$\mathbf{S}\mathbf{v}_1 = \sigma_1 \mathbf{u}_1 \tag{11.22}$$

or

$$\mathbf{S} \begin{pmatrix} \frac{1}{\sqrt{2}} \\ -\frac{1}{\sqrt{2}} \end{pmatrix} = 2 \begin{pmatrix} -\frac{1}{\sqrt{2}} \\ \frac{1}{\sqrt{2}} \end{pmatrix} \text{ or } \mathbf{S} \begin{pmatrix} 1 \\ -1 \end{pmatrix} = 2 \begin{pmatrix} -1 \\ 1 \end{pmatrix} \tag{11.23}$$

The row space is spanned by $\mathbf{v}_1^T = (1, -1)/\sqrt{2}$, meaning that for a net flux through the reaction, the time derivatives in the column space are moved in the opposite direction multiplied by a factor of two. There are three possible outcomes:

Net reverse reaction ($v_1 < v_2$): as shown in Figure 11.6, if (v_1, v_2) is located above the 45 degree line (i.e., the green dot), the distance from the 45 degree line is doubled and projected in the opposite direction in the time derivative space.

Net forward reaction ($v_1 > v_2$): the opposite is true for a point located below the 45 degree line (i.e., the blue square).

No net reaction: if the numerical values of v_1 and v_2 are the same, there is no net reaction and the time derivatives are zero (i.e., the red triangle). This corresponds to the flux vector being in the null space, i.e., Equation (11.21).

These considerations tell us that: (i) the null space has a combination of fluxes that correspond to *pathways* (i.e., Equation (11.21)), (ii) the left null space has a combination of concentrations that correspond to *pools* (i.e., Equation (11.20)) and, (iii) that the row and column spaces have a driver–direction relationship (i.e., Equation (11.22)).

Numerical example We can trace these mappings using a specific numerical example. If we pick $\mathbf{v} = (2\sqrt{2}, \sqrt{2})^T$ then

$$\mathbf{V}^T \cdot \mathbf{v} = \frac{1}{\sqrt{2}} \begin{pmatrix} 1 & -1 \\ 1 & 1 \end{pmatrix} \begin{pmatrix} 2\sqrt{2} \\ \sqrt{2} \end{pmatrix} = \begin{pmatrix} 1 \\ 3 \end{pmatrix} \tag{11.24}$$

which corresponds to the projection of $\mathbf{v}$ onto the two right singular vectors. The first one is the net flux in the row space, while the latter is the corresponding steady-state flux (equilibrium in this case as the system is closed). Then

$$\mathbf{\Sigma} \mathbf{V}^T \mathbf{v} = \begin{pmatrix} 2 \\ 0 \end{pmatrix} \tag{11.25}$$

shows that the row space content maps but the null space has no singular value, thus finally mapping by the left singular vectors

$$\dot{x} - \frac{1}{\sqrt{2}} \begin{pmatrix} -1 & 1 \\ 1 & 1 \end{pmatrix} \begin{pmatrix} 2 \\ 0 \end{pmatrix} = 2 \cdot \mathbf{u}_1 + 0 \cdot \mathbf{u}_2 = 2\mathbf{u}_1 = \begin{pmatrix} -\sqrt{2} \\ \sqrt{2} \end{pmatrix} \tag{11.26}$$

to generate the time derivatives of the concentrations. We see that the motion is along the reaction vector, and there is no change in the pools size of $x_1 + x_2$ during the motion.

11.3.2   The bi-linear association reaction

Next we consider the reaction:

$$x_1 + x_2 \underset{v_2}{\overset{v_1}{\rightleftharpoons}} x_3 \tag{11.27}$$

where we can consider $x_1 = C$, $x_2 = P$ and $x_3 = CP$ (e.g., Equation (9.3)) to show that the reaction involves the association of C with P.

The SVD   The stoichiometric matrix is decomposed as

$$
\begin{pmatrix}
-1 & 1 \\
-1 & 1 \\
1 & -1
\end{pmatrix} =
$$

$$
\begin{pmatrix}
-1/\sqrt{3} & 2/\sqrt{6} & 0 \\
-1/\sqrt{3} & -1/\sqrt{6} & 1/\sqrt{2} \\
1/\sqrt{3} & 1/\sqrt{6} & 1/\sqrt{2}
\end{pmatrix}
\begin{pmatrix}
\sqrt{6} & 0 \\
0 & 0 \\
0 & 0
\end{pmatrix}
\begin{pmatrix}
\frac{1}{\sqrt{2}} & -\frac{1}{\sqrt{2}} \\
-\frac{1}{\sqrt{2}} & -\frac{1}{\sqrt{2}}
\end{pmatrix}
$$

Note that the row and null spaces are spanned by the same right singular vectors, v_1^T and v_2^T, respectively, as for the reversible conversion, leading to an analogous interpretation as above. However, in this case, the left null space is two-dimensional. The orthonormal basis vectors for the column and left null spaces are shown in Figure 11.7A. The second and third left singular vectors, u_2 and u_3, that span the left null space can be interpreted chemically, as discussed below, while the column space is simply spanned by the reaction vector normalized to its length. The row and column spaces are related by

$$ Sv_1 = \sigma_1 u_1 \tag{11.28} $$

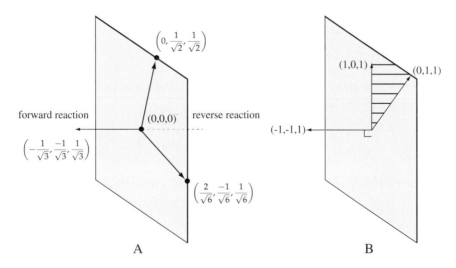

Figure 11.7   A depiction of the column and left null spaces for a simple bi-linear association reaction. Panel A: orthonormal basis. The plane is the left null space and the line is the column space. The vectors shown are an orthonormal set. If the flux vector is on the left-hand side of the plane as indicated, then the reaction is proceeding in the forward direction, and vice versa. Panel B: convex basis. The two basis vectors, (1,0,1) and (0,1,1), for the left null space are not orthogonal to each other, but both are orthogonal to the basis vector for the column space. The wedge shown corresponds to non-negative combinations of the two basis vectors.

or

$$
\begin{pmatrix} -1 & 1 \\ -1 & 1 \\ 1 & -1 \end{pmatrix} \begin{pmatrix} 1/\sqrt{2} \\ -1/\sqrt{2} \end{pmatrix} = \sqrt{6} \begin{pmatrix} -1/\sqrt{3} \\ -1/\sqrt{3} \\ 1/\sqrt{3} \end{pmatrix}
\tag{11.29}
$$

Formulating linear combinations of fluxes and concentrations We can familiarize ourselves with the details of these transformations by writing them out explicitly. The flux vector in the dynamic equations can be transformed using $\mathbf{V}^T$ as

$$
\frac{d\mathbf{x}}{dt} = \mathbf{S}\mathbf{V}\mathbf{V}^T\mathbf{v}
\tag{11.30}
$$

or

$$
\frac{d}{dt} \begin{pmatrix} x_1 \\ x_2 \\ x_3 \end{pmatrix} = \begin{pmatrix} -1 \\ -1 \\ 1 \end{pmatrix} (v_1 - v_2) - \begin{pmatrix} 0 \\ 0 \\ 0 \end{pmatrix} (v_1 + v_2)
\tag{11.31}
$$

forming two groupings of the fluxes:

- the first term corresponds to the row space and the grouping $v_1 - v_2$ is the net flux through the reaction, and it is orthogonal to $v_1 + v_2$, analogous to what is shown for the flux space in Figure 11.6; and
- the second term corresponds to the null space, and the combination $v_1 + v_2$ is a Type III extreme pathway that we will discuss in Chapter 12. It creates no motion and corresponds to the equilibrium state.

Multiplying Equation (11.31) by $\mathbf{U}^T$ leads to

$$
\frac{d}{dt} \begin{pmatrix} (-x_1 - x_2 + x_3)/\sqrt{3} \\ (2x_1 - x_2 + x_3)/\sqrt{6} \\ (x_2 + x_3)/\sqrt{2} \end{pmatrix} = \begin{pmatrix} \sqrt{6} \\ 0 \\ 0 \end{pmatrix} \frac{(v_1 - v_2)}{\sqrt{2}} - \begin{pmatrix} 0 \\ 0 \\ 0 \end{pmatrix} \frac{(v_1 + v_2)}{\sqrt{2}}
$$

Note that the singular value of $\sqrt{6}$ shows up and that the two column vectors on the right-hand side of the equation are the two columns of Σ. This system is now fully decomposed, showing how independent groupings of concentrations are moved by independent groupings of the fluxes.

Non-orthonormal basis vectors The two left singular vectors of $\mathbf{U}$ that span the left null space

$$
\begin{pmatrix} 2/\sqrt{6} & 0 \\ -1/\sqrt{6} & 1/\sqrt{2} \\ 1/\sqrt{6} & 1/\sqrt{2} \end{pmatrix}
\tag{11.32}
$$

are not easy to interpret chemically. Because they are not changed by the groupings of fluxes (zeros in the last two rows of the vectors on the right-hand side), they can be combined without changing the dynamic solution.

The second and third left singular vectors can be combined as

$$
\begin{pmatrix} 2/\sqrt{6} & 0 \\ -1/\sqrt{6} & 1/\sqrt{2} \\ 1/\sqrt{6} & 1/\sqrt{2} \end{pmatrix} \begin{pmatrix} \sqrt{6}/2 & 0 \\ \sqrt{2}/2 & \sqrt{2} \end{pmatrix} = \begin{pmatrix} 1 & 0 \\ 0 & 1 \\ 1 & 1 \end{pmatrix} \tag{11.33}
$$

to give $\mathbf{l}_1 = (1, 0, 1)$ and $\mathbf{l}_2 = (0, 1, 1)$ that also span the left null space. However, these are not orthonormal vectors spanning the left null space, i.e.,

$$
\langle \mathbf{l}_1^T, \mathbf{l}_2 \rangle = 1 \neq 0 \tag{11.34}
$$

but they represent chemical conservation moieties, or pools, that are

$$
\langle \mathbf{l}_1^T, \mathbf{x} \rangle = x_1 + x_3 \tag{11.35}
$$

(the moiety C based on Equation (11.27)) and

$$
\langle \mathbf{l}_2^T, \mathbf{x} \rangle = x_2 + x_3 \tag{11.36}
$$

(the moiety P based on Equation (11.27)), respectively. This basis for the left null space is shown in Figure 11.7B.

Note that the segment of the left null space that is chemically meaningful lies in the wedge spanned by these two vectors as only a non-negative combination of them is chemically possible. This basis is called a *convex basis* for the left null space. Thus, although mathematically convenient and useful, the use of the orthonormal bases obtained by SVD may not be well suited for chemical and biological interpretation of the left null space. We will also see in Chapter 12 that convex bases for the null space are biologically meaningful.

11.4   Studying Network Structure Using SVD

The simple examples in the previous section describing simple chemical reactions are useful conceptually. We now use SVD to decompose stoichiometric matrices for biochemically meaningful networks. We focus on the core *E. coli* metabolic network.

The metabolic network   The SVD of matrices for the core metabolic pathways in *E. coli* shows a singular value spectra where 16 of the 72 modes represent 50% of the spectrum, and 49 modes give 90% (Figure 11.8). The dominant modes take the form of systems reactions that correspond to the major co-factors found in the network (Figure 11.9):

- The first mode describes the translocation of a proton and represents 8.6% of the spectrum. The main forward drivers for the motion are the ETS associated enzymes that translocate protons (NADH16, CYTBD), while the reverse drivers are the ATP synthase and the transport reactions coupled to the proton gradient.

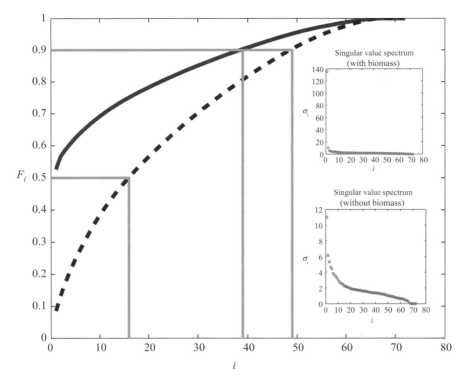

Figure 11.8 The cumulative singular value spectrum for the core *E. coli* metabolic model. The spectra for the model with (solid line) and without (dashed line) the biomass function are shown. The inserts are the two corresponding fractional singular value spectrum. Prepared by Harish Nagarajan.

- The second mode describes ATP synthesis and represents 4.8% of the spectrum. The main forward driver for the motion is the ATP synthase (ATPS4r), while the reverse drivers are the reactions that use ATP such as the ATP load (ATPM) and import of glutamine (GLNabc).
- The third mode describes both the formation of AcCoA and the transhydrogenation of NADPH to NADH. It represents 4.2% of the spectrum. The main forward drivers for the motion are the pyruvate dehydrogenase, that are associated with the AcCoA formation, and α-keto glutarate dehydrogenase (AKGDH) and acetaldehyde dehydrogenase (ACALD) while the main reverse driver is the transhydrogenase (THD2).

One can continue to look through the list of modes and interpret them similarly. Sometimes the modes are closely related to particular chemical transformations, and sometimes they are combinations of significant metabolic reactions.

The metabolic network with biomass formation   The biomass function can be added to the matrix to represent an integrated metabolic demand and the SVD can be performed again. In this case, the singular value spectrum is of lower effective dimension

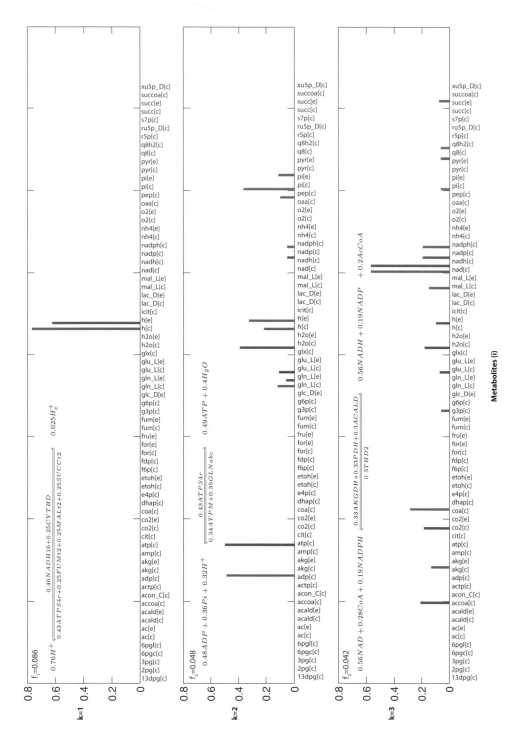

Figure 11.9 The dominant modes of the core metabolic model for *E. coli*. The panels show the magnitudes of the elements of $\mathbf{u}_i$ and the corresponding eigen reaction is shown. Blue indicates negative values and green positive. Prepared by Harish Nagarajan.

(Figure 11.8). The network has to satisfy a demand function that ties together many parts of the network.

The first mode, representing 52% of the spectrum, describes the use of ATP for biomass formation that basically reflects the biomass synthesis requirement (Figure 11.10). Note that this dominant feature is 'clean' in the sense that it has a clear biochemical form and interpretation. The second mode, representing 4.1% of the spectrum, describes the translocation of a proton. The third mode, representing 2.1% of the spectrum, describes the synthesis of the NADH co-factor as above. Note that the third mode is the same as that without the biomass function being included, and the same is true for the subsequent modes.

Observations   In general, the SVD calls out the co-factor uses as the major features of this network map. The fractional loadings on the modes correspond to systems reactions that are not proper chemical reactions. Thus, the SVD is a mathematically defined transformation that does not respect the chemical properties of the network. Sometimes the modes become clear combinations of dominant metabolic reactions, and when multiple simultaneous demands are imposed on the network (such as the growth function) they can become a dominant feature of the mapping that $\mathbf{S}$ represents. Rotation methods [149] might be developed to disentangle modes that represent interacting chemical reactions.

Genome-scale matrices   Genome-scale matrices can also be studied using SVD [111]. As in the smaller networks, the dominant modes that show the main features of their stoichiometric matrices are co-factor uses. The proton, ATP, NADPH, and NADH are main features of the maps and represent on the order of 25–30% of the singular value spectrum.

11.5   Drivers and Directions

We have seen that the column and row spaces of the stoichiometric matrix contain the concentration time derivatives and the thermodynamic driving forces, respectively, and that there is a mapping between the two. A full analysis of the mapping between these spaces leads to analysis of dynamic states of networks and requires knowledge of kinetics. Such analysis is found in the companion book [317].

11.5.1   Directions: the column space

The reaction vectors form the basis for the column space   The column space contains the derivatives of the concentrations of the compounds contained in a network. It is spanned by the reaction vectors ($\mathbf{s}_i$) as

$$\frac{d\mathbf{x}}{dt} = \mathbf{s}_1^v v_1 + \mathbf{s}_2^v v_2 \cdots + \mathbf{s}_n^v v_n \tag{11.37}$$

as weighted by the fluxes through each of the reactions at any given instant in time. We note, therefore, that the different flux levels or the changes in the flux levels determine the location of the vector of derivatives in the column space.

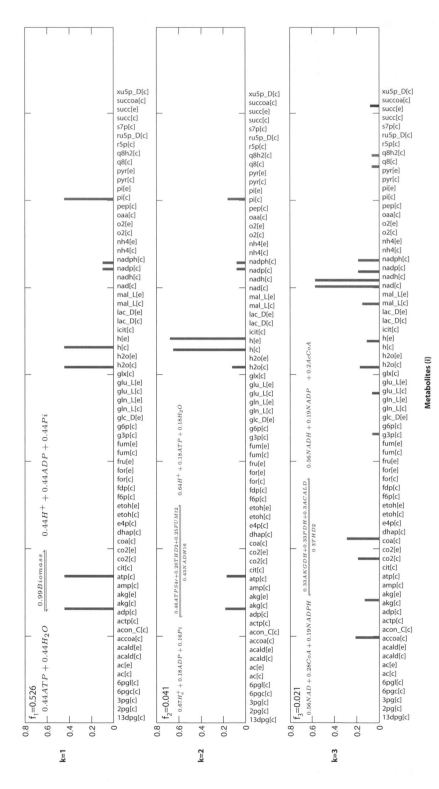

Figure 11.10 The dominant modes of the core metabolic model for *E. coli* when the biomass function is included in the network. The panels show the magnitudes of the elements of $\mathbf{u}_i$ and the corresponding eigen reaction is shown. Blue indicates negative values and green positive. Prepared by Harish Nagarajan.

Dynamic states   Reactions with high flux levels and those changing much in time will thus generate large motion along the corresponding reaction vector. The reaction vectors can be organized by the expected size and responsiveness of the reactions. Although network dynamics are outside the scope of this book, we note that reactions that are fast and quickly come to some sort of a quasi-steady state effectively reduce the column space's dimension on slower time scales. Time-scale separation results, and for every dimension that the column space is reduced in this fashion, an effective additional dimension in the left null space is created.

11.5.2   Drivers: the row space

Thermodynamic driving forces   If the fluxes are imbalanced, there will be a net generation or elimination of compounds in the network. If we denote a row in $\mathbf{S}$ by $\mathbf{s}_i^x$ then the corresponding dynamic mass balance is:

$$\frac{dx_i}{dt} = \langle \mathbf{s}_i^x, \mathbf{v} \rangle \tag{11.38}$$

thus, the time derivative is the inner product of that row vector and the flux vector:

$$\langle \mathbf{s}_i^x, \mathbf{v} \rangle = \|\mathbf{s}_i^x\| \|\mathbf{v}\| \cos(\theta_i) \tag{11.39}$$

where θ_i is the angle between the two vectors. If this inner product is zero, then the flux vector is orthogonal to the row vector. If not, this inner product sets the magnitude of the time derivative. Geometrically, the magnitude of this inner product may be viewed as a projection of the flux vector on the row vector. When all these projections become zero, the flux vector is orthogonal to the row vector and it resides solely in the null space, and the system is in a steady state.

Constraints on flux values   The magnitude of the individual fluxes is constrained. These constraints are derived from the limitation on the concentrations of the reactants and upper limits on the numerical values of the kinetic constants. The turnover rate of an enzyme–substrate intermediate complex (x)

$$v = k_1 x \leq k_1(x + e) = k_1 e_{\text{total}} \tag{11.40}$$

where the total amount of enzyme (e_{total}) present is limited to $x + e$. Note that this combination is a time invariant pool that lies in the left null space unless the enzyme (e) is synthesized or degraded.

11.5.3   The fundamental subspaces are of a finite size

We now revisit the reaction from Equation (11.17). Fluxes of elementary reactions are positive and have a maximal rate. Thus, all flux values fall into a range

$$0 \leq v_i \leq v_{i,\text{max}} \tag{11.41}$$

This range for v_1 and v_2 for the elementary reaction $x_1 \rightleftharpoons x_2$ is shown in Figure 11.11. The null space is on a diagonal line, while a perpendicular line spans the row space.

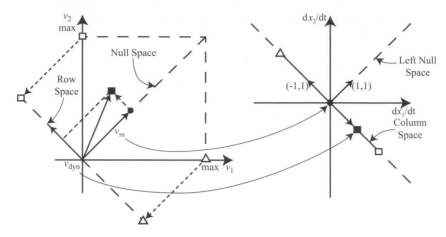

Figure 11.11   A graphical depiction of the null, row, and column spaces for $x_1 \rightleftharpoons x_2$. Because the fluxes v_1 and v_2 are finite, all these three spaces are finite.

Every point in the square of allowable flux states can be decomposed into a steady-state ($\mathbf{v}_{ss}$) and a dynamic ($\mathbf{v}_{dyn}$) component. These two components are represented by the $(1,1)/\sqrt{2}$ and $(1,-1)/\sqrt{2}$ vectors. They are orthogonal and span the null and row spaces, respectively; $\mathbf{v}_{ss}$ is mapped onto the origin by $\mathbf{S}$, whereas $\mathbf{v}_{dyn}$ is mapped onto $\mathbf{u}_1 = (-1,1)/\sqrt{2}$ and stretched by the singular value (see Equation (11.23)).

Note that the bounded range of the fluxes also set the bounds of the column space. The extreme points of the row space (the open triangle and square) correspond to the maximum allowable values on the time derivatives of x_1 and x_2. Thus, the extreme points of the row space lead to extreme points in the column space.

The constraints on the flux values confine the null, row and the column space. The consequences of this important limitation on the size of the null space will be discussed in the next chapter.

11.6   Summary

- SVD provides unbiased information about the four fundamental subspaces of $\mathbf{S}$.
- The first r columns of the left singular matrix $\mathbf{U}$ contain a basis for the column space of $\mathbf{S}$, and the remaining $m - r$ columns contain a basis for the left null space.
- The first r columns of the right singular matrix $\mathbf{V}$ contain a basis for the row space of $\mathbf{S}$ and the remaining $n - r$ columns contain a basis for the null space.
- The sets of basis vectors in $\mathbf{U}$ and $\mathbf{V}$ are orthonormal.
- The first r columns of $\mathbf{U}$ give systems reactions, analogous to a single column of $\mathbf{S}$, representing a single reaction.
- The corresponding column of $\mathbf{V}$ gives the combination of the fluxes that drive a systems reaction.

- Orthonormal basis vectors are mathematically convenient, but not necessarily biologically or chemically meaningful. If they are not, the basis vectors can be rotated onto a new set of basis vectors that we deem to be more biochemically appropriate.
- The column space contains the direction of motion and the row space the drivers for motion. To compute the resulting dynamic states, one needs kinetic constants. Dynamic states are discussed in the companion book [317].
- The magnitude of the individual fluxes is limited by kinetics and caps on concentration values. This limitation constrains the possible values of the time derivatives and thus the column space. Thus, the column and row spaces are finite closed spaces.

12 Pathways

Pathways are concepts, networks are reality – Uwe Sauer

The right null space of the stoichiometric matrix contains the steady-state flux distributions through the network that the matrix represents. This space is typically just called the *null space*. The choice of basis for the null space is important in describing its contents in meaningful chemical and biological terms. We will show that the basis vectors that span the null space correspond to continuous paths through the network that put it into a balanced state. As there is only a finite part of the null space that is of interest, convex representation has proven useful. The convex basis vectors correspond to a set of unique continuous and balanced pathways through a reconstructed network that represent one of its properties. We will introduce the basic concepts in this chapter.

12.1   Network-based Pathway Definitions

Definition of the null space   The right null space of $\mathbf{S}$ is defined by:

$$\mathbf{S}\mathbf{v}_{ss} = \mathbf{0} \tag{12.1}$$

thus, all the steady-state flux distributions, $\mathbf{v}_{ss}$, are found in the null space. The null space has a dimension of $n - r$. Note that $\mathbf{v}_{ss}$ must be orthogonal to all the rows of $\mathbf{S}$ simultaneously, and thus represents a linear combination of flux values on the reaction map that sum to zero. The null space is orthogonal to the row space of $\mathbf{S}$ (recall Figure 9.7).

Spanning the null space with pathway vectors   The null space is spanned by a set of $n - r$ basis vectors, $\mathbf{b}_i, i \in [1, n - r]$. The set of basis vectors form columns of a matrix $\mathbf{B}$ that satisfy:

$$\mathbf{S}\mathbf{B} = \mathbf{0} \tag{12.2}$$

and any steady state is represented as

$$\mathbf{v}_{ss} = \sum_{i=1}^{n-r} w_i \mathbf{b}_i \quad \text{where} \quad -\infty \leq w_i \leq \infty \tag{12.3}$$

A set of linear basis vectors is not unique, but once the set is chosen, the weights (w_i) for a particular $\mathbf{v}_{ss}$ are unique. A basis vector, $\mathbf{b}_i$, makes nodes in the flux map balanced as it is orthogonal to all the rows of $\mathbf{S}$ simultaneously. This balancing requirement leads to *network-based* pathway definitions and the use of basis vectors that represent such pathways.

Internal and external reactions   If a reacting system is closed, only internal cycles are possible. These cycles are at a thermodynamic equilibrium and thus of little biological interest. Living systems are open to the environment and thus it is the trafficking of substances in and out of a cell that becomes a key interest. This feature leads to the segregation of the flux vector into two parts, internal and external fluxes, as discussed in Chapter 9. The internal fluxes can be defined by elementary reactions that have non-negative flux values and thus are naturally considered to be uni-directional. The exchange fluxes can involve diffusive processes and may not have a natural representation as chemical reactions. The exchange fluxes are therefore naturally considered to be bi-directional.

12.2   Choice of a Basis

A linear basis for the null space can be computed using a number of standard methods, including SVD (see Chapter 11). An infinite number of different bases exist for a linear space. We are interested in finding a basis that is biochemically meaningful, and thus useful for biological interpretation.

Linear basis   A simple linear reaction network is shown in Figure 12.1A. The null space is defined by:

$$\begin{pmatrix} 1 & -1 & 0 & 0 & -1 & 0 \\ 0 & 1 & -1 & 0 & 0 & 0 \\ 0 & 0 & 1 & -1 & 0 & 1 \\ 0 & 0 & 0 & 0 & 1 & -1 \end{pmatrix} \begin{pmatrix} v_1 \\ v_2 \\ v_3 \\ v_4 \\ v_5 \\ v_6 \end{pmatrix} = \begin{pmatrix} 0 \\ 0 \\ 0 \\ 0 \end{pmatrix} \tag{12.4}$$

The matrix is full rank, and thus the dimension of the null space is 2 $(= 6 - 4)$. Because columns 4 and 6 do not contain pivots, this set of linear equations can be solved using v_4 and v_6 as the free variables to give

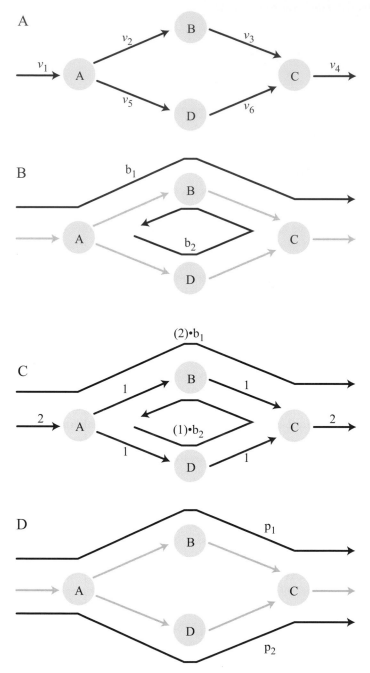

Figure 12.1 Simple network and basis vectors for the null space of the corresponding **S**. Panel A: the reaction map; panel B: the basis vectors from Equation (12.5) drawn as flux distributions on the reaction map; panel C: representation of flux distributions as a combination of basis vectors from Equation (12.6); and panel D: a set of non-negative basis vectors (from Equation (12.7)), in panel B. Adapted from Christophe Schilling.

$$
\begin{pmatrix} v_1 \\ v_2 \\ v_3 \\ v_4 \\ v_5 \\ v_6 \end{pmatrix} = \begin{pmatrix} v_4 \\ v_4 - v_6 \\ v_4 - v_6 \\ v_4 \\ v_5 \\ v_6 \end{pmatrix} = v_4 \begin{pmatrix} 1 \\ 1 \\ 1 \\ 1 \\ 0 \\ 0 \end{pmatrix} + v_6 \begin{pmatrix} 0 \\ -1 \\ -1 \\ 0 \\ 1 \\ 1 \end{pmatrix} = w_1 \mathbf{b}_1 + w_2 \mathbf{b}_2 \qquad (12.5)
$$

where $\mathbf{b}_1$ and $\mathbf{b}_2$ form a basis. They are both in the null space and are linearly independent. For any numerical values of v_4 and v_6, a flux vector will be computed that lies in the null space. These basis vectors can be shown schematically by graphing them onto the reaction map (Figure 12.1B).

Any steady-state flux distribution is a unique linear combination of the two basis vectors. For example:

$$
\mathbf{v} = \begin{pmatrix} 2 \\ 1 \\ 1 \\ 2 \\ 1 \\ 1 \end{pmatrix} = w_1 \mathbf{b}_1 + w_2 \mathbf{b}_2 = (2) \begin{pmatrix} 1 \\ 1 \\ 1 \\ 1 \\ 0 \\ 0 \end{pmatrix} + (1) \begin{pmatrix} 0 \\ -1 \\ -1 \\ 0 \\ 1 \\ 1 \end{pmatrix} = (2)\mathbf{b}_1 + (1)\mathbf{b}_2
$$

$$(12.6)$$

This combination can also be drawn on the reaction map (Figure 12.1C). This set of basis vectors, although valid mathematically, is unsatisfactory chemically. The reason is that the second basis vector, $\mathbf{b}_2$, represents fluxes through irreversible elementary reactions, v_2 and v_3, in the reverse direction, and it thus represents a chemically unrealistic event.

Non-negative linear basis The problem with the acceptability of the basis above stems from the fact that the flux through an elementary reaction can only be positive, i.e., $v_i \geq 0$. A negative coefficient in the corresponding row in the basis vector that multiplies the flux is thus undesirable. This consideration leads to the need to have non-negative basis vectors for the null space. In the example above, we can combine the basis vectors to eliminate all negative elements in them. This combination is achieved by transforming the set of basis vectors by:

$$
(\mathbf{b}_1, \mathbf{b}_2) = \begin{pmatrix} 1 & 0 \\ 1 & -1 \\ 1 & -1 \\ 1 & 0 \\ 0 & 1 \\ 0 & 1 \end{pmatrix} \begin{pmatrix} 1 & 1 \\ 0 & 1 \end{pmatrix} = \begin{pmatrix} 1 & 1 \\ 1 & 0 \\ 1 & 0 \\ 1 & 1 \\ 0 & 1 \\ 0 & 1 \end{pmatrix} = (\mathbf{p}_1, \mathbf{p}_2) \qquad (12.7)
$$

In this new basis, $(\mathbf{p}_1, \mathbf{p}_2)$, the first basis vector is the same as in the old basis, whereas the second basis vector in the new set is an addition of the two basis vectors in the old basis. These two new basis vectors are shown on the reaction map in Figure 12.1D, and they contain no fluxes that operate in the incorrect direction. Notice that these non-negative basis vectors look like pathways through this simple system. We point out to the reader that this toy system, although conceptually useful to make a key point, is biochemically irrelevant as there are no carrier or co-factor exchange reactions.

12.3   Confining the Steady-state Flux Vector

At the end of the previous chapter, we showed that the fundamental subspaces are finite in size. In particular, because the fluxes take on a limited range of values, the null space is finite and the steady-state flux vector, $\mathbf{v}_{ss}$, is confined to a closed space. The confining of the steady-state flux vector leads to special considerations associated with forming a set of basis vectors for finite spaces.

12.3.1   Finite or closed spaces

A polytope: an intersection of a hyperplane and a hypercube   Because all physical variables are finite, the flux and concentration variables are found in limited numerical ranges. The elementary reactions have non-negative fluxes, $v_i \geq 0$. In addition, they have an upper bound, $v_i \leq v_{i,max}$, giving them a finite range of allowable numerical values,

$$0 \leq v_i \leq v_{i,max} \tag{12.8}$$

thus, the allowable flux vectors are in a rectangular hyperbox in the positive orthant of the flux space (that can be a high-dimensional space), bounded by planes parallel to each axis as defined by $v_{i,max}$. This hyperbox contains all allowable flux states, both steady state and dynamic.

   If we are interested in only the steady states, then we must add the flux balance equation

$$\mathbf{Sv}_{ss} = 0 \tag{12.9}$$

that is a hyperplane that intersects the hyperbox forming a finite segment of a hyperplane (see Figure 12.4B). This intersection is a polytope in which all the steady-state flux distributions lie. This polytope can be spanned by the convex basis vectors that are *edges* of the polytope with restricted ranges on the weights. In other words,

$$\mathbf{v}_{ss} = \sum_{i=1}^{q} \alpha_i \mathbf{p}_i \quad \text{where} \quad 0 \leq \alpha_i \leq \alpha_{i,max} \tag{12.10}$$

where the $\mathbf{p}_i$ turn out to be edges, or extreme states in the polytope, and α_i are the weights which are positive and bounded. The number of such basis vectors, q, can be greater than the dimension of the null space. The $\alpha_{i,max}$ value will be related to a combination of the most restricting $v_{i,max}$.

Convex vs. linear bases   The introduction of non-negative basis vectors leads us to convex analysis. Convex analysis is based on equalities (i.e., $\mathbf{Sv}_{ss} = 0$) and inequalities

Table 12.1 Comparing the properties of linear and convex bases.

Linear spaces	Convex spaces
Described by linear equations	Described by linear equations and inequalities
$\mathbf{Sv}_{ss} = 0$	$\mathbf{Sv}_{ss} = 0$ and $0 \leq v_i \leq v_{i,max}$
Vector spaces defined by a set of linearly independent basis vectors $(\mathbf{b}_i)$	Convex polyhedral cone defined by a set of conically independent vectors $(\mathbf{p}_i)$
$\mathbf{v} = \sum w_i \mathbf{b}_i \quad -\infty \leq w_i \leq +\infty$	$\mathbf{v} = \sum \alpha_i \mathbf{p}_i \quad 0 \leq \alpha_i \leq +\infty$
Every point in the vector space is uniquely described by a linear combination of basis vectors (unique representation for a given basis)	Every point in the vector space is described as a non-negative linear combination of the convex basis vectors (non-unique representation)
Number of basis vectors equals dimension of the null space	Number of convex basis vectors may exceed dimension of the null space
Infinite number of bases that can be used to span the space	The set of convex basis vectors is unique

(i.e., $0 \leq v_i \leq v_{i,max}$). We will not discuss the mathematics underlying the formation of convex basis vectors, but list their properties in Table 12.1, contrasting them with the properties of the basis vectors for a linear space. Important features of the convex basis are that it is *unique* and determined based on network topology. However, the number of convex basis vectors can be *greater* than the dimension of the null space, leading to multiple ways to represent a flux distribution with the unique set of convex basis vectors.

Illustrating two key properties Simple geometries can be used to illustrate these properties:

The number of convex basis vectors can exceed the dimension of the space. The four-sided pyramid, seen in Figure 12.2, is a 3D object described by four convex basis vectors. A non-negative combination of these four vectors leads to the generation of a point inside the pyramid.

The set of convex basis vectors is unique and they lie on the extremities of the space. The flux vector lies in the positive orthant further constrained by the flux balance equations (Figure 12.3). The general shape of this space is one of a cone or a triangle in a two-dimensional representation. Two pairs of non-negative basis vectors are shown. Only the vectors that lie on the edges can give a non-negative representation of all the points in the cone. It is thus a unique convex basis. The convex basis vectors lie on the edges of the finite space and are thus sometimes called extreme vectors.

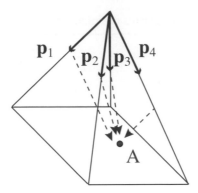

Figure 12.2   A four-sided pyramid. This object is a bounded three-dimensional convex space, but there are four convex basis vectors. A non-negative combination of these four vectors will generate a point inside the pyramid. The dashed lines show how a point 'A' is generated by a non-negative combination of $\mathbf{p}_1$, $\mathbf{p}_2$, and $\mathbf{p}_4$.

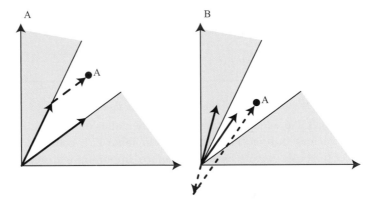

Figure 12.3   Schematic illustration of a flux cone in the positive orthant. Panel A: the entire cone can be spanned with $\alpha_i > 0$. Point 'A' can be represented in such a way. Panel B: using any other set of basis vectors may require $\alpha_i < 0$. For instance, representing the point 'A' would require a negative weighting on one of the basis vectors.

12.3.2   Importance of constraints

Redundant and dominant constraints   A node with three reaction links forms a simple flux split (Figure 12.4A). The minimum and maximum constraints on the reactions form a three-dimensional box that is intersected by the plane formed by the flux balance equation

$$0 = -v_1 - v_2 + v_3 = \langle (-1, -1, 1) \cdot (v_1, v_2, v_3) \rangle \tag{12.11}$$

This intersection forms a segment of a plane (Figure 12.4B). This two-dimensional polytope is spanned by two convex basis vectors (Figure 12.4C). These basis vectors are $\mathbf{b}_1 = (1, 0, 1)$ and $\mathbf{b}_2 = (0, 1, 1)$.

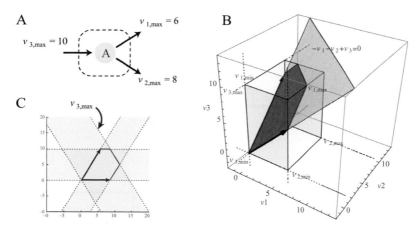

Figure 12.4 Forming a 2D polytope in 3D. (A) A simple flux split. (B) The two-dimensional null space is constrained by the v_{max} planes corresponding to the three reactions in the network. (C) A 2D representation of the finite null space (green segment). The blue parallelepipeds enclosing the null space chosen form parallel edges of v_{maxes} and v_{mins} for the three reactions. Figure 12.5 further discusses the consequences of changing $v_{3,max}$. Modified from [347]. Prepared by Niko Sonnenschein.

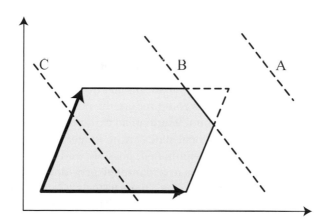

Figure 12.5 Bounding the flux cone; redundant and dominant constraints. Three constraints on $v_{3,max}$ in Figure 12.4C are illustrated; (A) $v_{3,max} > 14$ ($= v_{2,max} + v_{1,max}$); (B) $6 < v_{3,max} < 14$; (C) $v_{3,max} < 6$ ($= min\{v_{1,max}, v_{2,max}\}$). In case A, the $v_{3,max}$ is redundant; in case B, all $v_{i,max}$ are relevant; and in case C, $v_{3,max}$ is relevant, while $v_{1,max}$ and $v_{2,max}$ are redundant.

The length of these basis vectors is thus limited by the $v_{i,max}$ values. The relative magnitudes of the $v_{i,max}$ values lead to *dominant* and *redundant* constraints, as illustrated in Figure 12.5. The upper bounds on the fluxes close the flux cone to form a polytope. This finite set has *extreme points* that can be used to define the space.

Biological interpretation of varying constraints The constraints offer a mechanism to determine the effects of various parameters on the achievable functional states of a network. There are several practical consequences of altering governing constraints.

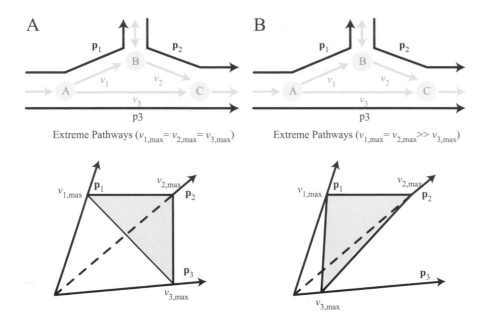

Figure 12.6   Changing constraints and shrinking polytopes. If a $v_{\max}$ constraint is reduced, a portion of the solution space may become inaccessible (shaded region). Enzymopathies can have this effect [183]. Taken from [453].

- Enzymopathies can reduce the maximum flux through a particular reaction in a network. Such constraints could reduce the numerical value for $\alpha_{i,\max}$ and thus shorten the maximum length of an extreme vector (see Figure 12.6). In such a case, several functional states are no longer possible. A desirable functional state may thus be eliminated, possibly leading to a pathological condition [183]. The consequences of complete gene deletions can be analyzed in this fashion as well (see Chapter 23).
- Regulation leads to shrinking or elimination of pathway vectors. Thus, regulation can shrink the solution space around a desirable state by eliminating all other candidate states. Regulation can thus be thought of as *restraint* (see Section 13.5).
- As will be discussed in Chapter 14, a global assessment of the consequences of altering constraints can be obtained through randomized sampling of points in the solution space.

12.4   Pathways as Basis Vectors

12.4.1   Some perspective

From reactions to pathways   Early in the history of biochemistry, enzymes isolated from cells were shown to be able to carry out specific chemical reactions (Figure 12.7A). It was then recognized that the products of one reaction were substrates of another. Thus, one could link different chemical transformations to form a series of reactions

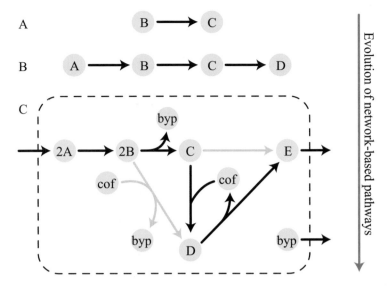

Figure 12.7 The historical development of the network-based pathway diagram: from reactions to pathways to networks. Taken from [324].

(Figure 12.7B) to form basic metabolic pathways, such as glycolysis, the TCA cycle, and so on. The definition and biochemical functions of such pathways have been taught to generations of life scientists. With the advent of whole-genome sequencing and the development of network reconstruction methods (i.e., Part I), we can now piece together entire networks. The integrated properties of such networks can and must be studied.

Network-based pathway definitions The stoichiometric matrix is a data matrix that results from the reconstruction process. Interestingly, as illustrated above, the finite null space of the stoichiometric matrix has a natural set of basis vectors that can be used to span all allowable network states. Thus, network-based definitions of pathways emerged (see Figure 12.7C) that account for the function of the network as a whole. These pathways are defined mathematically and thus free of human bias in their definition, and are useful for studying the systems properties of networks.

Some key concepts: mathematics vs. biology The finite null space contains all the allowable steady-state flux distributions through the network. It provides a clear link between mathematical and biological concepts.

- The null space represents all the possible functional, or phenotypic, states of a network.
- A particular point in the null space represents one network function, or one particular phenotypic state.
- As we will see in Chapter 20, there are equivalent points in the null space that lead to the same overall functional state of a network. Biologically, such conditions are called *silent phenotypes*.

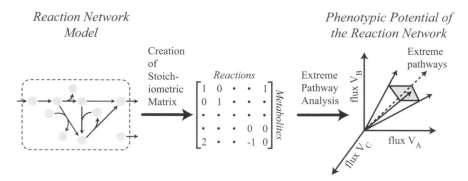

Figure 12.8 Network reconstruction and its mathematical representation leads to the definition of the null space. Physical constraints bound the null space. The bounded null space of the stoichiometric matrix is a convex cone. The edges of this cone are the extreme pathways. Redrawn from [324].

- The edges of the null space are the unique extreme vectors. Any flux state in the cone can be decomposed into the extreme vectors.

The unique set of extreme vectors thus gives a mathematical description of the range of flux levels that are allowed. It is important to note that the unique set of convex basis vectors is computed from the flux balance and inequalities. Both of these equations are rooted in the information that went into the network reconstruction (i.e., forming S), and the data used to specify $v_{i,\max}$. The convex basis vectors are thus *properties of the data* (Figure 12.8). The result from this structured data analysis represents the deeper understanding of the network and is a result of the formation of a knowledge base and deciphering its characteristics.

12.4.2   Extreme pathways

Biochemically meaningful steady-state flux solutions can be represented by a non-negative linear combination of convex basis vectors, as defined in equation (12.10). The vectors $\mathbf{p}_i$ are a unique set of convex basis vectors, but α_i may not be unique for a given $\mathbf{v}_{\mathrm{ss}}$. The number of the convex basis vectors can exceed $n - r$. The $\mathbf{p}_i$ have been studied extensively [325, 382]. They correspond to the edges of a polytope in an $(n - r)$-dimensional space (see Figure 12.8) and correspond to pathways when represented on a flux map, and are called *extreme pathways*, as they lie at the edges of the bounded null space in its conical representation.

Computing extreme pathways   Algorithms have been developed to compute the extreme pathways [379]. The details of such algorithms are not important here, but open source code has been provided to compute them [38]. The computation of extreme pathways for small systems is relatively easy. However, as the size of a network grows, the number of extreme pathways typically grows much faster. The combinatorial nature of their computation leads to an N-P hard problem. Thus, it has only been possible to compute extreme pathways at the genome scale for small matrices and under a limited set of conditions [323, 346]. The development of robust

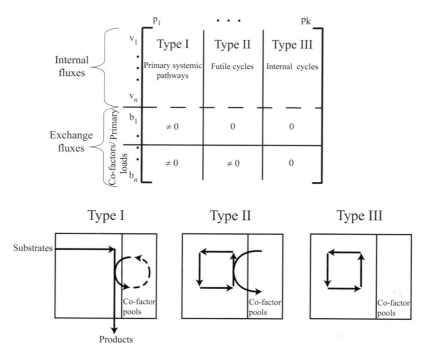

Figure 12.9 Classification of pathways based on the exchange fluxes that they contain. Top: the structure of the pathway matrix **P** for the three types of pathways. Bottom: the corresponding flux maps. Redrawn from [342].

algorithms for their computation is needed and so are analysis methods to interpret the results. In general, a full set of extreme pathways may not be of importance, but a subset of them will be. Fortunately, linear programming can be used to find particular extreme pathways of interest which typically is a fast computation (see Chapter 16). A linear basis based on sparsity of **P** can compute a minimal set of pathways (see Section 12.4.4).

12.4.3   Classifying extreme pathways

Classification of extreme pathways relates to the exchange fluxes   Extreme pathways have been classified into three groups according to their use of exchange metabolites (Figure 12.9). The pathways are computed based on the null space of S_{exch} (see Figure 9.9). The exchange fluxes are grouped into two categories: external fluxes, and fluxes external to metabolism but internal to the cell. Thus, the first set of exchange fluxes are with the surroundings of the cell, while the second set represents a virtual boundary separating metabolism from other cellular functions. The second category primarily contains the use of co-factor molecules or to meet metabolic loads.

Types I, II and III extreme pathways   With this definition of exchange fluxes, the pathways are classified into three categories based on which exchange fluxes they contain. The *pathway matrix* **P** and the classification of extreme pathways is illustrated in Figure 12.9.

- Type I pathways involve the conversion of primary inputs into primary outputs and thus contain exchange fluxes with the environment.
- Type II pathways involve the internal exchange of co-factors (or carrier) molecules only, such as ATP and NADH.
- Type III pathways are solely internal cycles; there is zero flux across all system boundaries.

Eliminating type II pathways: a matter of regulation   Type II pathways have no net exchanges with the environment, but are coupled to a co-factor pair. As the co-factor pair can only give up energy, type II pathways can only proceed in the direction of dissipating the charged co-factor. By either including currency metabolites as internal to the system or considering currency metabolites as primary exchange metabolites, pathway analysis yields zero type II pathways.

Historically, such a possibility has been termed a *futile cycle* and the first such situation that was discussed extensively was the cycle formed by the opposing reactions of PFK and FDPase in glycolysis and the gluconeogenic pathway (Figure 12.10). If both of these reactions are active they serve to convert ATP into ADP and P_i, dissipating the energy stored on ATP. Although such futile cycles are implicated in thermoregulation, they most likely are undesirable in most physiological states. The elimination of their occurrence by regulation of enzyme activity or simply by gene expression has been postulated [42, 430]. Thus, it is likely that most type II pathways are not relevant physiologically, and their activity is eliminated, or severely reduced, by regulation. Those that cannot be eliminated would represent non-specific loads on a network.

Eliminating type III pathways: a matter of basic physics   Type III pathways are infeasible thermodynamically. They correspond to internal loops, around which there can be no net flux. They are analogous to Kirchhoff's second law for electrical circuits [345]. Algorithms to eliminate such loops exist and one example is given in [377].

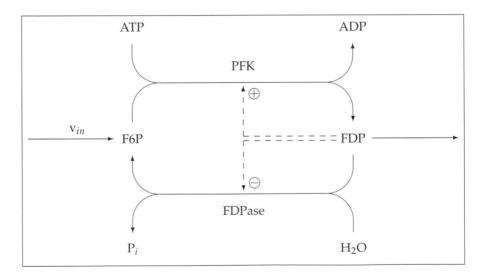

Figure 12.10   Proposed regulation of a futile cycle glycolysis by FDP.

12.4.4 The simplest set of linearly independent basis vectors

There is an infinite number of choices for a linear basis for a linear space. As we have already seen, some choices are better than others. One way of choosing a basis is to choose the simplest set. This means that the matrix $\mathbf{P}$, where the $\mathbf{p}_i$ are the columns, has to have as few non-zero elements as possible. Choosing such a basis will come down to solving an optimization problem. This approach chooses the minimal number of non-zero entries in the pathway matrix. Choosing such a basis will come down to solving an mixed-integer linear programming (MILP) problem [47].

12.4.5 Examples of pathway computation

Glycolysis The stoichiometric matrix for the glycolytic system discussed in Chapter 9 had a null space of dimension 3. A set of three null space vectors are shown in Figure 12.11. This set was initially chosen based on biochemical intuition. The lower portion of the figure shows flux map images of these basis vectors and it can be seen that they are continuous pathways through the network. A minimal basis can be computed using the MINSPAN algorithm. It is the same as that formed by intuitive reasoning.

The convex basis for this system has 4 type I extreme pathways, one more than the linear basis. The first pathway represents glycolysis from glucose to lactate, the second represents the conversion of lactate to pyruvate, and the third represents AMP metabolism. The fourth pathway is essentially a combination of the first and second pathways representing the conversion of glucose to pyruvate.

Pathways in red blood cell metabolism The glycolysis example above is a simple one. The core network of red cell metabolism is represented by a 40×45 matrix

	Glycolytic reactions											AMP metabolism		Primary export		Cofactors		Primary inputs		Inorganic	
	v_{hk}	v_{pgi}	v_{pfk}	v_{tpi}	v_{ald}	v_{gapdh}	v_{pgk}	v_{pglm}	v_{eno}	v_{pk}	v_{ldh}	v_{amp}	v_{apk}	v_{pyr}	v_{lac}	v_{atp}	v_{nadh}	v_{gluin}	v_{ampin}	v_{H^+}	v_{H_2O}
$\mathbf{p}_1$	1	1	1	1	1	2	2	2	2	2	2	0	0	0	2	2	0	1	0	2	0
$\mathbf{p}_2$	0	0	0	0	0	0	0	0	0	0	-1	0	0	1	-1	0	1	0	0	2	0
$\mathbf{p}_3$	0	0	0	0	0	0	0	0	0	0	0	1	0	0	0	0	0	0	1	0	0
v_{stst}	1.12	1.12	1.12	1.12	1.12	2.24	2.24	2.24	2.24	2.24	2.016	0.014	0	0.224	2.016	2.24	0.224	1.12	0.014	2.69	0

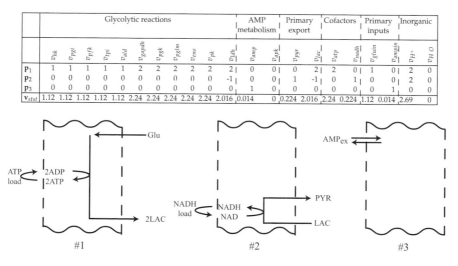

Figure 12.11 The pathway vectors for the glycolytic system in Figure 9.3. These vectors are shown graphically at the bottom of the figure. The last line of the table shows the steady-state fluxes (in mM/h) in the human red blood cell [317], and it is a linear combination of the pathway vectors.

A

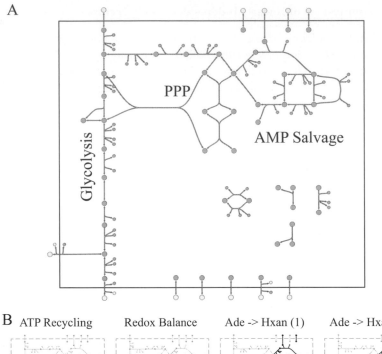

B

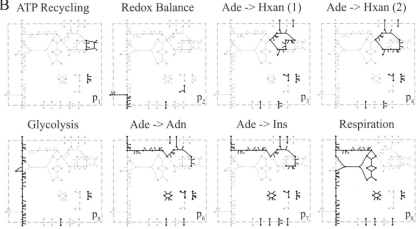

Figure 12.12   (A) A metabolic network for the human red blood cell encompassing glycolysis, pentose phosphate pathway, and AMP salvage metabolism. (B) The red blood cell network, and any accompanying flux distributions, can be decomposed into eight linearly independent MINSPAN pathways. These eight pathways are interpreted to carry biochemical functions including: (i) ATP recycling, (ii) redox balance through pyruvate to lactate conversion, (iii) adenine to hypoxanthine conversion through IMP, (iv) adenine to hypoxanthine conversion through adenosine, (v) glycolysis ending in lactate production, (vi) glucose and adenine to adenosine production, (vii) glucose and adenine to inosine production, and (viii) cellular respiration of glucose to CO_2 through the pentose phosphate pathway. Ade, adenine; Hxan, hypoxanthine; Adn, adenosine; Ins, inosine.

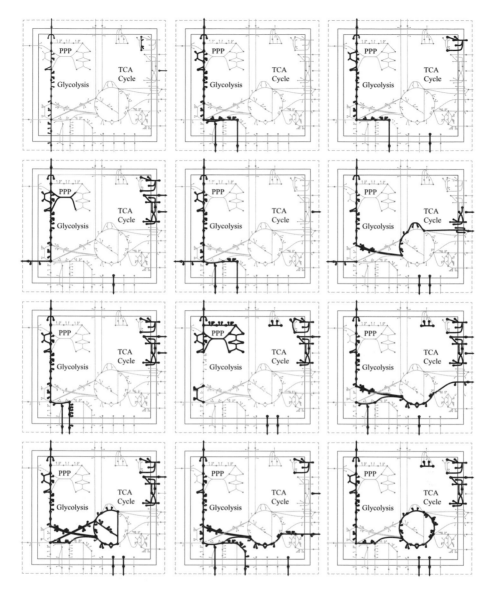

Figure 12.13 The 12 MINSPAN pathways type I for the core metabolic network for *E. coli*. Prepared by Aarash Bordbar.

(see chapter 12 of [317]). The MINSPAN algorithm computes the eight pathway vectors shown graphically in Figure 12.12. These vectors have a clear biochemical interpretation.

The number of type I extreme pathway vectors for this network is 91, notably larger than 8. This large increase in convex pathway vectors is due to a combinatorial explosion, which is even more pronounced for other definitions of convex basis vectors [320].

Pathways in core *E. coli* metabolism The dimension of null space of the core *E. coli* metabolic model is 23. This space is too complicated to attempt an intuitive construction of a basis vector. The number of convex basis vectors is very large (16,672 type I extreme pathways) and thus of limited conceptual value.

However, the MINSPAN algorithm computes a useful basis. The number of type I pathways is 12, the number of type II pathways is 10 and there is one type III pathway. Thus, the most biochemically relevant part of the null space is 12-dimensional. These 12 pathways are shown graphically in Figure 12.13.

Biochemical vs. mathematical dimensions Mathematically speaking, the null space is of high dimension, but its biochemical dimensionality may effectively be much lower. Understanding the difference between the two ways of viewing the dimensionality of the null space results from biochemical analysis of the basis vectors of null (S).

12.5 Summary

- The stoichiometric matrix has a null space that corresponds to a linear combination of the reaction vectors that add up to zero and result in a steady state for the network.
- The orthonormal basis given by SVD or other linear bases do not yield a useful biochemical interpretation of the null space of the stoichiometric matrix. There are alternative ways to define the bases.
- Convex basis vectors for the null space can be formulated by considering elemental reactions only. As they are irreversible ($0 \leq v_i$) and finite in magnitude ($v_i \leq v_{i,max}$), it only makes sense to add them in a non-negative fashion leading to a convex representation of the null space.
- The convex representation has edges that represent a set of vectors that span the convex space in a non-negative fashion. These edges correspond to biochemical pathways and are the extreme functions that a network can have, and are therefore called *extreme pathways*.
- Extreme pathway vectors are classified based on the exchange reactions that they contain. Type I involve the transfer of properties to currency molecules (carriers or co-factors), type II have no external exchange reaction and represent irreversible futile cycles, and type III have no exchange reactions and due to thermodynamic restrictions must have a zero flux.
- The number of extreme pathways grows faster than the number of components in a network, giving rise to the need to study them in large numbers. Thus, an approach like MINSPAN may be a more practical way to get meaningful pathways for larger networks.

13 Use of Pathway Vectors

Out of intense complexities, intense simplicities emerge – Winston Churchill

There are two approaches that we can take to understand the contents of the space that contains all possible networks states. First, we can characterize all the allowable states by using a set of basis vectors that can be used to span all the allowable solutions, and second, we can sample the solution spaces uniformly and determine the statistical properties of a large number of candidate solutions. These two approaches amount to an *unbiased* assessment of the properties of all the allowable states of a biological network. We discuss the use of pathway vectors in this chapter and the uniform random sampling approach in Chapter 14.

13.1   The Matrix of Pathway Vectors

The pathway matrix   The extreme pathways (Chapter 12) are convex basis vectors and can represent all the functional states of a network. The *pathway matrix*, $\mathbf{P}$, is formed using the extreme pathways ($\mathbf{p}_i$) as its columns:

$$\mathbf{P} = (\mathbf{p}_1,\ \mathbf{p}_2,\ \mathbf{p}_3,\ \ldots) = (||| \ \ldots) \tag{13.1}$$

Thus, entries in row i of $\mathbf{P}$ indicate whether reaction i (v_i) is used in an extreme pathway and the columns indicate which reactions are involved in the makeup of a particular pathway. $\mathbf{P}$ can be written in a binary form, $\hat{\mathbf{P}}$, whose elements are defined by

$$\hat{p}_{ij} = 1,\ \text{if}\ p_{ij} \neq 0\ \ \text{and}\ \ \hat{p}_{ij} = 0,\ \text{if}\ p_{ij} = 0 \tag{13.2}$$

This binary form of $\mathbf{P}$ is analogous to the binary form of $\mathbf{S}$ discussed in Chapter 10. As with $\hat{\mathbf{S}}$, $\hat{\mathbf{P}}$ can be used to compute interesting features of $\mathbf{P}$.

Systems properties of interest   Because the entire contents of a solution space can be represented by its basis, the pathway vectors can be used to obtain a global viewpoint of network capabilities and characteristics. Many such properties have been studied. We will discuss the following properties.

Pathway length.   Classically, we think of a pathway as a linear sequence of events and the length of the pathway is the number of steps in this sequence. Extreme pathways can be much more complicated, and 'length' becomes the number of reactions that participate in the pathway. This number is computed from an adjacency matrix of $\hat{\mathbf{P}}$.

Reaction participation.   The number of pathways in which a reaction participates is an important quantity. It indicates how many pathways are affected if a reaction is removed from the network. The knock-out of a gene can lead to the removal of a reaction, or a reaction can be down-regulated. Reaction participation is computed from an adjacency matrix of $\hat{\mathbf{P}}$. More importantly, one can calculate *correlated subsets* of reactions that always appear together in the extreme pathways, effectively forming a network *module*.

Input–output relationships.   Type I extreme pathways contain primary exchange reactions (recall Figure 12.9A). These exchange reactions can be used to determine which output can be achieved from a given input. Conversely, one can determine all the input combinations that lead to a given output. Furthermore, one can determine which pathways have an identical set of inputs and outputs. The number of pathways with identical inputs and outputs give a measure of network, or *pathway redundancy*, by counting the number of different internal states that give the same external state. Finally, based on the input–output relationships, one can mathematically define *crosstalk* using the overlap between the inputs and outputs in a set of pathways.

Consequences of regulatory rules.   The reconstruction of regulatory networks can lead to a set of causal relationships that describe regulatory interactions. These logical statements can be such that an extreme pathway can never be expressed due to conflicts with the regulatory rules, or only expressed under certain environmental conditions. One can therefore get an assessment of how regulation reduces the allowable functional states.

Example network We will use a simple, easy to understand network shown in Figure 13.1 to illustrate some of these properties. Each of the extreme pathways results in the production of E and a by-product, but their internal states differ.

13.2   Pathway Length and Flux Maps

A Pathway Length Matrix ($\mathbf{P}_{LM}$) can be calculated directly from the binary form of the extreme pathway matrix ($\hat{\mathbf{P}}$). $\mathbf{P}_{LM}$ is computed by pre-multiplying $\hat{\mathbf{P}}$ by its own transpose,

$$\mathbf{P}_{LM} = \hat{\mathbf{P}}^T \hat{\mathbf{P}} \tag{13.3}$$

resulting in a symmetric matrix. $\mathbf{P}_{LM}$ is an adjacency matrix of $\hat{\mathbf{P}}$.

- The diagonal values of $\mathbf{P}_{LM}$ correspond to the number of reactions in an extreme pathway. In the simple example system (Figure 13.2), the first value

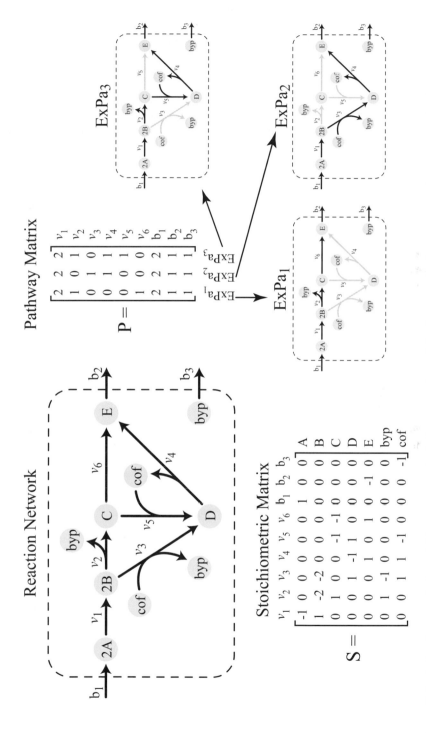

Figure 13.1 A simple network, its stoichiometric matrix, pathway matrix, and the flux map for the extreme pathways. From [323].

$$P = \begin{bmatrix} 2 & 2 & 2 \\ 1 & 0 & 1 \\ 0 & 1 & 0 \\ 0 & 1 & 1 \\ 0 & 0 & 1 \\ 1 & 0 & 0 \\ 2 & 2 & 2 \\ 1 & 1 & 1 \\ 1 & 1 & 1 \end{bmatrix} \longrightarrow \tilde{P} = \begin{bmatrix} 1 & 1 & 1 \\ 1 & 0 & 1 \\ 0 & 1 & 0 \\ 0 & 1 & 1 \\ 0 & 0 & 1 \\ 1 & 0 & 0 \\ 1 & 1 & 1 \\ 1 & 1 & 1 \\ 1 & 1 & 1 \end{bmatrix} \longrightarrow \hat{P}^T \hat{P} = \begin{bmatrix} 6 & 4 & \circled{5} \\ & 6 & 5 \\ & & 7 \end{bmatrix} \begin{matrix} \text{ExPa}_1 \\ \text{ExPa}_2 \\ \text{ExPa}_3 \end{matrix}$$

Figure 13.2 The pathway length matrix $\mathbf{P}_{LM}$ for the simple network in Figure 13.1. The lengths of ExPa$_1$, ExPa$_2$, and ExPa$_3$ are 6, 6, and 7, respectively, and are the highlighted diagonal elements of the final matrix. ExPa$_2$ and ExPa$_3$ have a shared length of 5, indicated by the circle. From [323].

along the diagonal is 6, meaning that six reactions participate in ExPa$_1$. A quick count of the fluxes shown in ExPa$_1$ (Figure 13.1) shows that there are indeed six reactions participating in the first extreme pathway.
- The off-diagonal terms of $\mathbf{P}_{LM}$ are the number of reactions that a pair of extreme pathways have in common. For example, notice the circled off-diagonal term in Figure 13.2, which is a comparison of ExPa$_3$ (the column) and ExPa$_1$ (the row) and contains a value of 5. ExPa$_1$ and ExPa$_3$ have five reactions in common. Upon examining ExPa$_1$ and ExPa$_3$ in Figure 13.1, one can readily see that the five reactions shared are b_1, v_1, v_2, b_2, and b_3.

Thus, the diagonal and off-diagonal terms of $\mathbf{P}_{LM}$ are the number of reactions in an extreme pathway and the number of reactions common to the two pathways, respectively. They can be used to obtain a measure of pathway 'length.'

Genome-scale example Pathway lengths have been computed for genome-scale matrices [323]. For *Helicobacter pylori*, all the extreme pathways that lead to protein synthesis from a set of substrates have been computed (Figure 13.3). These pathways are all composed of about 100–110 reactions even though the protein yields vary significantly. This demonstrates that basically the same set of reactions can be used to generate quite different overall protein synthesis rates.

13.3   Reaction Participation and Correlated Subsets

The Reaction Participation Matrix ($\mathbf{R}_{PM}$) is calculated by post-multiplying $\hat{P}$ by its own transpose,

$$\mathbf{R}_{PM} = \hat{P}\hat{P}^T \tag{13.4}$$

forming a symmetric matrix. $\mathbf{R}_{PM}$ is an adjacency matrix of $\hat{P}$. The computed $\mathbf{R}_{PM}$ for the simple example system is shown in Figure 13.4.

- The diagonal terms in $\mathbf{R}_{PM}$ give the number of pathways in which a particular reaction participates. For example, the first diagonal term in $\mathbf{R}_{PM}$ for the simple example system, corresponding to reaction v_1, has a value of 3. Thus, reaction v_1 participates in all three extreme pathways. An examination

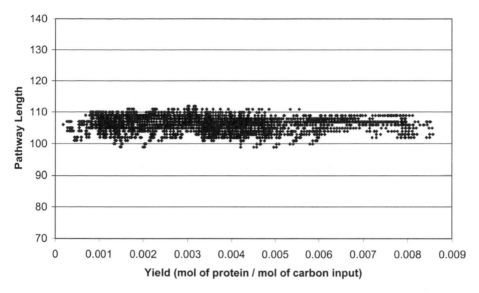

Figure 13.3 Correlation of extreme pathway length and yield (mol of protein/mol of carbon input) in *Helicobacter pylori* protein synthesis. There was essentially zero correlation between target product yield and extreme pathway length. From [323].

$$
P = \begin{bmatrix} 2 & 2 & 2 \\ 1 & 0 & 1 \\ 0 & 1 & 0 \\ 0 & 1 & 1 \\ 0 & 0 & 1 \\ 1 & 0 & 0 \\ 2 & 2 & 2 \\ 1 & 1 & 1 \\ 1 & 1 & 1 \end{bmatrix} \longrightarrow \tilde{P} = \begin{bmatrix} 1 & 1 & 1 \\ 1 & 0 & 1 \\ 0 & 1 & 0 \\ 0 & 1 & 1 \\ 0 & 0 & 1 \\ 1 & 0 & 0 \\ 1 & 1 & 1 \\ 1 & 1 & 1 \\ 1 & 1 & 1 \end{bmatrix} \longrightarrow \hat{P}^T\hat{P} =
$$

	v_1	v_2	v_3	v_4	v_5	v_6	b_1	b_2	b_3	
	③	2	1	?	1	1	③	③	③	v_1
		2	0	1	1	1	2	2	2	v_2
			1	1	0	0	1	1	1	v_3
				2	1	0	2	②︎	2	v_4
					1	0	1	1	1	v_5
						1	1	1	1	v_6
							③	③	③	b_1
								③	③	b_2
									③	b_3

Figure 13.4 The reaction participation matrix $\mathbf{R}_{PM}$ for the simple network in Figure 13.1. The number of extreme pathways in which each reaction participates is indicated in the diagonal elements, as highlighted in the final matrix. Redrawn from [323].

of ExPa$_1$, ExPa$_2$, and ExPa$_3$ in Figure 13.1 shows that reaction v_1, which converts A to B, is in fact utilized in all three extreme pathways.

- The off-diagonal terms give the number of extreme pathways that contain the pair of corresponding reactions. For example, the off-diagonal element boxed in Figure 13.4 has a value of 2. This element refers to the number of pathways that contain both reaction b_2 (the column) and reaction v_4 (the row). Both of these reactions are utilized in ExPa$_2$ and ExPa$_3$, while only b_1 is utilized in ExPa$_1$ (Figure 13.1).

Genome-scale example Reaction participation numbers have been computed for all the extreme pathways that lead to protein synthesis in *H. pylori*. The participation

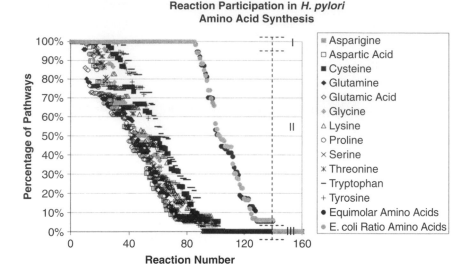

Figure 13.5   Reaction participation in the extreme pathways of the genome-scale *H. pylori* metabolic network. From [323].

numbers for the individual reactions can be rank-ordered (Figure 13.5) leading to the definition of three categories of reactions: (1) reactions that participate in all of the extreme pathways; (2) reactions that participate in varying amounts of extreme pathways; and (3) reactions that do not participate in any of the extreme pathways. The first group represents essential reactions for protein synthesis, and the last represents reactions irrelevant to protein synthesis. The second group represents a set of reactions that can be used for protein synthesis, but are not essential because there are pathways that lead to protein synthesis without them.

Correlated subsets   The off-diagonal elements of $\mathbf{R}_{PM}$ can be used to define correlated subsets of reactions [323, 332]. The circled elements in Figure 13.4 show reaction pairs that participate in exactly the same extreme pathways. In this particular case, each of these reaction pairs participates together in all of the extreme pathways. Thus, reactions v_1, b_1, b_2, and b_3 are always present. They form a correlated reaction subset, meaning that if one of them is utilized, the others must also be utilized.

Regulation of expression and co-sets   One would expect that a common transcriptional regulatory mechanism would develop for the genes that encode the members of a co-set. This is the case for rhamnose metabolism in *E. coli* (Figure 13.6). A larger study found a correlation between the expression of genes in co-sets and transcription units based on expression data [359]. With more comprehensive knowledge about transcription unit (TU) architecture, one might expect that such correlations will improve.

Co-sets and single nucleotide polymorphisms (SNPs)   One can use co-sets to seek dependencies among SNPs with causal implications on network function, by grouping

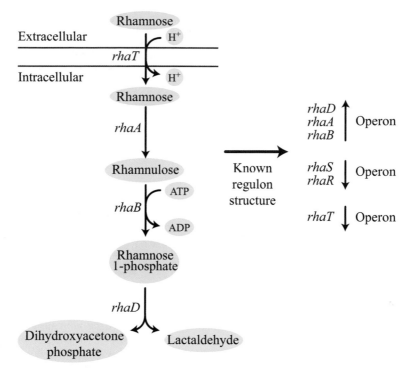

Figure 13.6   A co-set containing four reactions involved in the utilization of rhamnose in *E. coli*. These reactions are catalyzed by the products of four genes (shown in italics next to the reactions), which are organized into two operons. These operons form a regulon that is regulated by rhaS. From [325].

SNPs in proteins that catalyze different reactions within the same co-set. SNPs that adversely affect the function of an enzyme in a co-set will have the same effect on network state. Thus co-sets represent a way to correlate and analyze SNPs.

One can classify a group of genes that encode members of a co-set into three fundamental types; Type A, Type B, and Type C (Figure 13.7). Type A describes a multi-meric enzyme, where an SNP in any subunit of the multimer can thus result in the same phenotype. Type B represents a co-set of reactions in a contiguous pathway, and Type C co-sets are formed by non-contiguous reactions. Examples of such co-sets are found in the human mitochondria [181].

Flux-coupling assessment through optimization   A more thorough assessment of the relationships between the use of fluxes in reconstructed networks has been developed [56], called the *Flux Coupling Finder* (FCF). FCF is based on a linear programming approach to minimize and maximize the ratio between all pairwise combinations of fluxes in a reaction network. The computations classify reaction pairs to be either:

directionally coupled,   if a non-zero flux for v_i implies a non-zero flux for v_j, but not necessarily the reverse;

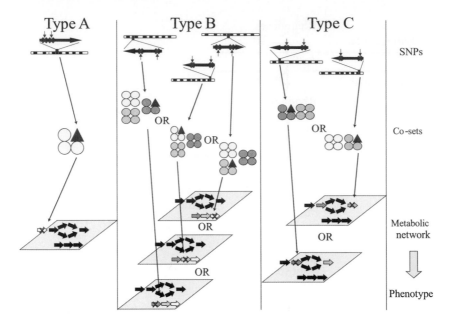

Figure 13.7   Relating SNPs, diseases, and correlated reaction sets. Each subpanel exemplifies a type of co-set represented by colored arrows and circles (non-black). The circles represent the protein subunits of a particular enzyme. Each enzyme is the same color. Red triangles represent altered protein subunits due to causal SNPs. Each group of colored circles (enzyme) corresponds to the same colored reaction flux in the pathway. SNPs in the gene coding for the protein subunits that catalyze the transformation from one metabolite to another will result in a phenotype that is unable to produce that particular product (Type A co-set). Type B co-sets function in an analogous manner, where causal SNPs affecting a reaction that is part of a correlated reaction set (a linear chain of reactions) will result in a phenotype characterized by an inability to produce the end point of the chain. More complex schemes may also occur in which the correlated reaction set is not a linear pathway (Type C co-set). From [181].

partially coupled,   if a non-zero flux for v_i implies a non-zero, though variable, flux for
   v_j, and vice versa; or

fully coupled,   if a non-zero flux for v_i implies not only a non-zero but also a fixed flux
   for v_j, and vice versa.

The FCF was utilized to analyze the genome-scale networks of *E. coli*, *Saccharomyces cerevisiae*, and *H. pylori* [56]. The percentage of reactions in the networks for each microorganism that were found in a coupled set was 60% for *H. pylori*, 30% for *E. coli*, and 20% for *S. cerevisiae*. This percentage is indicative of the flexibility of a network and the degrees of freedom available in a network.

13.4   Input–output Relationships and Crosstalk

Extreme pathways type I have an input–output signature that is made up of the exchange reactions found in the pathway. These signatures can be used to generate an *Input/Output Feasibility Array* (IOFA).

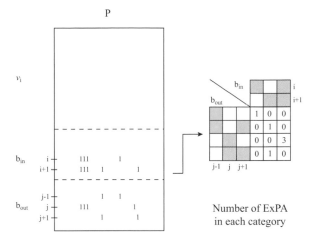

Figure 13.8 The input–output feasibility array.

The IOFA The primary exchange reactions of an extreme pathway will contain input and output reactions (Figure 13.8). The columns of $\hat{P}$ can be ordered based on the input and output status of a pathway. The extreme pathways with identical inputs and outputs will be placed in an adjacent position, and the number of such pathways enumerated.

The part of $\hat{P}$ that contains the primary inputs and outputs can be segmented out of the matrix. This part of the matrix can be represented in a different format. An array can be formed where the columns represent a unique set of inputs (the *input signature*) and the rows represent a unique set of outputs (the *output signature*). If there is a pathway that connects an input signature to an output signature, the corresponding entry in the array can be assigned a value. This value can be a color so that one can visualize all the possible matches between a set of inputs and outputs. One can also put a numerical value in the array that corresponds to the number of pathways that connect a particular set of inputs to a particular set of outputs. The IOFA is a concise representation of a set of input–output properties of extreme pathways.

Computing the number of identical input/output states in a genome-scale metabolic network Sets of extreme pathways for genome-scale metabolic networks for *H. pylori* and *Haemophilus influenzae* have been computed for a number of growth environments and for a number of required outputs (such as the production of individual amino acids). The average number of extreme pathways for all these different functional states that have identical input/output signatures have been computed [322]. The results show that for *H. pylori*, the number of pathways with identical inputs and outputs is 2, whereas the corresponding number for *H. influenzae* is 46. Thus, even though the metabolic networks appear similar, *H. influenzae* has much more flexibility in the choice of an internal state for a given overall network function.

Defining crosstalk Extreme pathway analysis can be used to analyze the interconnection of multiple inputs and multiple outputs of signaling pathways, often called

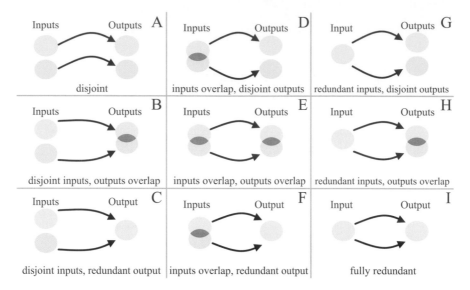

Figure 13.9 Classification scheme for crosstalk in the extreme pathways of a signaling network. From left to right, each pair of input sets is disjoint, overlapping, and identical. From top to bottom, each pair of output sets is disjoint, overlapping, and identical. From [322].

crosstalk. As the extreme pathways are fundamental and irreducible functional states of a signaling network, crosstalk can be defined as the non-negative linear combination of extreme pathways of a signaling network. The pairwise combination of extreme pathways is thus the simplest form of crosstalk.

Classifying crosstalk   With this definition, crosstalk can be classified into nine different categories as shown in Figure 13.9. Each circle in Figure 13.9 represents a set of pathway inputs or a set of pathway outputs. From left to right, each pair of pathway input sets is classified as disjoint, overlapping, or identical. From top to bottom, each pair of pathway output sets is classified as disjoint, overlapping, or identical. For example, the representation in the middle of the figure corresponds to two extreme pathways with shared (but not identical) sets of inputs, and shared (but not identical) sets of outputs. Thus, the two independent extreme pathways in this instance have overlapping but not identical functionality. Panel B represents the circumstance most commonly referred to as crosstalk.

13.5   Regulation Eliminates Active Pathways

All the allowable functional states of a reconstructed network are described by the corresponding set of extreme pathways. However, regulatory networks may prevent some of these functional states. Thus, regulation may be viewed as a way to shrink the steady-state flux solution space. This principle is illustrated in Figure 13.10. Thus, regulatory networks can be viewed as a mechanism to generate self-imposed constraints that restrict the allowable functional states of a network. These constraints may be thought of as *restraints*.

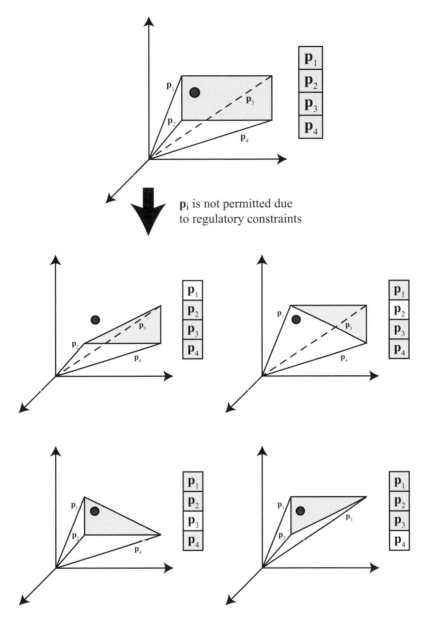

Figure 13.10 Regulatory constraints reduce the steady-state solution space of a network by eliminating operational pathways. In the top panel, all of the pathways are considered operational in the network (denoted by the highlighted green boxes at right). Under certain environments, however, regulatory constraints may cause one or more of the extreme pathways to be inoperable, p_1 through p_4 in the cases shown, resulting in a more restricted space corresponding to a metabolic network with fewer available behaviors (bottom panels). Note that if p_1 is rendered inoperative, the physiological state represented by the red dot cannot be attained. Modified from [86].

13.6   Summary

- Pathway vectors can be used to characterize all functional states of networks. Because they are basis vectors that characterize the solution space, the properties that they describe are global, i.e., throughout the space.
- The extreme pathways can be used as columns in a matrix to form a pathway matrix. This matrix can be represented in a binary form ($\hat{P}$), where an entry of '1' indicates that a reaction participates in a pathway.
- The adjacency matrices of $\hat{P}$ give the number of reactions that make up extreme pathways (pathway length), and the number of pathways in which a particular reaction participates (pathway participation).
- Analysis of pathway participation leads to the identification of correlated reaction sets (co-sets), which are sets of reactions that always appear together in the functional states of a network. The reactions that make up a co-set may be regulated coordinately.
- The input/output status of pathways can be used to determine network redundancy and its crosstalk characteristics.
- Regulatory rules can be used to reduce the number of allowable pathways and thus shrink the solution space.

14 Randomized Sampling

Everything we care about lies somewhere in the middle, where pattern and randomness interlace – James Gleick

Pathways as basis vectors for the null space are useful for studying the capabilities of a network and to determine network properties. An alternative approach to characterizing the contents of solution spaces is *uniform random sampling*. This approach involves obtaining a statistically meaningful number of solutions uniformly distributed throughout the entire solution space and then studying their properties. Randomized sampling of candidate states throughout an entire solution space gives an unbiased assessment of its properties and can result in significant biological insights and understanding.

14.1 The Basics

A simple flux split Uniform random sampling can be illustrated by looking at a simple flux split (see Figure 14.1). The null space can be shown in two dimensions using its two conical basis vectors:

$$\mathbf{b}_1 = (1, 0, 1) \quad \text{and} \quad \mathbf{b}_2 = (0, 1, 1) \tag{14.1}$$

Note that the maximum weight that can be placed on these vectors is $\alpha_{1,\max} = 6$ and $\alpha_{2,\max} = 8$, whereas the minimum values in both cases is zero. There is an additional constraint association with reaction 3 that states that

$$\alpha_{1,\max} + \alpha_{2,\max} \leq \alpha_{3,\max} = 14 \tag{14.2}$$

thus both $\alpha_{1,\max}$ and $\alpha_{2,\max}$ cannot be attained simultaneously.

The null space can be enclosed with a parallelepiped by ignoring the constraint in Equation (14.2), which is a parallelogram in two-dimensions. The parallelepiped can then be sampled uniformly. Then a set of candidate flux solutions is formed by selecting only points in the null space; basically by excluding points in the sampled set that violate the constraint of Equation (14.2). The probability distributions for the

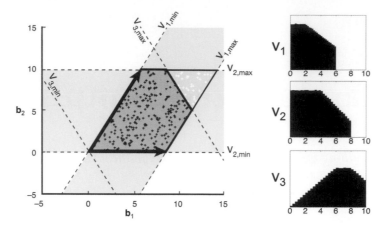

Figure 14.1 The outcome and representation of randomized sampling for the simple flux split shown in Figure 12.4. Note that the 'white' points are excluded from the sampled points if the $v_{3,max}$ constraint is applied. Modified from [347].

flux through each individual reaction can then be graphed. The algorithms used to perform the randomized sampling are discussed in the subsequent sections.

The overall procedure The process of obtaining a uniform set of candidate solutions and studying their properties consists of three basic steps:

1 defining the space to be sampled based on imposed constraints,
2 randomly sampling it based on uniform statistical criteria, and
3 further segmenting the solution space based on additional post-sampling criteria as necessary.

It is particularly convenient to use only linear constraints in the first step. Linear equalities and inequalities lead to the formation of a polytope. Then, following the random sampling, candidate solutions can be eliminated based on non-linear criteria in the third step, or based on additional experimental information, as illustrated in Figure 14.2. A large number of candidate solutions can then be characterized using statistical measures.

14.2   Sampling Low-dimensional Spaces

To compute the size and contents of a solution space, Monte Carlo integration is generally implemented by defining a range of the variables that encompasses the solution space and then uniformly and randomly sampling points within this region. The solution space size is then calculated by determining the fraction of the uniformly distributed points that lie within the solution space and multiplying by the volume of the enclosing region (see Figure 14.1).

It is easy to sample regularly shaped geometric objects. A solution space can be enclosed by a geometric object in which uniformly distributed random points can be readily generated. Ideally, the shape of the chosen geometric object needs to fit

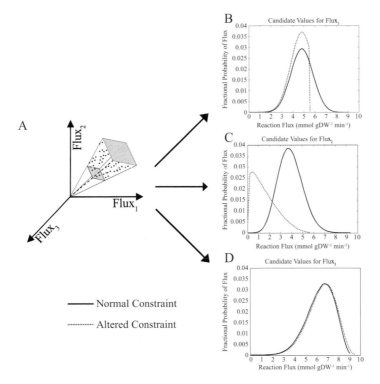

Figure 14.2 Sampling a solution space followed by tightening constraints in it that leads to the exclusion of a set of candidate solutions. The statistical properties of candidate solutions with and without the tighter constraints can be compared. For some fluxes the constraints may not matter (Panel D), while for others they may make a big difference (Panel C). Taken from [235].

as tightly as possible around the solution space. This would lead to a high fraction of points that are in the geometric object and in the enclosed solution space. This approach has been described in detail and the computational methods have been developed [347].

Parallelepipeds   A parallelepiped with the same dimension as the rank, r, of the null space of $\mathbf{S}$, is a geometric object that can enclose polytopes and can be sampled readily. A parallelepiped can be represented as a matrix, $\mathbf{B}$, where the columns of $\mathbf{B}$ represent a set of spanning edges of the parallelepiped,

$$\mathbf{B} = (\mathbf{b}_1, \ldots, \mathbf{b}_i, \ldots, \mathbf{b}_r) \tag{14.3}$$

Its volume is simple to compute [407],

$$\text{Volume} = \sqrt{Det(\mathbf{B}^T\mathbf{B})} \tag{14.4}$$

Sampling parallelepipeds   Uniform random samples of points can be generated readily within a parallelepiped. Uniform random weightings, α_i, are generated on all of the spanning edges, $\mathbf{b}_i$. A random point inside the space is generated by:

$$\mathbf{v} = \sum_i \alpha_i \mathbf{b}_i, \ \alpha_{i,\min} \leq \alpha_i \leq \alpha_{i,\max} \tag{14.5}$$

where $\mathbf{v}$ is a point within the space.

Elimination of redundant constraints in determining $\alpha_{\max}$ Many reaction $v_{i,\max}$ levels cannot be reached in a steady-state as the saturation of other reactions can be more constraining for v_i than its own $v_{i,\max}$ (recall Figure 12.5). Thus, many of the $v_{i,\max}$ constraints may be systemically redundant. Redundant $v_{i,\min}$ and $v_{i,\max}$ constraints are not needed to define the solution space. One can readily determine if a particular $v_{i,\max}$ is redundant by determining if

$$\max v_i < v_{i,\max} \quad \text{redundant} \tag{14.6}$$

or

$$\max v_i = v_{i,\max} \quad \text{not redundant} \tag{14.7}$$

along a spanning edge using the optimization methods described in Chapter 18. Determining redundant $v_{i,\min}$ constraints is done similarly.

Choice of enclosing parallelepiped   Because each pair of $v_{i,\min}$ and $v_{i,\max}$ constraints form parallel hyper-planes, the shape of the null space leads naturally to the choice of a high-dimensional parallelepiped in which it will be enclosed. The set of possible parallelepipeds that can enclose the steady-state flux space is chosen by forming the faces of the parallelepiped along the directions defined by these $v_{i,\min}$ and $v_{i,\max}$ constraints. As each parallelepiped is defined by r planes which are chosen from the set of m $v_{i,\min}$ and $v_{i,\max}$ planes, the number of such parallelepipeds that could be used to enclose the space is

$$\text{Number of possible parallelepipeds} = \frac{m!}{r!(m-r)!} \tag{14.8}$$

where m is the number of $v_{\max}$ constraints and r is the dimension of the null space. For the simple split, $m = 3$ and $r = 1$, thus the number of possible parallelepipeds is 3. All three can be seen in Figure 14.1 using the dashed lines.

Uniform random sampling   A set of uniform random points can be generated within a solution space by randomly sampling within the enclosing parallelepiped. Sampling can be performed by choosing a weight on each of the edges uniformly, b_i, of the parallelepiped. Each point in the space is uniquely defined by weightings on the edges spanning the parallelepiped (Equation (14.5)). The weighting, α_i, on each basis vector, b_i, can be selected uniformly by generating a random number, f, between 0 and 1 and computing each weight as

$$\alpha_i = \alpha_{i,\min} + f(\alpha_{i,\max} - \alpha_{i,\min}) \tag{14.9}$$

Points generated uniformly within the parallelepiped were then compared to the set of $v_{i,\max}$ and $v_{i,\min}$ constraints to verify whether or not the point falls in the solution space. If the point satisfies all constraints, it is a valid solution and is kept in the random set. If the point does not satisfy all the constraints, it is excluded.

Computing the 'volume' of the solution space We can determine the *hit fraction* ($\bar{p}$) as the ratio of sampled points that fall inside the solution space and n as the total number of sample points generated. The volume of the solution space can be calculated by multiplying the volume of the enclosing parallelepiped by the fraction of generated points that fall within the solution space:

$$estimated\ volume\ of\ solution\ space \approx \bar{p} \cdot volume\ of\ enclosing\ parallelepiped$$

Note that the notion of a volume here is not tied to our usual definition of volume as the size of a three-dimensional solution space. For instance, the so-calculated volume of a 2D solution space is actually an area (i.e., see Figure 14.1).

Error in computed volume estimate The variance in the volume estimate is [341]:

$$\sigma^2 = \frac{\bar{p}(1-\bar{p})}{n} \leq \frac{1}{4n} \tag{14.10}$$

where n is the total number of sample points generated. The maximum variance is at $\bar{p} = 1/2$. The estimated relative error in the volume calculation, ϵ, can be calculated as the ratio of the standard deviation (σ) to the mean (μ) as

$$\epsilon = \frac{\sigma}{\mu} = \sqrt{\frac{\bar{p}(1-\bar{p})/n}{\bar{p}^2}} = \sqrt{\frac{1/\bar{p}-1}{n}} \tag{14.11}$$

14.3 Sampling High-dimensional Spaces

The sampling methods described above have proven to work for polytopes of dimensions of up to 12. In spaces of higher dimension, more sophisticated sampling methods are required. The detailed description of these methods requires sophisticated statistics and is beyond the scope of this text. The interested reader can consult with some of the primary references [199, 244].

There are many applications of sampling for use in studying network states and to address important biological issues [378]. This section describes three insightful approaches to solution space characterization.

Growth of *E. coli* and a high-flux backbone The steady-state flux solution space of a genome-scale reconstruction of *E. coli* metabolism has been randomly sampled [14]. The distribution of all the steady-state flux levels through all of the individual reactions in the sampled solutions shows an approximate power-law distribution (see Figure 14.3). This feature is a global property of the functional states of the network.

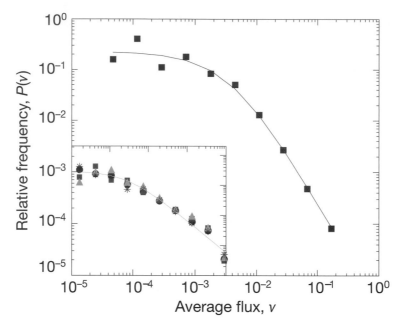

Figure 14.3   Distribution of the flux levels through reactions in a genome-scale metabolic network of *E. coli*. The inset shows the flux distribution in four randomly chosen sample points. From [13].

Recall from Chapter 10 that the distribution of the number of reactions that a compound participates in also follows an approximate power-law distribution.

This study also looked at the correlated use of reactions in the *E. coli* metabolic network. It was found that a large fraction of the candidate solutions had common network-scale patterns in the used sets of reactions. This large co-set was termed the *high-flux backbone* of the flux map, because it was similar in all solutions. The individual solutions then have distinct deviations from this state.

Analyzing the exo-metabolome   Metabolomics is a data type that holds great potential for the analysis of metabolic networks. The metabolome is divided into the endo-metabolome (EnMe, metabolites inside a cell) and the exo-metabolome (ExMe, metabolites in the medium outside the cell). The ExMe typically contains fewer compounds than the EnMe and they are stable. The EnMe is complex and has labile components creating challenges with experimental sampling procedures that capture the EnMe in a physiologically representative state [350, 432].

A sampling approach on the flux state of the internal network, on a genome-scale, can be linked to ExMe measurements forming an input/output (I/O) type analysis. By using a randomized sampling approach (Figure 14.4), ExMe changes can be linked to intracellular flux perturbations in an unbiased manner without relying on defining any optimal flux distributions. The so-inferred perturbations in intracellular reaction fluxes can be analyzed further using reporter metabolite and subsystem approaches [30] in order to identify dominant metabolic features that are collectively perturbed. The sampling-based approach also has the additional benefit of being less sensitive

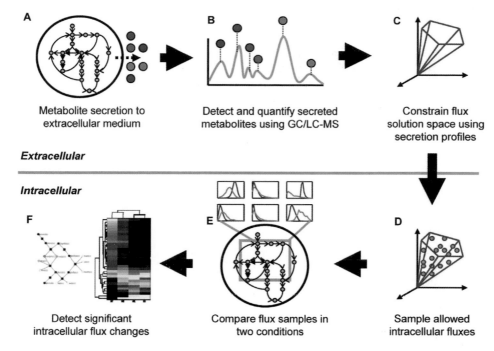

Figure 14.4 Schematic illustrating the integration of exo-metabolomic (ExMe) data with a constraint-based framework to perform an input/output type analysis. (A) Cells are subjected to genetic and/or environmental perturbations to secrete metabolite patterns unique to that condition. (B) ExMe is detected, identified, and quantified. (C) ExMe data are used to generate secretion flux constraints to define allowable solution space. (D) Random sampling of solution space yields the range of feasible flux distributions for intracellular reactions. (E) Sampled fluxes are compared to sampled fluxes of another condition to determine which metabolic regions were altered between the two conditions. (F) Significantly altered metabolic regions are identified. From [266].

to inaccuracies in metabolite secretion profiles than optimization-based methods and thus can more readily be used in settings such as biofluid metabolome analysis.

Classifying dominant modes of regulation The solution space can be constrained further by adding transcriptomic data to fluxomic data obtained [54]. The most likely flux change through a reaction can thus be compared with the corresponding change in the expression of the proteins that make up the enzyme. The comparison of flux change and gene expression leads to the identification of enzymes showing a signifi-cant correlation between flux change and expression change (interpreted as transcrip-tional regulatory mode) as well as reactions whose flux change is not correlated with changes in gene expression, and thus likely to be driven by changes in the metabolite concentrations (metabolic regulation). Lastly, changes in gene expression that are not correlated to changes in rate are interpreted as post-transcriptional regulation. The result of the analysis is illustrated by the table in the middle of Figure 14.5.

The utility of the methods was examined using data from *S. cerevisiae* [54]. Data were collected from growth on four different carbon sources (glucose, maltose,

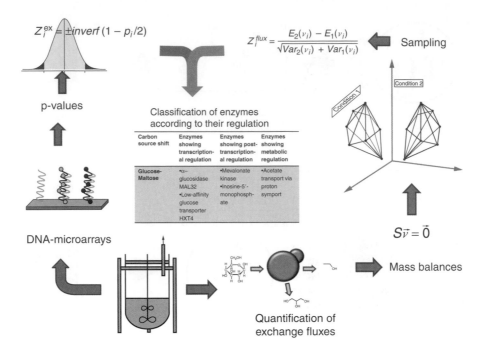

Figure 14.5 A workflow that uses fluxomic and transcriptomic data to constrain possible flux states and to classify regulatory modes for particular gene products. Gene expression data are transformed into significance scores and p-values for the expression change between conditions examined. Fluxomic data are obtained under the two conditions. Novel sampling method of allowable flux states under two conditions is performed and the results are compared. The analysis infers the dominant regulatory action for a particular gene product. From [54].

ethanol, and acetate) in chemostat. Five deletion mutants (Δ-*grr1*, Δ-*hxk2*, Δ-*mig1*, Δ-*mig1*Δ-*mig2*, and Δ-*gdh1*) grown in batch cultures were used. The regulatory modes for a series of metabolic enzymes are estimated. The enzymes exhibiting transcriptional regulation showed enrichment in known transcription factors.

Rotating basis vectors to elucidate regulatory needs Monte Carlo sampling of the steady-state flux space of a large-scale metabolic system in conjunction with Principal Component Analysis (PCA) and eigenvector rotation results in a low-dimensional and biochemically interpretable decomposition of the steady flux states of a network [28] (Figure 14.6). The solution space is randomly sampled. Additional flux constraints can be obtained from an mRNA expression profile. The flux states that are inconsistent with the mRNA expression profiles are eliminated. A PCA decomposition (similar to SVD) is performed on the permissible flux states. The singular vectors are rotated onto the stoichiometric structure of the network to identify the pathways whose flux levels create the largest dimensions in the allowable solution space.

This analysis method allows for the identification of the key post-translational regulatory needs to confine the solution close to a single point and also the number of regulatory steps that are needed to control the phenotypic state of a network. This approach is essentially a network-level analysis of regulation and regulatory needs;

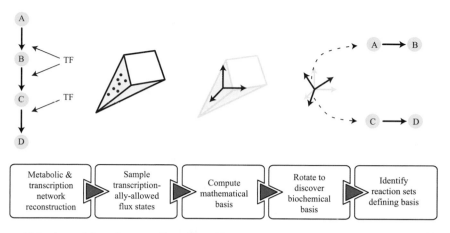

Figure 14.6 A workflow for sampling allowable states and to rotate the results from PCA decomposition of the sampled states onto biochemically meaningful pathways. The possible steady-state flux states of the transcriptionally allowed regions of the *E. coli* metabolic network are sampled and analyzed to reveal a small number of reaction sets that account for nearly all of the flux variation in a dynamic growth environment. From [28].

a systems biology approach that contrasts that of molecular biology that focuses on molecular mechanisms associated with individual regulatory events.

14.4   Sampling Network States in Human Metabolism

A growing number of studies have appeared that use sampling of networks derived from Recon 1. These studies represent a new approach to analyze metabolic states in human cells and tissues.

Mitochondria in the heart An early human cardiac mitochondrial network reconstructed from proteomic data [429] was analyzed using the sampling approach. Network capabilities were characterized under different conditions, including various diets and diabetic conditions. It was found that the pyruvate dehydrogenase flux in diabetic patients is constrained to be lower than in non-diabetic patients due to mass conservation constraints alone. This result was surprising because although it had been known that pyruvate dehydrogenase has a decreased flux in diabetics, it was thought to be a consequence of unknown regulatory mechanisms. The sampling approach thus demonstrated how bottom-up reconstructions can describe real biochemical and physiological conditions and provide mechanistic insights into cause and effect relationships.

Enzymopathies in red blood cell metabolism The steady-state flux solution space of a historical model describing metabolism in the human red blood cell has been studied through randomized sampling [347, 452]. The probability distributions for each flux in the network can be shown on the reaction map (Figure 14.7). This representation allows one to visualize all the allowable flux values for all the reactions in the network

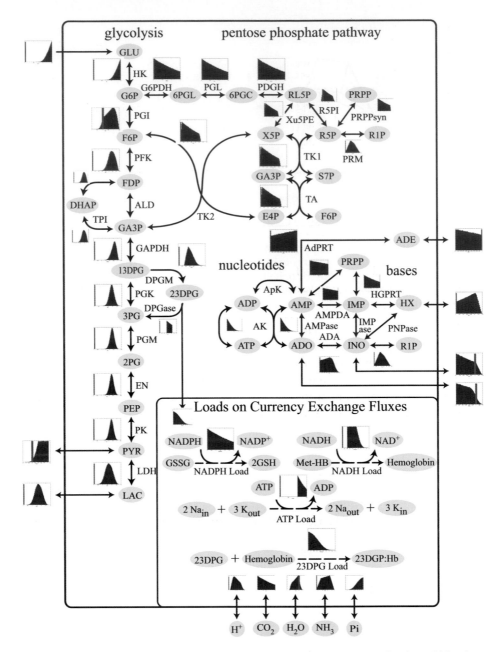

Figure 14.7 Monte Carlo sampling of the steady-state flux solution space for the red blood cell. The red boxes in the histograms denote zero flux. From [347].

simultaneously. These studies have led to several notable results, three of which are briefly described here.

How well can fluxes be estimated? The histograms provide information about the 'shape' of the solution space and how likely the fluxes are to fall into certain numerical values. For instance, some of the histograms are flat, implying that

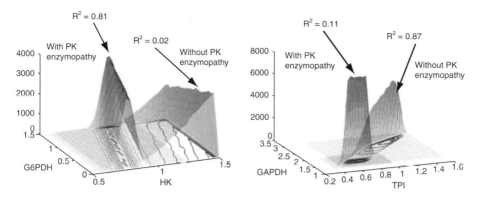

Figure 14.8 Simulated enzymopathies in the human red blood cell using segmentation of sampled solution spaces. From [347].

every numerical value for a flux through a particular reaction is equally likely, and thus more data are needed to estimate what these fluxes actually are under physiological conditions.

Should measurements be biased towards uncorrelated fluxes?   The cross-correlations between every pair of flux values can be computed. Such computations lead to identifying co-sets if the correlation coefficient is unity ($r^2 = 1.0$). The co-sets are also computable from the extreme pathways (Chapter 13). Poor correlations can be found between pairs of flux variables. Such correlations can be used to guide experimental design. The measurement of poorly correlated fluxes is likely to be more informative than measuring highly correlated fluxes.

Are fluxes correlated differently under different physiological conditions?   There may be regions in the solution space where fluxes are correlated strongly and regions where they are correlated weakly. Such regions can be found by segmenting the solution space and computing the correlations in each segment. Solution spaces have been segmented based on inborn errors in metabolism, i.e., lowered $v_{i,\max}$ values due to genetic variation. Figure 14.8 shows an example of the analysis of enzymopathies in red cell metabolism. Pyruvate kinase (PK) and glucose-6-phosphate dehydrogenase (G6PDH) have many genetic variations in the human population; some of which are related to pathological conditions [183]. Figure 14.8 shows how the correlation between fluxes can be very different between the full solution space and the segment created by the imposition of a $v_{i,\max}$ constraint for PK and G6PDH. This suggests that the characteristics of metabolic states could be significantly different in individuals with these enzymopathies as compared to a normal state. Uncorrelated fluxes in a normal physiological state may be subject to independent modes of regulation. An enzymopathy that leads to stoichiometric coupling of such pairs of fluxes would confuse such regulatory mechanisms, thus exacerbating the effects of the enzymopathy.

Intracellular pathogens   Metabolic coupling of *Mycobacterium tuberculosis* to its host is foundational to its pathogenesis. Computational genome-scale metabolic models

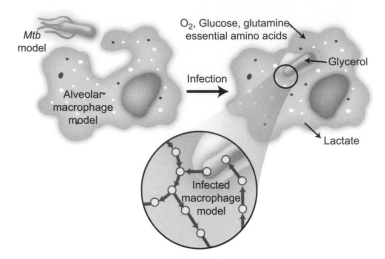

Figure 14.9 Host–pathogen models are formed by linking together separate models of a host and the pathogen. Prepared by Nathan Lewis.

can be used for integrating omic as well as physiologic data for systemic, mechanistic analysis of metabolism [48]. First, a cell-specific alveolar macrophage model from the global human metabolic reconstruction, Recon 1, was constructed and used to successfully predict experimentally verified ATP and nitric oxide production rates in macrophages. This validated model was then integrated with an *M. tuberculosis* H37Rv model, to build an integrated host–pathogen genome-scale reconstruction (Figure 14.9).

The integrated host–pathogen network enables simulation of the metabolic changes during infection. Expression profiling data from infected macrophages were mapped onto the host–pathogen network and were able to describe three distinct pathological states. Integrated host–pathogen reconstructions thus form a foundation upon which understanding the biology and pathophysiology of infections can be developed.

The changes in flux states of the *M. tuberculosis* portion of the host–pathogen model can be analyzed using randomized sampling. The sampled candidate flux states show a shift in carbon uptake and overall usage. There are major shifts in flux states in central metabolism. Glycolysis is suppressed (Figure 14.10, ENO) with the production of acetyl-CoA coming from fatty acids through the glyoxylate shunt (Figure 14.10, ICL). In addition, glucose is generated from gluconeogenesis (Figure 14.10, FBP). Beyond central metabolism, there are several changes. For example, there is an up-regulation of fatty acid oxidation pathways, as fatty acids become a major carbon source. In addition, there is a shift toward mycobactin and mycolic acid synthesis. Production of nucleotides, peptidoglycans, and phenolic glycolipids is also reduced. Thus, randomized sampling of allowable flux states enables us to develop understanding of host–pathogen interactions.

Metabolic subtasking in the human brain and Alzheimer's disease   Metabolic subtasking and coordination of metabolic functions between specialized cell types in the

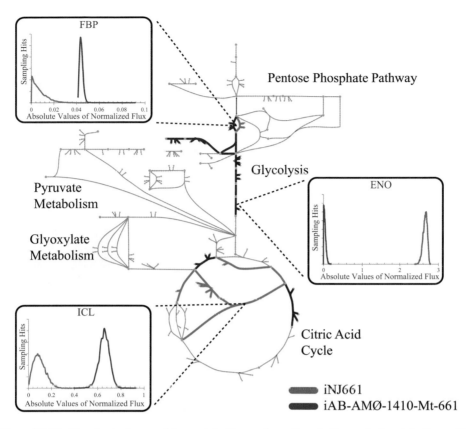

Figure 14.10   Topological map of the metabolites and reactions in the central metabolism of *M. tuberculosis*. The map shows predicted change of expression states when comparing *in vivo* and *in vitro* conditions. Blue represents the metabolic network for the pathogen alone. Red represents the integrated model. The expression states were determined from the change in the solution space by randomized sampling. From [48].

tissue microenvironment are of great interest. Metabolic network reconstructions have been built for several cell types in the human brain [233]. *In silico* metabolic models for three different neuron types were each connected with an astrocyte model; the most common supporting cell in the brain (recall Figure 6.9). With the use of randomized sampling, all feasible flux states were identified, representing the internal capacities of the cells and demonstrating how metabolic functions are distributed across multiple cells in the tissue micro-environment.

These models of brain energy metabolism were used subsequently to seek insight into experimental observations in Alzheimer's disease (Figure 14.11). Known enzyme deficiencies were incorporated into the models, specifically, α-ketoglutarate dehydrogenase complex was inhibited to levels observed in Alzheimer's disease. Sampling methods demonstrated that the metabolic rates of different cell types were affected differently, as demonstrated by the predicted change in CO_2 release (Figure 14.11B). This difference was also reflected in decreased rates of oxidative phosphorylation in glutamatergic and cholinergic neurons (Figure 14.11C,D).

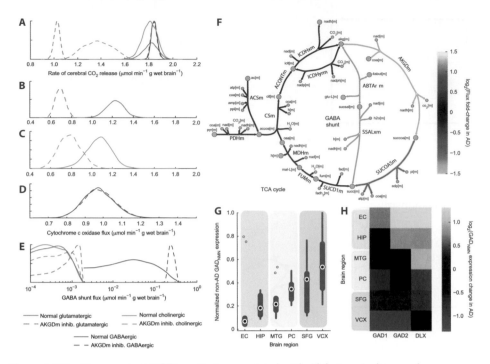

Figure 14.11 Decrease in AKGDm activity associated with Alzheimer's disease shows cell-type and regional effects *in silico* consistent with experimental data. (A–E) Kernel density plots show the distribution of feasible fluxes for various reactions. An *in silico* reduction of AKGDm flux from normal activity (solid lines) to Alzheimer's disease brain activity (dashed) decreases the oxidative metabolic rate for glutamatergic and cholinergic neurons, but not GABAergic neurons (A). This results from a decrease in the feasible fluxes for oxidative phosphorylation (e.g., cytochrome c oxidase) for both glutamatergic (B) and cholinergic neurons (C), but not GABAergic cells (D). (E,F) This cell-type-specific protection from the AKGDm deficiency results from an increased flux through the GABA shunt in GABAergic cells (E), by bypassing the damaged AKGDm (F). GABAergic cells maintain a higher GABA shunt flux because of the expression of glutamate decarboxylase (GAD). Neuroprotective properties of GAD are supported by gene expression. (G) Severely damaged brain regions in Alzheimer's disease patients have lower GAD_{NMN} expression in control brain, whereas high GAD_{NMN} regions (SFG and VCX) show little damage. (H) In Alzheimer's disease brain, severely affected regions (HIP and entorhinal cortex) show an increase in GAD_{NMN} and the GAD-inducing DLX family, suggesting that non-GAD-expressing neurons may be lost in Alzheimer's disease. GABA, gamma-aminobutyric acid; EC, entorhinal cortex; HIP, hippocampus; MTG, middle temporal gyrus; PC, posterior cingulate cortex; SFG, superior frontal gyrus; VCX, visual cortex; NMN, neuron marker normalized; inhib., inhibited. From [233].

Because neurons depend heavily on high rates of ATP production in order to maintain axonal ionic gradients, this drop in oxidative phosphorylation flux can induce apoptosis in these neuron types. However, GABAergic neurons (Figure 14.11E) and astrocytes do not show this drop in oxidative phosphorylation flux (GABA is the metabolite gamma-aminobutyric acid). This cell-type specificity is consistent with the known levels of sensitivity of different cell types in Alzheimer's disease; i.e., glutamatergic and cholinergic cells are damaged and lost earlier on in Alzheimer's disease, while

GABAergic neurons and astrocytes are relatively spared. The sampling results further demonstrated that the deficiency in the α-ketoglutarate dehydrogenase complex was absorbed by the GABA shunt (Figure 14.11F) in GABAergic neurons, and that this was made possible by a GABAergic neuron-specific enzyme, glutamate decarboxylase.

Thus, randomized sampling can be used to analyze complex multicellular metabolic functions and provide insights into disease states.

14.5   Summary

- Solution spaces can be studied by randomly sampling points contained within them. Large sets of candidate solutions can be analyzed statistically.
- A three-step sampling procedure can be implemented: (1) confinement of space by linear constraints, (2) randomized sampling, and (3) further confinement by non-linear and other additional constraints.
- Solution spaces can be enclosed by regularly shaped objects, such as parallelepipeds, making the sampling procedure easy and enabling the computation of the size (volume) of the solution space. Low-dimensional spaces (≤ 10 to 12) can be sampled in this fashion.
- Sampling high-dimensional spaces requires more sophisticated approaches.
- Metabolomic data, fluxomic data, and thermodynamic properties can be used to set additional constraints. The consequences of additional constraints can be assessed by determining the reduction in the size of the solution space.
- Numerous biological questions and problems have been addressed successfully using randomized sampling of the null space of the stoichiometric matrix.

PART III

Determining the Phenotypic Potential of Reconstructed Networks

How often have I said to you that when you have eliminated the impossible, whatever remains, however improbable, must be the truth? – Sherlock Holmes, A Study in Scarlet

The functional states of reconstructed networks are directly related to cellular phenotypes. With reconstructed networks represented formally, we can use mathematics to compute their candidate functional states. If one adopts the informatics point of view of the stoichiometric matrix, **S**, and its annotated information as a biochemically, genetically, and genomically structured knowledge base, then these *in silico* methods are viewed as *query* tools.

Whether viewed from an informatic or mathematical standpoint, the result of applying *in silico* analysis methods is the study of *network properties*, sometimes referred to as *emergent properties*. These properties represent functionalities of the whole network and are hard to decipher from a list of its individual components. In some sense, these properties are a reflection of the hierarchical nature of living systems.

A variety of methods have been developed to examine the properties of genome-scale networks. The third part of this text summarizes the COBRA approach and some of its methods. The first four chapters describe the conceptual framework of the COBRA approach and how optimal states are computed. The next two describe how general properties of optimal solutions can be studied. The last chapter then discusses the objective function, that to some is perhaps one of the most interesting and fundamental topics in constraint-based modeling. More details of the COBRA methods and their implementation are available [236, 299, 344].

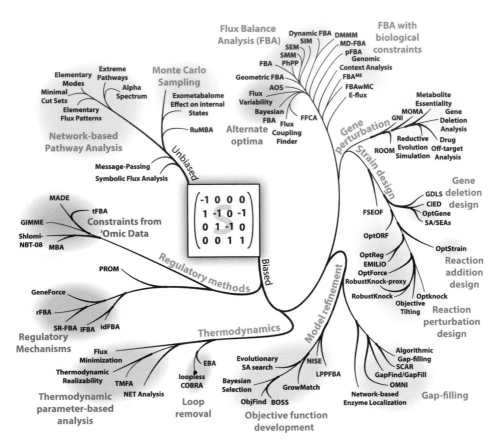

Reconstructed networks can be converted into a mathematical format that in turn can be translated into quantitative networks models. This ability has enabled the field of COnstraint Based Reconstruction and Analysis (COBRA) to develop. This figure illustrates the many COBRA methods and their interrelationships. The COBRA philosophy and basic methods are covered in Part III of this text. From [236].

15 Dual Causality

Nothing in biology makes sense, except in the light of evolution –
Theodosius Dobzhansky

The stoichiometric matrix and the information associated with it fundamentally represent a biochemically, genetically, and genomically structured knowledge base. It can be used to analyze network properties and to relate the components of a network and its genetic bases to network or phenotypic functions. Biology is subject to *dual causality*, or *dual causation* [261]. It is governed not only by the physical laws but also by genetic programs. Thus, while biological functions obey the physical laws, their functions are not predictable by the physical laws alone. Biological systems function and evolve under the confines of the physical laws and environmental constraints. How organisms operate within these constraints is a function of their evolutionary history and their survival strategy.

15.1 Causation in Physics and Biology

Physics Classically, 'cause and effect' is established by formulating mathematical descriptions of conceptual models of fundamental physical phenomena. One example is molecular diffusion (see Figure 15.1). The fundamental process underlying diffusion is the random walk process that a collection of molecules undergoes. The statistical properties of the random walk process can be assessed quantitatively, and its macroscopic consequences are described with Fick's law. This law is described by a simple equation that is used as the basis to describe mass transfer processes from regions of high concentration to regions of low concentration. The established causality is the basis for computations that reliably predict mass transfer processes. The Boltzman and Nernst equations provide other specific cases of causality in physics, and there are many more examples.

Engineering design can be based on such predictions. Thus, in engineering, "there is nothing more practical than a good theory," as the physical laws can be used for design, often with minimal experimentation and prototyping.

Cause and effect for physical phenomena are often well established and can be described mathematically. Mathematical descriptions are in the form of equations and inequalities. An interesting discussion of the character of physical law is found in [114].

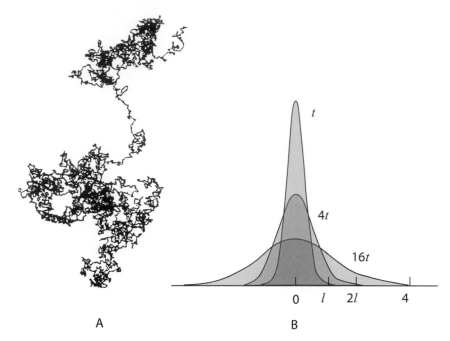

Figure 15.1 Causation in physics: an example of a random walk process and diffusion of molecules. Panel A: the simulated random walk trajectory of a single molecule. Panel B: the probability distribution for the molecule's location as a function of time when it was located at $l = 0$ at $t = 0$. The width of the distribution, l, increases with the square root of time, t. Modified from [318].

Biology Causation in biology is much different from physics. Biological causation originates fundamentally from the evolutionary process leading to genetic variation within a population. There are four key parts to the conception of an evolutionary process, shown in Figure 15.2:

1   initial phenotype resulting from a genotype;
2   natural selection of the new organism leading to the ability to produce offspring successfully;
3   successful mating leads to the possible formation of a new genotype; and
4   processes, such as mutation and recombination, that lead to the formation of a new genotype.

This process repeats itself. The result is diversity: a *biopopulation* of non-identical individuals. Therefore, living systems are time-variant: they evolve and change over time. In contrast, physical phenomena are time-invariant; e.g., oxygen, a homogeneous population of identical molecules, always diffuses the same way in water under a given set of circumstances, the unit charge on the electron does not change, and so on.

The outcome of the selection process is, in part, stochastic, and is influenced by environmental variables. The selection process in biology gives the appearance of

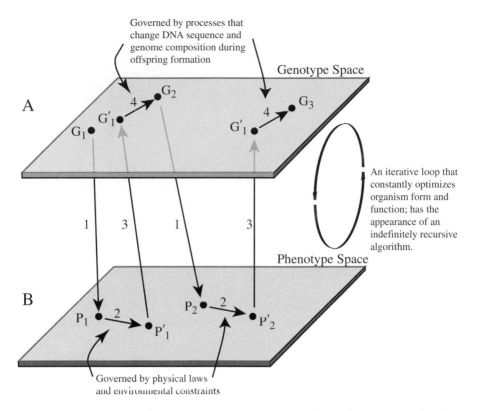

Figure 15.2   Causation in biology. The genotype–phenotype relationship conceptualized as an iterative mapping process. Panel A depicts the position of genotype (G) and Panel B a phenotype (P) in their respective spaces. Iteration over two generations is illustrated. Redrawn based on [173].

'sense of purpose' that, fundamentally, is survival. The sense of purpose for quantitative model-building is represented by an objective function (see Chapter 21) that is meant to describe the basis of selection. In general, it is hard to know what the detailed objective is underlying a selection process. Thus, the objective function becomes the focus of study as one seeks to understand the selection process and the distal causation it represents. The objective function itself is now subject to an experimental investigation through adaptive laboratory evolution in a controlled setting.

Causation in biology, therefore, is in some respect an endless iterative process that seeks to find an optimal 'solution' for survival. Alterations in the genetic program with each iteration have the potential to induce changes in phenotypic functions. This recursive process takes place within the constraints imposed by physics and chemistry under given environmental conditions. Necessary ingredients to understand distal causation mechanistically are thus constraints and optimality.

Systems biology   As described above, causation differs in physics and biology. However, both are relevant to systems biology and they represent opposite ends of a hierarchical process (Figure 15.3). Systems biology tries to bridge the two ends of

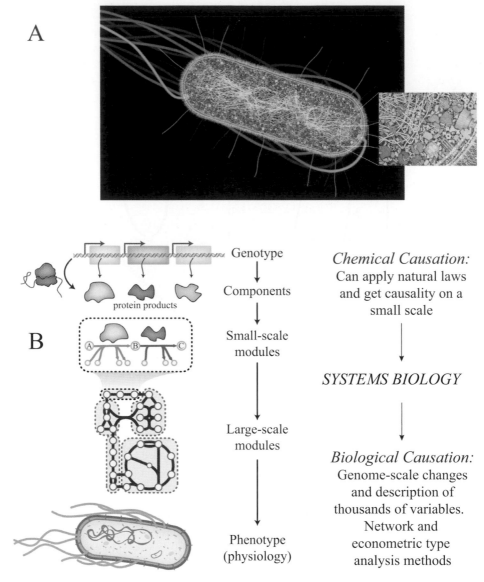

Figure 15.3 The hierarchical nature of living systems and multi-level causation. (A) Living systems are complex and hard to define precisely in biophysical terms. (B) Systems biology tries to provide structure to the hierarchical relationship between molecular and physiological events. Inserted image from [148] used with kind permission from Springer Science+Business Media B.V. Prepared by Nathan Lewis.

this spectrum and develop what amounts to a *quantitative and mechanistic genotype–phenotype relationship.*

The genotype–phenotype relationship has been the subject of argument and speculation. However, since the first genome sequence appeared in 1995, we have learned how to build such relationships with a mechanistic basis and have applied it to metabolism and growth of microorganisms and metabolism in mammals. Genome sequences provide comprehensive, albeit not yet complete, information about the genetic elements that create the form and function of an organism. A constraint-based analysis provides a framework within which these basic considerations can be accommodated. Before we describe this framework, we will discuss the challenges with applying theory-based approaches to large-scale model-building in biology.

15.2 Building Quantitative Models

15.2.1 The physical sciences

Proximal causation *In silico* model-building in the physico-chemical sciences starts with basic principles such as thermodynamics, chemical potential, the diffusion equation, mass conservation, or the Nernst and Boltzman equations. These equations are based on well-developed fundamental physical theories, and they typically contain a large number of parameters, most of which can be measured individually under defined conditions. These parameters, such as the diffusivity of oxygen or the unit charge on the electron, are time-invariant. These equations then form the basis for computer models and simulation.

Limitations of theory-based modeling approaches in biology Traditional theory-based models of large-scale biological processes are faced with fundamental challenges.

First, the intracellular chemical environment is complex (e.g., see Figure 15.3A) and hard to define in terms needed for the formulation of equations that describe the physics of the intracellular milieu.

Second, assuming that we had all the governing equations defined, we would have to find numerical values for all the parameters that appear in these equations. These values would have to be accurate for intracellular conditions.

Third, even if we could overcome the first two challenges, we have to face the fact that evolution changes the numerical values of kinetic constants over time. In addition, in a biopopulation, even if we had a perfect *in silico* model for one individual organism it would not apply perfectly to other individuals in the biopopulation due to genetic and epigenetic differences between individuals. Such time-dependency and diversity of parameter values are key distinguishing features between biological and physico-chemical systems.

15.2.2 The life sciences

Distal causation: the selection process The process of evolution is fundamental to the biological sciences. Organisms exist in particular environments and, as they replicate, they produce offspring that are not genetically identical to the parent, thus generating a biopopulation of individuals that are each slightly different from one another (Figure 15.2). Over time, natural selection favors those individuals in the biopopulation that have more *fit* functions than other members of the biopopulation.

To survive in a given environment, organisms must satisfy myriad constraints, which limits the range of available phenotypes. The better an organism can achieve a relatively fit function in a given environment, the more likely it is to survive.

Constraining behaviors   Because of dual causality, mathematical model-building in biology at the network and genome-scale will need to differ from that practiced in the physico-chemical sciences. The third limitation listed above is due to the dual causality that needs to be accounted for in realistic models of biological processes. An approach to the *in silico* analysis of cellular functions can be formulated based on the fact that cells are subject to governing constraints that limit their possible behaviors. Imposing these constraints can determine what functional states can and cannot be achieved by a reconstructed network. Imposing a series of successive constraints can limit allowable cellular behavior, but will never predict it precisely.

The imposition of constraints leads to the formulation of *solution spaces* rather than the computation of a *single solution* (or a discrete set of a few solutions), the hallmark of theory-based models. Cellular behaviors (i.e., functional states of networks) within the defined solution space can be attained, those outside cannot. Each allowable behavior basically represents a different candidate phenotype based on the component list, the biochemical properties of the components, their interconnectivity, and the imposed constraints. The constraint-based approach leads to *in silico* analysis procedures that are helpful in analyzing, interpreting, and even predicting the genotype–phenotype relationship.

Thinking about constraints   Cells are subject to a variety of constraints. There are both *non-adjustable* (i.e., invariant or hard) and *adjustable* constraints (Table 15.1). The former can be used to bracket the range of possible phenotypic functions. The latter can be used to further limit allowable behavior, but these constraints can adjust through an evolutionary process or through changing environmental conditions. In addition, the adjustable constraints may vary slightly from one individual to another in a biopopulation. Together, these constraints define a range of possible functions, described mathematically as a solution space, and direct the realization of phenotypic expression.

This distinction between adjustable and non-adjustable constraints and their role in understanding dual causation is illustrated schematically in Figure 15.4. The feasible space of steady-state reaction fluxes is determined by non-adjustable constraints (the outer octagon). A narrower subset of functional states within this space is defined by the regulation of kinetic properties, or adjustable constraints. These adjustable constraints can be modified further through evolution to alter the limits of the subspace of expressed functional states. The direction of such an evolutionary expansion of the regulated subspace is driven by a need for improved performance, which can be described by an objective function (see Chapter 21).

15.2.3   Genome-scale models

Phenotypic functions are the results of the interactions of multiple gene products. In principle, all expressed gene products under a given condition affect the phenotypic

Table 15.1 Constraints on the functions of biochemical reaction networks. Adapted from [311].

Factor	Type of constraint
Physico-chemical constraints	
Osmotic pressure, electroneutrality, solvent capacity, membrane space, molecular diffusion, thermodynamics	Hard, non-adjustable constraints
Connectivity	
Systemic stoichiometry Causal relationships	Hard, non-adjustable constraints, but can be adjusted by horizontal gene transfer
Capacity	
Maximum/minimum flux	Non-adjustable maximum based on maximum association rates Adjustable by transcriptional regulation
Rates	
Mass action, enzyme kinetics, regulation	Highly adjustable by an evolutionary process

state of an organism. Thus, to develop mechanistic genotype–phenotype relationships a genome-scale model is needed. The availability of whole-genome sequences made the construction of such models possible.

Three generations of genome-scale models The successive application of constraints lends itself to a step-wise development of increasingly refined data-driven *in silico* models [313]. These models broaden in scope with the establishment and imposition of additional constraints. Constraint-based models can address questions relating to determining the possible functions of a network, which of these functions the cell actually chooses, and how such choices are made (Table 15.2).

The *first generation* of constraint-based models for microbial metabolism appeared at the turn of the century [105, 106]. The 'omics' data type on which they are based is *genomic*. Literature (*bibliomic*) data are used as well as the formulation of hard physico-chemical constraints, such as mass, energy and redox balance, thermodynamic, and maximal reaction rates. These constraints collectively define all the possible functional states of a reconstructed network. They are confined mathematically to a solution space. The properties of this space can be studied by the methods described in the following chapters.

The *second generation* of constraint-based models includes the imposition of transcriptional regulatory networks, leading to the shrinking of the allowable states of metabolic networks (see Figure 15.4). In response to environmental queues

Table 15.2   Generations of genome-scale models. From [313].

Generation	Usage	Type of data used
First	What states are possible	Genomic and bibliomic
Second	What states are chosen	Condition-dependent transcriptomic and genome location data
Third	How states are chosen	Time-dependent proteomic and metabolomic

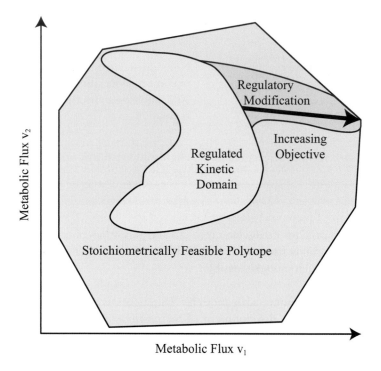

Figure 15.4   Early conceptualization of hard and adjustable constraints and how distal causation drives the choice of functional states within the hard constraints. Redrawn from [439].

and built-in regulation, the solution space is shrunk [85, 86] to contain network functions that the cell has chosen through an evolutionary process. The choices that a cell makes can then be identified and analyzed. Through the reconstruction of transcriptional regulatory networks, we can now begin to impose condition-dependent constraints, or restraints on reconstructed metabolic networks and formulate the second-generation models.

The *third generation* of constraint-based models will account for the abundance or concentration of the cellular components. Various 'omic' data types can now be obtained in a time-resolved fashion. Such data will help clarify just how the cell implements the choices it has made and how it evolves to find new choices. This approach is likely to lead to the definition of the rate constants of the network as a whole rather than constants for the individual underlying biochemical events. Initial efforts in this direction are represented by the MASS modeling procedure [317].

Some properties of genome-scale models   Biological networks have several fundamental properties that need to be considered when interpreting large-scale data sets and building models to describe their functions (see Figure 15.5).

Redundancy.   Biochemical reaction systems have redundancy built into them at many levels. Often, individual steps in a network can be carried out in more than one way. Isozymes represent different enzymes that carry out the same reaction. Similarly, some codons can be translated by more than one tRNA. There are also network-level redundancies. The overall function of a network to support a phenotype can be achieved in more than one way. Thus, there are multiple equivalent outcomes from the same biological selection process. The mathematical aspect of this feature, *equivalent optimal solutions*, is detailed in Chapter 20. Biologically, these may be called *silent phenotypes*.

Multi-functionality.   There are components in biochemical networks that can carry out more than one function. Examples include generalist of *promiscuous* enzymes that can catalyze many related chemical reactions. Similarly, some tRNA molecules can translate more than one codon. At the network level, there could be global network states that would give similar phenotypes even if the environments were different. This feature would be called a *generalist phenotype*. The notion of a high-flux backbone in metabolism [14] is composed of a set of reactions that lead to optimal growth on different substrates. A high-flux backbone is an example of a large correlated set of reactions that function together in optimal solutions [325].

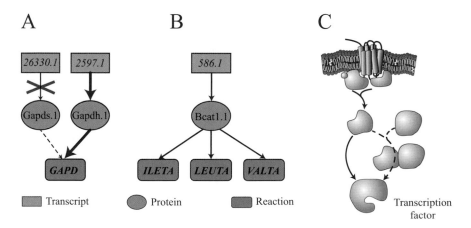

Figure 15.5   Some properties of genome-scale models. Panel A: redundancy; panel B: multi-functionality; panel C: non-causality. Prepared by Nathan Lewis.

Non-causality.   Due to the hierarchical organization of organisms, changes on one level may not percolate up to functions at a higher-level of organization, and would thus be non-causal. A well-known example of non-causality are *hitchhiker mutations* that co-select with a causal mutation located nearby on the genome. In the field of signal transduction, many are interested in knowing 'who-talks-to-whom,' meaning that one wants to know all possible chemical interactions between two components. Protein–protein interaction maps provide one example. In this case, however, the biologically meaningful question is 'who-*listens*-to-whom,' as we are only interested in knowing if chemical interactions are a part of a higher-order biological function. Thus, there can be many non-causal (biologically), but detectable, chemical interactions among macromolecules.

These three attributes are important considerations in studying the hierarchical nature of biological systems. Multi-scale, multi-parameter analysis methods will be needed to study this hierarchical organization. They will need to be able to deal with non-regular patterns, which will be a deviation from classical methods such as Fourier analysis that looks for repeated regular patterns. All of these features have appeared through the evolutionary process which abides by a series of constraints.

Higher-order properties   There are some other notable higher-order properties of biological networks, which will not be detailed here. Such properties include *self-assembly* of components to form a functioning network spontaneously, the *selection* that seems to be at work during both distal and proximal causation, the notion of *a self* in biology (namely, is a component a part of a network, or not?), and the notion of *awareness* (that ultimately is related to the mathematical concept of a functional state of a network). It is an interesting and fundamental challenge to the field to determine if such important biological properties can be defined mathematically in the context of genome-scale models. If possible, molecular systems biology will advance notably.

15.3   Constraints in Biology

All expressed phenotypes resulting from the selection process must satisfy the governing constraints. Therefore, clear identification and statement of constraints to define ranges of allowable phenotypic states provides a fundamental approach to understanding biological systems that is consistent with our understanding of the way in which organisms operate and evolve.

Different types of constraints limit cellular functions and several authors have discussed general constraints in biology [84,93,139,179,261]. Here we start this discussion by dividing constraints into four categories [344]: (1) fundamental physico-chemical, (2) spatial or topological, (3) condition-dependent environmental, and (4) regulatory, or self-imposed constraints.

Physico-chemical constraints   Many physico-chemical constraints govern cellular processes. These constraints are inviolable and thus represent *hard* constraints.

Conservation of mass, elements, energy, and momentum represent hard constraints. The interior of a cell is densely packed, forming an environment where the viscosity may be on the order of 100–1000 times that of water. Diffusion rates inside a cell may be slow, especially for macro-molecules. The confinement of a large number of molecules within a semi-permeable membrane causes high osmolarity. Thus, cells require mechanisms for dealing with the osmotic pressure generated, such as sodium–potassium pumps to balance osmolarity or a rigid cell wall to physically withstand it. Intracellular reaction rates are determined by local concentrations inside cells. Reactions have maximal reaction rates (denoted with v_{max}) estimated to be about a million molecules per μm^3 per second (see Equation (17.6)). Furthermore, biochemical reactions need to have a negative free energy charge in order to proceed in the forward direction. These are some of the many basic physico-chemical constraints under which cells must operate.

Spatial constraints The crowding of molecules inside cells leads to *spatial*, or three-dimensional, constraints. The linear dimension of the bacterial genome is on the order of 1000 times that of the length of the cell. DNA must therefore be tightly packed in the nuclear region in an accessible and functional configuration because DNA is only functional if it is accessible. Thus, at least two competing needs (to be tightly packed, yet accessible) constrain the physical arrangement of the bacterial genome. DNA in eukaryotes is organized in a highly hierarchical fashion.

As a further example, we note that the ratio between the total number of tRNA molecules and the number of ribosomes in a typical E. coli cell is approximately 10 to 1 [282]. With 43 different types of tRNA, there is less than one full set of tRNAs per ribosome. The genome, therefore, may have to be configured such that the location of rare codons is spatially close and translated by the same ribosome. Protein localization and crowding of space in membranes represent additional topological constraints.

Identification of these constraints and analysis of their consequences will be important for the understanding of the three-dimensional organization of cells. This challenge is hard, but progress is being made [477].

Environmental constraints Environmental constraints on cells, such as nutrient availability, pH, temperature, osmolarity, etc., are typically time- and condition-dependent. For example, H. pylori, a human gastric pathogen, lives in a relatively constant environment, but is constrained by its low pH. It produces ammonia to sufficiently neutralize the pH in its immediate surroundings in order to stay alive. The growth of a plant is limited by the flux of incident photons as well as nitrogen and phosphorous availability in the soil.

Conversely, the life cycle of E. coli is characterized by a series of sudden environmental changes. Outside of an animal it lives at ambient temperature and in the presence of ample oxygen. Then it experiences a heat shock when it enters the mouth of an animal, followed by an acid shock when it reaches the stomach. Following entry into the small intestine, another pH shock is experienced, followed by a nutritionally rich anaerobic environment where it can grow rapidly in the presence of other bacterial species. Then, finally, it experiences a cold shock and ample

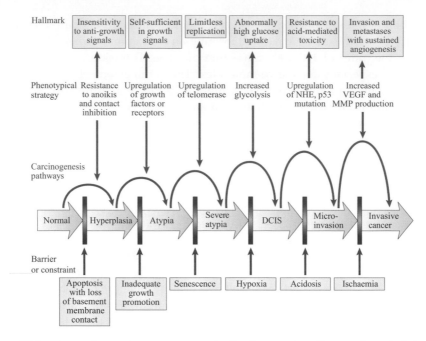

Figure 15.6   The carcinogenic process conceptualized as a series of losses of constraints. Taken from [136].

oxygen with diminishing nutrients surrounding it upon excretion. *E. coli* needs to be able to adjust its internal functional state to survive this series of environmental changes.

Knowing the environmental constraints is of fundamental importance for the quantitative analysis of microorganism functions; however, natural environments may be hard to define precisely. Conversely, in the laboratory, defined growth media can be used so that the environmental variables are known precisely.

Regulatory constraints   These constraints are fundamentally different than the three types discussed above. They are *self-imposed*, are subject to evolutionary change, and can thus be time-variant. As illustrated in Figure 15.4, they work within the outer constraints defined by physico-chemical processes. For this reason, these constraints may be thought of as regulatory *restraints*, in contrast to the physico-chemical constraints, the spatial constraints, and environmental constraints. Based on environmental conditions, regulatory constraints provide a mechanism to eliminate suboptimal phenotypic states and confine cellular functions to behaviors of high fitness. Regulatory constraints are produced in a variety of ways.

The loss of ability to impose constraints through regulation to maintain a certain phenotype would lead to a loss of the desired biological function. The carcinogenic process provides a serious example (Figure 15.6). This process can be understood as a series of losses of constraints that leads to a malignant phenotype and ultimately a metastatic state.

15.4   Summary

- Systems biology bridges multiple scales in biology.
- Dual causation in biology requires us to accommodate the physico-chemical constraints under which cells operate, as well as the fundamental biological processes of natural selection and generation of alternatives when building models in systems biology.
- Organisms have to abide by a series of constraints, including those arising from basic physical laws, spatial constraints, and the environment in which they operate.
- Many possible biological functions are achievable under these constraints, and organisms willfully impose constraints through various regulatory mechanisms to select useful functional states from all allowable states.
- A constraint-based approach that enables the simultaneous analysis of physico-chemical factors and biological properties emerges from these considerations.

16 Functional States

Life is a program written in DNA – Craig Venter

Chemical reactions link cellular components together to form a network. Although we can specify the chemical properties of links in biological networks, it is the way in which a multitude of such links form networks that determines phenotypic functions. These integrated network functions are also called *functional states*, and they correspond to the observed biological functions or phenotypic states that networks create. A functional state may be viewed as the outcome of the execution of the genetic program written in the DNA. In this chapter we detail the concept of a functional state of a genome-scale network and how it represents a physiologically observable state. The following chapters then describe the framework for computing functional states using the constraint-based approach.

16.1 Components vs. Systems

Components come and go Biological components all have a finite turnover time. Most metabolites turn over within a minute in a cell, mRNA molecules typically have two-hour half-lives in human cells [463], 3% of the extracellular matrix in cardiac muscle is turned over daily, and so forth. So a cell that you observe today, compared with the same cell yesterday, may only contain a small fraction of the same molecules.

Similarly, cells have finite lifetimes. The cellularity of the human bone marrow turns over every two to three days. The renewal rate of skin is on the order of five days to a couple of weeks. The lining of the gut epithelium has a turnover time of about five to seven days. Slower tissues, like the liver, turnover their cellularity approximately once a year. So a mammal that you observe today may only contain a small fraction of the same cells as the same mammal observed a year ago. Thus, the components of a biological system come and go, and their turnover takes place on multiple time scales.

However, the system remains Most of the cells that are contained in an individual today were not there just a few years ago. However, we consider the individual to be

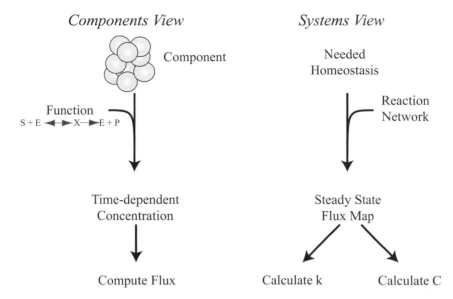

Figure 16.1   A contrast between the components view (on the left) and the systems view (on the right).

the same. Similarly, we consider one cell to be the same one a week later, even if most of its chemical components may have turned over. Thus, components come and go, but the system remains. Therefore, a key feature of living systems is how their components are connected. The interconnections between cells and cellular components define the essence of a living process.

Moving from viewing components in isolation to being a part of a system   The difference between the familiar components view of a cell and its molecular biology is different from the less familiar systems view in many subtle ways. Here, we illustrate this difference in Figure 16.1.

On the left side we see the classical component point of view. When we are looking at one gene product, in this case an enzyme carrying out its function, we study this component by placing it in a beaker with its substrates and then observe the time-dependent disappearance of a substrate and the appearance of a product. The component that we are studying is the centerpiece of this experiment, and it is responsible for concentration changing in a time-dependent manner.

The right side illustrates a systems viewpoint of a biochemical network. Contrary to the components view, it is not the components that matter, but it is the state of the whole system that is important. Any biological network will have a nominal state that we recognize as a homeostatic state. Thus, the fluxes that reflect the interactions among the components to form the functional state of the network are dominant variables, and the concentrations of the individual components are 'subordinate quantities.' The concentrations of the network components are determined to a first approximation by the flux map, or the state of the network, and then in detail by the kinetic properties of the links in the network.

Two key issues arise from the above considerations. The first deals with the nature of the links between components in a biological network, and the second deals with the functional states and the properties of a network that a set of links forms. At the genome-scale, the former is the process of network reconstruction discussed in Part I, and the latter is the subject of the current part of this book.

16.2   Properties of Links

Links between molecular components are given by chemical reactions or associations between chemical components. These links are therefore characterized and constrained by basic chemical rules. In tissue biology, the nature of links between cells is more complicated and often related to higher-order chemistry. We note that a T cell receptor, for instance, forms a complicated structure in the membrane of a cell. The properties of that structure, and how compatible it is with the complementary features of another cell, determine whether there is communication, or links, between these cells. Links between people in a social network have an even more complex basis. As we are focused on the characteristics of biochemical networks, we will discuss the chemical nature of links in molecular biology further.

Basic chemistry   The prototypical transformations in living systems at the molecular level are bi-linear. This association involves two compounds coming together to either be transformed chemically through the breakage or formation of a new covalent bond, as is typical of metabolic reactions or macromolecular synthesis:

$$X + Y \rightleftharpoons X\text{--}Y \quad \text{covalent bonds}$$

or two molecules coming together to form a complex through hydrogen bonds and/or other physical association forces, a complex that has different functionality from individual components:

$$X + Y \rightleftharpoons X : Y \quad \text{association of molecules}$$

Such association, for instance, could designate the binding of a transcription factor to DNA to form an activated site to which an activated polymerase binds. Such bi-linear association between two molecules might also involve the binding of an allosteric regulator to an enzyme that induces a conformational change in the enzyme.

Chemical transformations have certain key properties.

Stoichiometry.   The stoichiometry of chemical reactions is fixed, and is described by integers that count the molecules that react and that form as a consequence of the chemical reaction. Thus, stoichiometry represents 'digital information.' Chemical transformations obey elemental and charge balancing, as well as other features. Stoichiometry is invariant between organisms for the same reactions and does not change with pressure, temperature, or other conditions. Stoichiometry gives the fundamental topological properties of a biochemical reaction network.

Relative rates.   All reactions inside a cell are governed by thermodynamics. The relative rate of reactions, forward and reverse, are therefore fixed by basic thermodynamic properties. Unlike stoichiometry, thermodynamic properties do change with physico-chemical conditions such as pressure and temperature. The thermodynamics of transformation between small molecules in cells are fixed but condition-dependent. The thermodynamic properties of associations between macromolecules can be changed by altering the amino acid sequence of a protein.

Absolute rates.   In contrast to stoichiometry and thermodynamics, the absolute rates of chemical reactions inside cells are adjustable. Highly evolved enzymes can be very specific in catalyzing particular chemical transformations. Cells can thus extensively manipulate the rates of reactions through changes in DNA sequence. Enzymes evolve to bring molecules into particular orientation to control the rate of appropriately oriented collisions between two molecules that lead to a chemical reaction (see Figure 7.1). It should be noted that much of the chemistry that takes place in cells occurs on the surfaces of protein. Surfaces are encoded in the DNA sequence and they determine the catalytic properties, such as rate constants, binding specificity, subunit association, protein binding to the DNA, and so forth.

The formation of links is restricted   Links cannot just form between any two cellular components. The links that are formed are constrained by the nature of covalent bonds that are possible and by the thermodynamic nature of interacting macromolecular surfaces. The absolute rates are key biological design variables because they can evolve from a very low rate, as determined by the mass action kinetics based on collision frequencies, to a very high and specific reaction rate, as determined by appropriately evolved enzyme properties.

Information about links   We do not have detailed information about the nature of all links between molecules inside a cell. In fact, there is a range of levels of knowledge that we have available to us (see Figure 16.2). Full information about the links allows for a full description of how a genetic property is mechanistically related to a phenotypic one. Most of this book is focused on this level of information, although additional auxiliary information can be included in the form of logistical relationships.

16.3   Links to Networks to Biological Functions

Reaction bi-linearity and network topology   Most biochemical reactions are bi-linear. Bi-linearity gives the networks a hyper-graph property that is topologically non-linear. Consequently, biochemical reaction networks form a *tangle of cycles* [361] where different chemical properties and moieties are being transferred throughout the network from one carrier to the next. The coordinated movement of such *transferred*

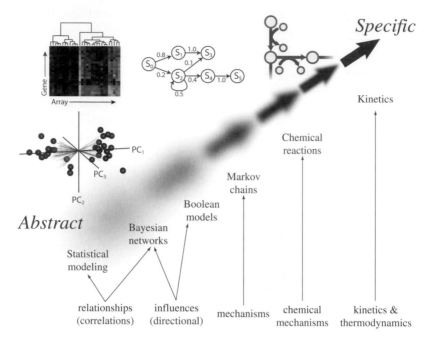

Figure 16.2   Illustration of different levels of knowledge that are available about a link in a network. Prepared by Nathan Lewis.

properties is determined by network topology and represents a key aspect of systems biology as they tie the whole system together.

We are familiar with the pathway maps that are used to describe cellular processes. We are less familiar with maps drawn around co-factors or carrier molecules that participate in multiple reactions. An example of the trafficking of redox equivalents in the core *E. coli* metabolic network is shown in Figure 16.3. This figure illustrates the two points of view, and shows how the carrier or co-factor molecules form a tangle of cycles that transmit the redox potential from one state to another. Another familiar example would be the movement of high-energy phosphate bonds between metabolites and proteins. ATP is the primary carrier of such high-energy bonds, and, for instance, a phosphate group is tied to glucose to form glucose-6-phosphate as the first step in glycolysis. The same feature is found in signaling networks whose components are in phosphorylated or dephosphorylated states. Other properties being transferred between molecules are one-carbon units, two-carbon units, ammonia groups, and so on.

Bi-linearity makes biochemical reaction networks highly interwoven and confers on them certain stoichiometric texture that affects their steady and dynamics states.

One network, many functional states   One interesting feature of biochemical networks as they grow in size is that due to combinatorics, the number of possible functional states that they can take can grow faster than the number of components in a network. Therefore, the number of phenotypic functions derivable from a genome

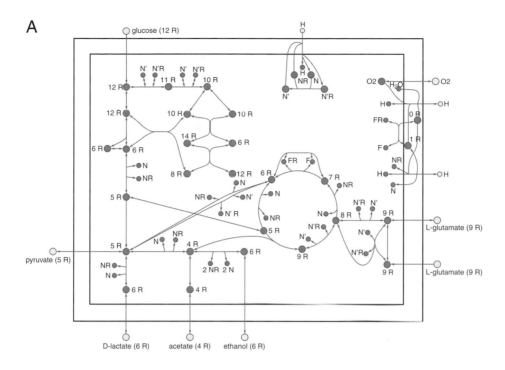

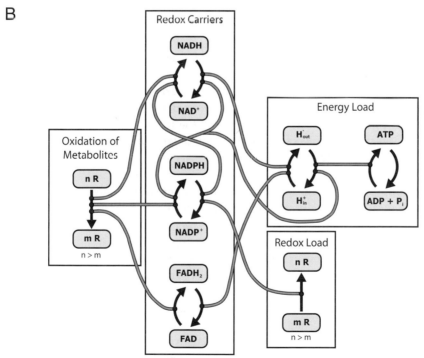

Figure 16.3 The tangle of cycles in trafficking of redox potential (R) in *E. coli* core metabolic pathways showing the redox equivalents (R) of the metabolites and carriers in the core *E. coli* metabolic model under aerobic conditions. (A) A reaction map organized around the core pathways. (B) A series of node maps organized around the molecules that carry redox potential. This map looks like a tangle of cycles. Taken from [317].

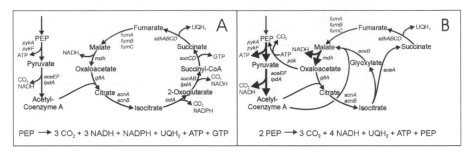

Figure 16.4 Stoichiometries of two alternative cycles for complete oxidation of PEP. The tricarboxylic acid cycle (A) and the PEP–glyoxylate cycle (B). Large solid arrows indicate reactions that are used twice per turn of the cycle. Gene names are shown in italics. Taken from [122].

does not linearly scale with the number of genes. For instance, the human genome may only have 50% more genes than the genome of *Caenorhabditis elegans*, a small worm, but nevertheless, human beings display much more complicated phenotypes and in greater variety. Thus, in general, it is hard to correlate organism complexity and functions to the number of genes its genome contains.

The fundamental property of biochemical networks having many possible functional states leads to the possibility of having the same network display many different phenotypic behaviors. A specific example in Figure 16.4 shows two experimentally determined alternative functional states of some of the metabolic pathways in the core pathways of *E. coli*. An organism does not fully exploit or use all possible functional states.

Many possible states will be useless to the organism in its struggle for survival. Therefore, a limited subset of these functional states needs to be selected and expressed by cells by imposing regulatory constraints. As we will discuss in Chapter 20, complex biological reaction networks can also have *equivalent* functional states, that is, there are identical overall functional states that differ in the ways in which they use the underlying links in the network.

Changing properties through evolution: distal causation Some of the key features of biological networks that distinguish them from other networks need to be accounted for in the analysis of their systemic properties. The first basic feature of biological networks is that they evolve; they change with time. They are *time-variant*. Principally, such changes occur through the kinetic properties of the links in the network and the changing of the available or active links in the network at any given point in time. The number of available links can be manipulated by regulation of gene expression, by horizontal gene transfer, and by other mechanisms.

The second feature that has to be taken into account is the fact that they have a sense of *purpose*. The fundamental purpose is survival. However, in complicated organisms that are fundamentally composed of many networks, some will have goals that are subtasks to the overall goal of survival. For instance, the goal of adipocytes would be to collect and store fat if there is an abundance of energy resources in its environment. A goal of the mitochondrion, being the powerhouse of the cell, seems to be to maximize ATP production from available resources. Therefore, the study of *objectives*, i.e., purpose, of biochemical reaction networks becomes a relevant and central issue (Chapter 21).

Temporal–spatial organization Thus, linking many biological components together forms a network. This network can have many functional states from which a subset is selected. Links, network topology, and functional states can all change with time or environmental conditions. It is important to be cognizant of the fact that biochemical reaction networks have to operate in the crowded interior of a cell (see Figures 15.3A and 17.1). Thus, the network view of the biological process has to be considered in the context of the three-dimensional physical arrangement of such networks. These considerations may limit the usefulness of analogies with other man-made networks such as electrical circuits.

16.4   Constraining Allowable Functional States

Disciplines differ in their approach to model-building The above considerations of the nature of links, how they form networks, and how networks form functional states, make it likely that *in silico* modeling and simulation of genome-scale biological systems is going to be different from that practiced in the physico-chemical sciences. First is the notion that a network can fundamentally have many different states or many different solutions. Which states (or solutions) are picked is up to the cell and such choices can change over time based on the selection pressure experienced. This difference from the physico-chemical sciences is illustrated in Figure 16.5.

Theory- versus constraint-based thinking All theory-based considerations in engineering and physics leads one to attempt to seek an 'exact' solution, typically computed based on the laws of physics and chemistry. However, in biology it appears that not only can a network have many different behaviors that are picked based on the evolutionary history of the organism, but also, as we shall see, these networks can carry out the same function in many different and equivalent ways. This leads to an interesting distinction in mathematical modeling philosophy between the key disciplines (Table 16.1).

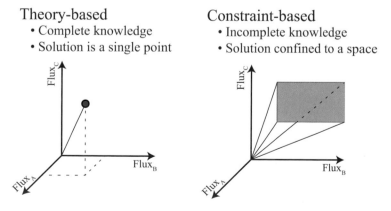

Figure 16.5 Theory- vs. constraint-based analysis. Illustration of finding an exact solution (a point) versus finding a range of allowable solutions (a solution space).

Table 16.1   Disciplinary differences in modeling philosophy.

	Equations	Boundary conditions	Nature of solutions
Physics	+++	+	Unique
Engineering	++	++	Design
Biology	+	+++	Multiple
			and changing

In physics, the emphasis has always been on deriving theory. Quantum mechanics developed about 100 years ago. Boltzman derived his famous equation prior to that. Theory, as expressed by mathematical equations representing our understanding of fundamental physical mechanisms, has been central to physics. If one wants to obtain particular solutions to these equations, one imposes boundary conditions that typically lead to the calculation of a unique solution.

Engineering takes a bit of a departure from this philosophy. The equations used in engineering do not need to be correct mechanistically, in a fundamental theoretic sense, as long as they describe the process at hand phenomenologically. Furthermore, the boundary conditions that need to be stated are very important and are often very specific to what an engineer is designing. In engineering, though, one is used to the fact that a problem can have multiple solutions, and that often comes down to the use of design variables to try to optimize a design.

In biology, based on the above consideration, we find that the equations needed to describe the physics of the intracellular environment may never be well-known, and furthermore, network functionalities evolve and change over time. Therefore, the fundamental equations describing biological functions may be hard to formulate and fully define. On the other hand, the boundary conditions or the constraints under which cells operate and evolve against are easier to identify, state, and use.

Constraint-based analysis methods   These considerations give a general conceptual background for functional states and that there are constraints on what functional states a cell can take on. There are many methods that have been developed under the constraint-based modeling approach. They can be used to address many network properties, functional states, and biological questions; some are summarized in a later chapter (Table 18.1). To complete this chapter, we will discuss the general types of constraints that biology operates under before we proceed to formalize and mathematically deploy them.

16.5   Biological Consequences of Constraints

The constraints under which a cell operates   Cells operate under myriad constraints. There are different ways to classify these constraints, and many authors have discussed them from different points of view. A few will be mentioned here.

- A statement of two very general categories of constraints imposed by natural selection have been described by F. Jacob [179]. They are basically (i) the requirement for reproduction and the genetic mechanisms required to produce offspring with non-identical genetic composition of the parent(s), and (ii) the permanent interaction with the environment that imposes thermodynamic constraints of constant flux of matter, energy, and information. The latter constraints are easier to describe in the language of the basic physical laws while the former describe distal causation.

- A. Danchin [93], in his insightful book about genomes, divides the cellular processes and their associated constraints into four general categories: (i) compartmentalization to segregate function in space and to differentiate the 'inside' from the 'outside;' (ii) metabolism that determines the flow of matter, energy, and redox potential within cells, and its relationship with the outside world; (iii) the transfer of memory to physico-chemical processes (i.e., 'actuating' inherited information); and (iv) memory transmitted from one generation to the next. This classification is similar to that of Jacob, with the first two describing the physico-chemical constraints that a cell deals with while the latter two are related to biological causation.

- In Chapter 15 we defined four categories of constraints that can be used to analyze the capabilities of reconstructed biochemical reaction networks: (i) physico-chemical constraints, (ii) spatial and topological constraints, (iii) environmental constraints, and (iv) regulatory constraints. This classification is operational and these constraints can be described mathematically and used to assess the capabilities of networks.

Picking candidate states: the role of regulation   Cells are subject to inviolable constraints such as those associated with mass and energy balances. Their underlying biochemical networks must obey these, and other spatial constraints. These constraints have been called *hard constraints* and, as illustrated by the pentagon in Figure 16.6, give a range of all allowable states of the network. One or more states may be deemed suitable by the cell, based on its evolutionary history and current challenges (i.e., the prevailing environmental constraints). A way to exclude all the unwanted states (i.e., those that are unsuitable, or selected against) is to implement a regulatory network that eliminates a large portion of the solution space (the pentagon), and by default forces the expression of the 'desired' phenotype.

If a state or phenotype is not the best one under given conditions, the solution can move within the allowable range. This change in the selection of a functional state can be accomplished by regulating the expression of the genes or by regulating the activity of the corresponding gene products. Such regulation has a relatively short time profile. Over longer times, of course, the components of the network can evolve and the properties change slightly, allowing a drift in the phenotypic function of the cell.

Hierarchical organization in biology   Many facets of cellular function and properties are organized hierarchically. The spatial organization of DNA is shown in Figure 16.7A. The linear dimension of the *E. coli* cell is about 1 mm while the length

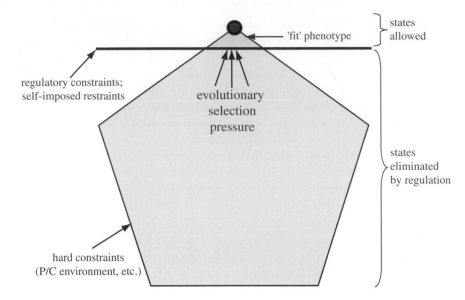

Figure 16.6 Illustration of the constraints on network functions. The pentagon illustrates the range of allowable functions based on hard physico-chemical and environmental constraints. The solid horizontal line illustrates self-imposed constraints (restraints) produced by regulatory networks, i.e., all the states below the line are ruled out by regulatory mechanisms (the blue segment). The red dot denotes the desired functional state, that is found among the admissible states (gray segment) after regulatory constraints have been imposed.

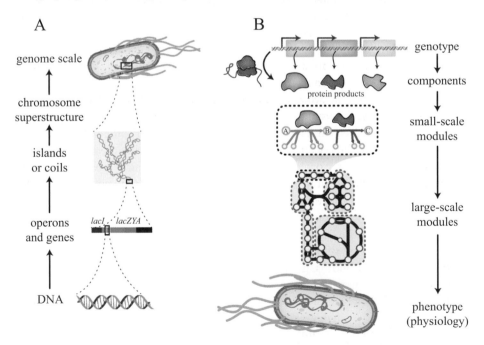

Figure 16.7 Illustration of hierarchical organization in biology: (A) of the DNA, (B) in network function. Prepared by Nathan Lewis.

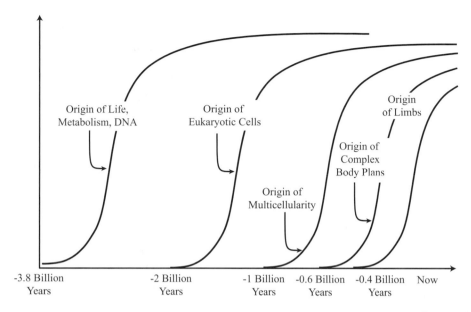

Figure 16.8 Timeline of the development and biological 'fraction' of major cellular and developmental processes. Inspired by Marc Kirschner.

of the cell is on the order of 1 μm, a 1000-fold difference. The bacterial genome is thus 'folded' a thousand times in a hierarchically organized fashion. Biochemical reaction networks can be similarly decomposed (Figure 16.7B). Reactions group together into coordinated units that may be co-localized in space, or even compartmentalized. Many such coordinated units can form a larger organized unit.

The constraints that apply to the lower levels of organization by necessity will constrain the subsequent higher-level functions. This upward application of constraints necessitates a *bottom-up* approach to the analysis of complex biological phenomena. Gödel's completeness theorem in mathematics that showed an axiomatic approach to proving mathematical theorems could not prove all properties of a system may in a general sense apply to biology. By analogy, we would expect that we cannot construct all higher-level functions from the elementary operations alone. Thus, observations and analyses of system-level functions will be needed to complement the bottom-up approach. Therefore, bottom-up and top-down approaches are complementary to the analysis of the hierarchical nature of complex biological phenomena.

Evolutionary adoption of constraints and formation of hierarchy The successive adoption of cellular functions over evolutionary times are illustrated in Figure 16.8. The basic biochemistry of cellular processes and the maintenance and expression of the information on the DNA molecule evolved early. This basic set of processes is found in all organisms today. The genetic code is essentially universal and most proteins are made up of about 20 amino acids. These are basic constraints under which all subsequent cellular processes must operate. The genetic code cannot be predicted from basic theory or physics [91], but is consistent with the basic laws of physics and chemistry. Once picked, it is essentially fixed over evolution. Similarly, most modern proteins are made up of a limited number of motifs, and the basic circuits that lay out the body plan are remarkably conserved. Thus, the constraints set at a lower level of biological hierarchy confine higher levels of organization, but may not explain or predict the more complex functions. Evolution is a tinkerer that combines the elements at hand together in new and unpredictable ways. The first 'wave' in Figure 16.8 is close to the underlying chemical principles and represents the focus of this text.

16.6   Summary

- Biological systems are defined by the interactions between their components.
- The links between molecular components are constrained by the basic laws of chemistry.
- Multiple links between components form a network, and the network can have functional states.
- Functional states of networks are constrained by various factors that are physico-chemical, environmental, and biological in nature.
- The number of possible functional states of networks typically grows much faster than the number of components in the network.
- The number of candidate functional states of a biological network far exceed the number of biologically useful states to an organism.
- Cells select useful functional states by elaborate regulatory mechanisms.
- One may view hierarchical organization and evolutionary change as biological consequences of dealing with constraints.

17 Constraints

Every good scientific theory is a prohibition: it forbids certain things to happen. The more a theory forbids, the better it is – Karl Popper

The importance of constraints and four general categories of constraints were described in Chapter 16. The formal imposition of constraints to form a solution space containing all candidate functional states of networks was discussed in Chapter 12. There are clearly a multitude of constraints that cellular functions have to abide by. In this chapter we discuss constraints in more detail, how they are defined, understood and described.

17.1 Genome-scale Viewpoints

Crowded and interconnected A cell is faced with myriad constraints on its integrated functions. The cell is a very crowded place that is packed with various types of molecules. These molecules have to have certain biochemical states (such as being properly phosphorylated or methylated), cellular location, and abundance to ensure proper overall function of a cell. The genome-scale biophysical and network view of an *E. coli* cell illustrates these features (Figure 17.1). Similar renderings of other cell types convey a similar picture [146–148]. At the biochemical level, these components interact chemically through chemical transformations or associations of molecules to form aggregates. The former represent the formation or breakage of covalent bonds, while the latter is based on intermolecular forces, such as those that lead to hydrogen bonds. The number of such biochemical interactions is large (Figure 17.1B). Thus the cell is crowded, that means all sorts of topological constraints and steric hindrances, and interconnected, that means all sorts of constraints on network functions.

Range of acceptable parameter values In the case of bacteria, all these processes take place simultaneously in a volume that is on the order of a cubic micron. Given the large number of constraints that govern cellular functions, it is surprising that numerical values exist for all the physical constants that allow coherent cellular functions. Thus, some scientists consider the living process to be highly improbable. Conversely, there are scientists that consider the parameter spaces that come with the governing

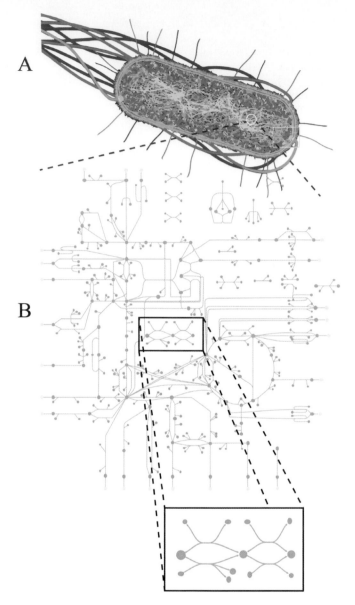

Figure 17.1   The cell is a crowded place and its components are interconnected. Panel A: biophysical 3D view of *E. coli*. From [148] used with kind permission from Springer Science+Business Media B.V. Panel B: the interconnections among cellular components are many.

equations to be very large and thus implicitly suggest that there are many ways to get the living process to work. The constraint-based view would argue that, although there are many parameters to consider in genome-scale models, such parameter spaces are effectively small as the numerical values that each parameter can take on is likely to be highly restricted. The challenge is to define, describe, and quantify these restrictions.

17.2   Stating and Imposing Constraints

Mathematical representation of constraints: balances and bounds   Following their definition, governing constraints need to be described mathematically in order to be used to constrain *in silico* models. Mathematically, constraints are represented as either *balances* or *bounds*.

A balance constraint is represented by an equation.   An example is the conservation of mass. In a steady state, there is no accumulation or depletion of compounds; thus, the rate of production equals the rate of consumption for each compound in the network. This balance is represented mathematically as $\mathbf{Sv} = 0$. Similar balance equations can be formulated for other quantities, such as osmotic pressure, electroneutrality, and free energy around biochemical loops.

A bound is represented by an inequality.   Bounds are constraints that limit the numerical ranges of individual variables and parameters such as concentrations, fluxes, or kinetic constants. Upper and lower limits can be applied to individual fluxes ($v_{min} \leq v \leq v_{max}$). For elementary (and irreversible) reactions, $v_{min} = 0$, and v_{max} is less than approximately one million molecules per μm^3 per second. Concentrations must always be non-negative, so $0 \leq x_i$. Upper bounds for concentrations can arise from solvent capacity constraints, represented as $x_i \leq x_{max}$. Kinetic constants are constrained to be positive and have an upper bound based on collision frequency ($0 \leq k \leq k_{max}$). Transmembrane potentials are limited to about 240–270 mV, above which lipid bi-layers destabilize.

Successive imposition of constraints   Constraints can be applied to the analysis of reconstructed networks to narrow the range of possible functional states (or attainable behaviors), and can be applied in a successive fashion, as illustrated in Figure 17.2. The first icon in Figure 17.2 shows a space where the axes represent fluxes through reactions in a network. Not all the points in this space are attainable, due to the inter-relatedness of the fluxes.

1   The flux balances ($\mathbf{Sv} = 0$) limit the steady-state fluxes to a subspace that is a hyperplane (step 1).

2   If the reactions are defined so that all the fluxes are positive, this plane is converted into a semi-finite space, called a cone (step 2). The edges of this cone become a set of unique, systemically defined extreme pathways, as detailed in Chapter 12. All the points inside the cone can be represented as non-negative combinations of these extreme pathways.

3   Because there are capacity constraints on the individual reactions (v_{max}) in the extreme pathways, the length of each edge is limited. This closes the cone (step 3) and forms a closed solution space in which all allowable network states lie.

4   The properties of this space can then be studied using the methods described in the subsequent three chapters. Further segmentation of this space can be achieved if additional constraints, such as kinetic constants or thermodynamic quantities, are available (step 4).

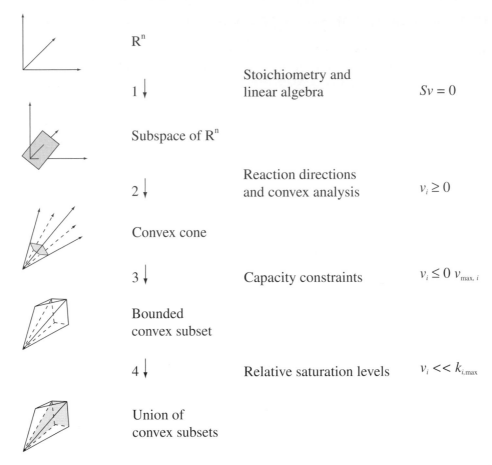

R^n

Stoichiometry and
linear algebra $Sv = 0$

$1 \downarrow$

Subspace of R^n

Reaction directions
and convex analysis $v_i \geq 0$

$2 \downarrow$

Convex cone

$3 \downarrow$ Capacity constraints $v_i \leq 0 \, v_{\text{max}, \, i}$

Bounded
convex subset

$4 \downarrow$ Relative saturation levels $v_i << k_{i, \text{max}}$

Union of
convex subsets

Figure 17.2   Narrowing down alternatives. Conceptual illustration of successive application of constraints to narrow down the range of attainable functional states. Redrawn from [311].

The first three issues were also described in Chapter 12, while the fourth was introduced in Chapter 14.

Example of successive imposition of constraints   The details of the successive imposition of constraints can be illustrated with a simple example. A small network with only two chemical reactions (flux $v_1 : A \rightarrow B$ and flux $v_2 : A \rightarrow C$) as well as two transport processes (metabolite A enters the cell with flux v_{in} and B and C exit together via flux v_{out} through a symporter) is depicted in Figure 17.3. The candidate functional states of this toy network can be defined through the imposition of successive constraints.

1   Just knowing the network topology imposes no constraints on flux values.
2   If the exchange fluxes are irreversible, then v_{in} and v_{out} become non-negative variables (i.e., they cannot take on negative values that would restrict the numerical values that they can take).

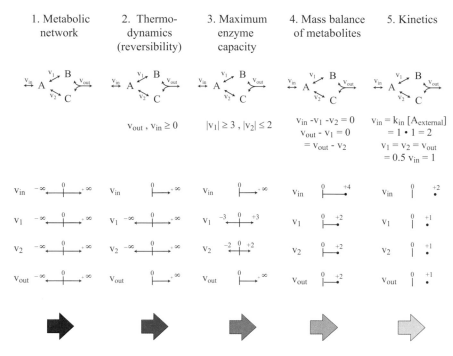

Figure 17.3 Narrowing down alternatives; a simple example. Taken from [84].

3   If the capacities of the internal enzymes are known (i.e., 3 for v_1 and 2 for v_2 in Figure 17.3), the numerical range for their flux values becomes finite.

4   If the network is in a steady-state, a flux balance is imposed, further restricting the allowable ranges of the flux values.

5   If the system is characterized completely, and all numerical values for the physical parameters are known, then the allowable numerical ranges are reduced to a single point, and we have a unique solution. In this simple example it may be enough to know the external concentration of A and the kinetics of its transporter, thus fixing v_{in}.

This simple example illustrates the implementation of constraints. There are many methods that have been developed under that constraint-based umbrella (see Figure on page 250). These methods have been deployed for a number of applications as described in Part IV.

Dominant vs. redundant constraints   As illustrated in Chapter 12, constraints can be redundant or dominant (also called governing). The implications of this distinction are important.

First, the number of parameters that one needs to know accurately may be a small fraction of the total. Those that are associated with governing constraints need to be known. Those that are associated with redundant constraints can be approximated or even left out in some cases. For many growth experiments, accurately knowing the metabolic inputs and outputs from the cell is important. Fortunately, these are measured easily.

Second, selection pressure will be placed on governing constraints in a given environment and thus an organism would be expected to have mutations related to such constraints. Experience with adaptive laboratory evolution (Chapter 26) bears this expectation out. It should be noted that governing constraints are condition-dependent.

17.3   Capacity Constraints

There are capacity constraints that govern cellular processes. These constraints exist because limited numbers of a particular molecule can be present in the cell and it has a maximal activity. Some examples of such constraints are given below.

Membrane surface and metabolism   For metabolic networks, the inputs and outputs are transporters. They reside in the membrane. Some of these protein complexes are large and there is a finite area of cellular membrane. Thus, if many transporters are needed, then the surface area of the membrane can represent a constraint.

Example: management of space in the E. coli *inner membrane.* The membrane capacity constraints associated with aerobic growth and metabolism have been determined [477]. The estimates of these constraints are more detailed than those above as they include the measured uptake rates and the turnover rates of the transporters. The study shows that glucose transporters alone occupy 48% of the membrane area available for protein expression. Similar estimates suggest that at the highest oxygen uptake rate of 18 mmol/g DW/h, about 15% of *E. coli*'s cytoplasmic proteins are cytochromes, while under microaerobic conditions it is about 11% (Figure 17.4). In addition, estimates based on ATP requirement suggest that about 13% of cytoplasmic proteins are ATP synthase during optimal aerobic growth. These major metabolic processes are limited by the space available in the membrane.

Example: protein multi-functionality to increase membrane area. A common way to deal with membrane capacity constraints is to fold the membranes to achieve a high surface-to-volume ratio. For energy-transducing membranes in the mitochondria and the chloroplast, such membrane packing is evident (Figure 17.5). While the ATPase in humans and *E. coli* have much in common, they do differ in a number of additional new protein subunits as members of the complex that are not involved in catalytic functions. These additional subunits serve to fold the membrane where the ATPase is placed, showing how membrane folding can be achieved through the structure of the transmembrane protein and the multiple functions of protein complexes.

Membrane surface and signaling   In addition to metabolic transporters, channels, pores, and other functional proteins, mammalian cells need an array of receptors for signaling molecules. It has been estimated that the maximum number of receptor proteins on a given human cell is on the order of 4 million molecules per cell [321]. This estimate is based on cell size, size of the receptor complex, and the protein carrying capacity of the membrane.

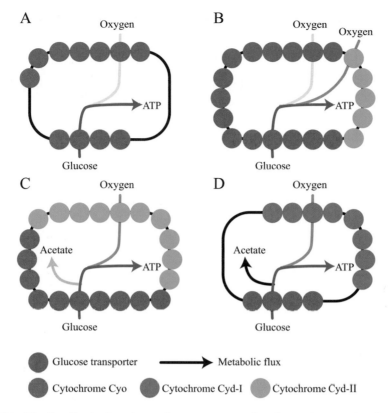

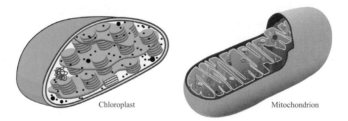

Figure 17.4   The *E. coli* cytoplasmic membrane occupancy by glucose transporters (brown), Cyo (blue), Cyd-I (orange), and Cyd-II (green) under aerobic low glucose uptake rate (GUR) (A), aerobic medium GUR (B), aerobic high GUR (C), and microaerobic conditions (D). In this figure, the spheres represent the membrane enzymes, and the black line underneath the spheres represents the cytoplasmic membrane. The arrows represent the metabolic flow. Taken from [477].

Figure 17.5   Energy transducing membranes are often highly folded to create a high surface-to-volume ratio, partially overcoming membrane surface constraints. Images from http://commons.wikipedia.org

This value seems a reasonable absolute maximum for the total number of receptor proteins considering published *in vivo* data. For example, 10^4 to 10^5 receptor proteins of one type have been observed for some cell types. The estimated maximum of 4 million receptor proteins implies that there could be a maximum of 40 different receptor types with 10^5 receptor proteins each. Certainly, these estimated constraints might change for a given cell that has a greater surface area, larger-sized receptor proteins, or a greater percentage of its plasma membrane devoted to receptor proteins.

Water solvent capacity is limited, leading to constraints on concentrations The approximate number of different metabolites present in a given cell is on the order of 1000 [115]. Assuming a metabolite has a median molecular weight of about 312 g/mol [317] and that the fraction of metabolites of the wet weight is 0.01, we can estimate a typical metabolite concentration of:

$$x_{ave} \approx \frac{1\,g/cm^3 \times 0.01}{1000 \times 312\,g/mole} \approx 32\ \mu M \qquad (17.1)$$

The volume of a bacterial cell is about one cubic micron, or about one femto-liter $(= 10^{-15}$ liter). As a cubic micron is a logical reference volume, we convert the concentration unit as follows:

$$1\ \mu M = \frac{10^{-6}\,mole}{1\,l} \times \frac{10^{-15}\,l}{1\,\mu m^3} \times \frac{6 \times 10^{23}\,molecules}{mole}$$

$$= 600\ \frac{molecules}{\mu m^3} \qquad (17.2)$$

This number is remarkably small. A typical metabolite concentration of 32 μM then translates into a mere 19,000 molecules per cubic micron. Thus, the space in which cellular functions take place, and the large number of different compounds required, leads to a limitation on how many copies of each molecule can be present in a cell.

Enzyme concentrations are limited Cells represent a dense solution of proteins. In the case of a typical bacterium that is perhaps on the order of a μm^3 in volume, the total number of protein molecules present is about 2.5 million.

One can estimate the concentration ranges for individual enzymes in cells. If we assume that the cell expresses about 1000 different types of proteins with an average molecular weight of 34.7 kDa, as is typical for an *E. coli* cell, and given the fact that the cellular biomass is about 15% protein, we get:

$$\frac{1\,g/cm^3 \times 0.15}{1000 \times 34,700g/mole} \approx 4.32\ \mu M \qquad (17.3)$$

This estimate is, indeed, the range into which the *in vivo* concentration of most proteins fall. It corresponds to about 2500 molecules of a particular protein molecule per cubic micron, about 2.5 million total protein molecules. This number is an average and there is a distribution around this mean. Important proteins such as the enzymes catalyzing major catabolic reactions tend to be present in higher concentrations, and pathways with smaller fluxes have their enzymes in lower concentrations. For instance, the 20 most abundant proteins in *Mycoplasma pneumoniae* account for 44% of the total cellular protein mass [257].

Estimating maximal reaction rates   Reaction rates in cells are limited by the achievable kinetics. The bimolecular association rate constant, k_1, for a substrate (S) to an enzyme (E);

$$S + E \quad \xrightarrow{k_1} \tag{17.4}$$

is on the order of $10^8\,\mathrm{M}^{-1}\,\mathrm{s}^{-1}$ [369, 387]. This numerical value corresponds to the estimated theoretical limit, due to diffusional constraints [153]. The corresponding number for macromolecules is about three orders of magnitude lower.

Using the order of magnitude values for concentrations of metabolites and enzymes given above, we find the representative association rate of substrate to enzymes to be on the order of

$$k_1 se = 10^8 \, (\mathrm{M} \times \mathrm{s})^{-1} \times 10^{-4}\,\mathrm{M} \times 10^{-6}\,\mathrm{M} = 0.01 \,\mathrm{M/s} \tag{17.5}$$

that translates into about

$$k_1 se = 10^6 \,\mathrm{molecules}/\mu\mathrm{m}^3\,\mathrm{s} \tag{17.6}$$

that is only one million molecules per cubic micron per second. However, the binding of the substrate to the enzyme is typically reversible, and a better order of magnitude estimate for *net* reaction rates is obtained by considering the release rate of the product from the substrate–enzyme complex, X. This release step tends to be the slowest step in enzyme catalysis [10,11,77], and thus most enzymatic rates are expected to be lower than this maximum estimate.

Measuring typical reaction rates   Large amounts of kinetic information about enzymes has been collected over many decades, and such information exists in an organized format. Analysis of such collections of thousands of k_{cat} and K_M values has been performed (Figure 17.6). The study found that the 'average enzyme' exhibits a k_{cat} of approximately $10/s$ and a k_{cat}/K_M of approximately $10^5/\mathrm{M/s}$, much below the diffusion limit described above. The study found that enzymes operating in secondary metabolism are, on average, approximately 30-fold slower than those in central metabolism. Physico-chemical properties of substrates, such as low molecular mass and hydrophobicity, were found to affect the numerical values of the kinetic parameters.

Thus, for metabolite and protein abundance, the values of kinetic parameters can be bracketed by basic estimates and then the distribution of numerical values can be assessed from data collections.

17.4   Constraints from Chemistry

The interactions between cellular components are described by chemistry. There are many restrictions on how chemical reactions take place. The conservations associated

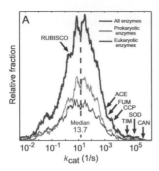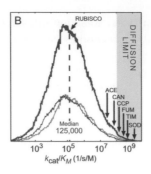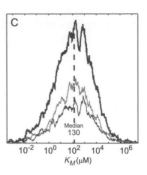

Figure 17.6   Distributions of kinetic parameters: (A) k_{cat} values ($N = 1942$), (B) k_{cat}/K_M values ($N = 1882$), and (C) K_M values ($N = 5194$). Only values referring to natural substrates were included in the distributions. Green and magenta lines correspond to the distributions of the kinetic values of prokaryotic and eukaryotic enzymes, respectively. The location of several well-studied enzymes is highlighted: ACE, acetylcholine esterase; CAN, carbonic anhydrase; CCP, cytochrome c peroxidase; FUM, fumarase; RUBISCO, ribulose-1,5-bisphosphate carboxylase oxygenase; SOD, superoxide dismutase; TIM, triosephosphate isomerase. Taken from [26].

with charge and elemental balancing, for example, were discussed in Chapter 9. In addition, there are thermodynamic constraints that are associated with the energetic states of molecules. These constraints can be described formally. The details of the chemical reactions that specify where atoms from the reactants end up in the products of a reaction can be traced, leading to fluxomic constraints. We will describe these two sets of constraints in this section. Other constraints, such as those related to numbers of molecules and osmotic pressure, are not described.

17.4.1   Mass conservation

Flux balances   Mass cannot be created or destroyed. Thus, when a network is in a steady state, all the fluxes into a node in the network must be balanced by the flows leaving the node. This leads to the flux balance equations.

$$\mathbf{Sv} = \mathbf{0} \tag{17.7}$$

A flux balance is analogous to Kirchhoff's first law of electrical circuits.[1]

The steady-state assumption   Although no network is strictly in a steady state, many homeostatic states are close to being steady, and flux balances are quite useful for analyzing functional states of the network. Although a flux balance may not be obeyed strictly at all times, over reasonably long times there is no accumulation of mass in the network and a time-averaged flux balance applies. By analogy, the reader may be surprised to learn that equilibrium thermodynamics are used to analyze many

---

[1] Kirchhoff's First Law, also known as Kirchhoff's current law, states that at any node (junction) in an electrical circuit, the sum of currents flowing into that node is equal to the sum of currents flowing out of that node.

engineering systems in our daily lives (such as the combustion engine), even though nothing is strictly at equilibrium. However, for practical purposes, this assumption works.

Time-scale hierarchy and fast subsystems: quasi-steady states In engineering systems science there are commonly used procedures to put fast subsystems into a quasi-steady state. This means that the subsystem is in a steady state internally but has input and output into a dynamically changing environment. An example of this would be a production organism in a fermenter. Its internal metabolism is approximately in a steady state, even if the chemical composition of the medium is changing over time.

A familiar example A classical example in the life sciences of the use of a quasi-steady-state assumption is the formulation of the Michaelis–Menten rate law. The reaction mechanism is:

$$S \; + \; E \; \underset{k_{-1}}{\overset{k_1}{\rightleftharpoons}} \; X \; \overset{k_2}{\rightarrow} \; E \; + \; P \tag{17.8}$$

where a substrate, S, binds reversibly to the enzyme, E, to form the intermediate, X, which can break down to give the product, P, and regenerate the enzyme. By assuming a quasi-steady state for the intermediate concentration (i.e., $dX/dt = 0$) one derives the famous and universally used rate equation:

$$\frac{dS}{dt} = \frac{-k_2 E_0 S}{K_m + S} = \frac{-V_m s}{K_m + S} \tag{17.9}$$

where $K_m = (k_{-1} + k_2)/k_1$ is the well-known Michaelis constant, $V_m = k_2 E_0$ is the maximal reaction velocity, and E_0 is the initial (and total) enzyme concentration. As discussed in chapter 5 of the companion book [317], the concentration of X does change in time, but it is a fast variable that adjusts to the time changes of S rapidly under certain numerical values of the kinetic parameters. Thus, underlying this universally used rate law in the life sciences is a steady state assumption, that essentially corresponds to a quasi-steady-state assumption that enables simplified but accurate descriptions under the appropriate numerical values of the rate constants.

17.4.2   Thermodynamics

Thermodynamics are universally applicable and inviolable laws of nature. They limit cellular functions in many ways, from an overall energetic perspective, to prevention of flux loops in networks, to the activities of individual reactions.

Gibbs free energy The net change in the Gibbs free energy in the reactants and products of a reaction has to be negative for the reaction to proceed in the forward direction. This is a statement of constraints and is stated as

$$\Delta G < 0 \tag{17.10}$$

where ΔG is computed from

$$\Delta G = \Delta G^o + RT \ln \left(\frac{\Pi_{\text{products}} \, x_i}{\Pi_{\text{reactants}} \, x_j} \right) \tag{17.11}$$

where ΔG^o is the free energy change under standard conditions. At equilibrium we have

$$\Delta G^o = -RT \ln K_{\text{eq}} \tag{17.12}$$

where K_{eq} is the equilibrium constant for the reaction.

Although this is a universal law, its use in network biology is hampered by the lack of available numbers for the standard free energies and the difficulty of measuring the concentrations accurately [9].

Eliminating loops A closed path inside a network that has no inputs and outputs cannot have any net flux. As discussed in Chapter 11, these are type III extreme pathways that carry no net flux. A no net flux constraint through an internal loop is analogous to Kirchhoff's second law of electrical circuits.[2] There are several ways in which net fluxes through internal loops can be eliminated in biochemical reaction networks [377].

17.4.3   Fluxomics

Detailing the chemistry using isotopomers Direct estimation of fluxes inside the cell are desirable. One widely used experimental method for estimating *in vivo* reaction fluxes is steady-state substrate ^{13}C isotope labeling [267,371]. This method relies on the fact that most carbon nuclei are of atomic mass 12, but an atomic mass 13 isotopomer exists. By placing a ^{13}C-labeled carbon atom in a particular place in a molecule, one can obtain a fine-grained resolution of metabolic fluxes by observing how this label travels through the various metabolites over time.

An overview of the general ^{13}C methods is described in Figure 17.7. By measuring the enrichment for ^{13}C in macromolecule pools after growing on a ^{13}C-labeled substrate, inferences about the internal flux state can be made. The approach can be summarized as a data-fitting problem between simulated and experimentally measured ^{13}C-labeled metabolite concentrations. ^{15}N-labeling is also deployed.

The procedure An isotopomer model, which describes the positional transfer of carbon atoms for all or a subset of reactions in the network, is used to simulate data (Figure 17.7A). For a specified carbon input label, an isotopomer model enables the calculation of an Isotopomer Distribution Vector (IDV) corresponding to a particular simulated steady-state flux distribution (Figure 17.7B). Mass spectrometry (MS) or nuclear magnetic resonance (NMR) experiments on ^{13}C-labeled cell biomass generate

---

[2] Kirchhoff's second law, also known as Kirchhoff's voltage law, states that the directed sum of the electrical potential differences (voltage) around any closed network is zero.

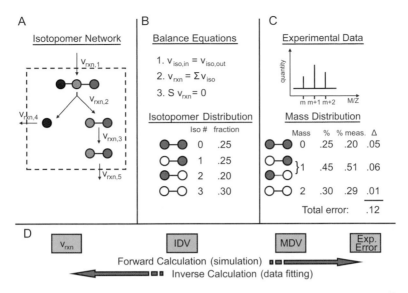

Figure 17.7 Overview of fluxomic computations illustrating the overall flux balance equation and the balance equations on each isotopomer. The Isotopomer Distribution Vector (IDV) and Mass Distribution Vector (MDV) are different because some of the isomers have the same mass and they cannot be distinguished experimentally. More details are provided in the text. Taken from [376].

fractional ^{13}C enrichments from fragmented macromolecules, forming a Mass Distribution Vector (MDV) (Figure 17.7C). The error between the measured MDV and the MDV corresponding to the simulated IDV summarizes how well the presumed flux distribution fits the ^{13}C experiment. The flux distribution v that minimizes this error can be computed by solving a global optimization problem.

While simulating ^{13}C-enrichment given a flux distribution is deterministic and inexpensive computationally, the inverse problem of calculating the flux distribution that best fits a ^{13}C-tracing experiment is of greater interest, yet is significantly more difficult computationally (Figure 17.7D). A review of these methods and associated challenges can be found in [408, 454, 470].

Complicated mathematics As indicated in Figure 17.7B, there are additional equations to solve over and above the regular flux balances. One has to trace the fate of each isotopomer. This leads to a vast increase in the number of equality constraints that one has to satisfy. The mathematics are too involved to be detailed here and the primary references should be consulted. The net result is that the solution space is now more constrained and the consequences of these additional constraints can be evaluated [19, 384, 433, 445, 454, 455].

17.5 Regulatory Constraints

Cells are subject to both adjustable and non-adjustable constraints. The latter are discussed above and are physico-chemical in origin, including stoichiometric, capacity, and thermodynamic constraints. As described above, they can be used to bracket the range of possible functional states of networks. Adjustable constraints are biological in origin, and they can be used to further limit allowable behavior. These constraints will change in a condition-dependent manner.

In general, regulation affects the amount and activity level of a gene product in a cell. The former comes with regulation of transcription and translation, while the latter involves chemical modification (i.e., phosphorylation, acetylation, etc.) of a protein or the binding of a ligand (i.e., allosteric regulator, needed co-factor, etc.). The description of transcriptional regulation at the level of absence or presence calls of protein has been achieved in the form of digital constraints, i.e., zero or one.

Regulation imposes additional constraints on cellular functions Regulatory events impose temporary, adjustable constraints on the solution space, as shown in Figure 17.8. This figure depicts a solution space and optimal solution (red circle) for a target network. A larger solution space defined by non-adjustable constraints is shown in Figure 17.8A. If the flux through a certain reaction is regulated, then the volume of the space (i.e., the range of allowable cellular behaviors) is reduced.

As illustrated in Figure 17.8B and C, respectively, the solution space is restricted to a smaller space. This restricted space is analogous to a cell with fewer behavioral possibilities. Note that the optimal solution remains in the subspace shown in Figure 17.8B, but not in the subspace shown in Figure 17.8C. If the optimal solution is no longer in the previously confined space, the phenotype that it corresponds to cannot be expressed and a different phenotype within the reduced space needs to be expressed.

Representing transcriptional regulatory constraints The transcriptional regulatory structure can be described using Boolean logic equations. This approach involves

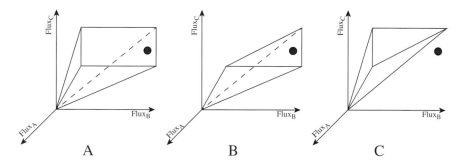

Figure 17.8 Regulatory constraints change the shape of the solution space. A hypothetical solution space defined by various non-adjustable constraints is illustrated in panel A. An optimal solution is indicated by a red circle. The flux through a certain reaction may be constrained, leading to a reduction in the size of the solution space. After regulatory constraints have been applied, the original solution may either remain in the smaller solution space (B), or may no longer be located inside the space (C), in which case, a new solution will be determined by the cell – that is, a different phenotype. Redrawn based on [88].

restricting expression of a transcription unit to the value of 1 if the transcription unit is transcribed and 0 if it is not. Similarly, the presence of an enzyme or regulatory protein, or the presence of certain conditions inside or outside of the cell, may be expressed as 1 if the enzyme, protein, or a certain condition is present, and 0 if it is not.

Example: results from imposing regulatory constraints [85]   A regulated flux balance model was used to simulate growth of *E. coli* quantitatively over the course of growth experiments. The resulting time courses of growth, substrate uptake, and by-product secretion were then compared with experimental data. *E. coli* has been observed to secrete acetate when grown aerobically on glucose in batch cultures; when glucose is depleted from the environment, the acetate is then reutilized as a substrate.

Using the regulated and non-regulated flux balance (rFBA and FBA, respectively) models, an aerobic batch culture of *E. coli* on glucose minimal medium was simulated; the calculations are shown together with experimental data [438] (Figure 17.9). The major difference between the rFBA and FBA simulations is in the delayed reaction of the system to depletion of glucose in the growth medium. The FBA model without regulation does not account for the delays associated with protein synthesis.

In addition, an *in silico* expression array and a regulatory protein activity array can be computed (Figure 17.9C). The *in silico* array predicted the up-regulation of four gene products, *aceA, aceB, acs*, and *ppsA*, as well as the down-regulation of three gene products, *adhE, ptsGHI-crr*, and *pykF*. DNA microarray technology has been used to detect differential transcription profiles on a collection of 111 genes in *E. coli* [291]. The difference in gene expression for aerobic growth on acetate versus growth on glucose as reported in [291] is included in Figure 17.9C. The calculated expression of the eight genes included in the rFBA model for which expression data were published was in qualitative agreement with the predictions of the rFBA model. The ability of the rFBA model to reutilize acetate depends on the up-regulation of the glyoxylate shunt genes, *aceA* and *aceB*, which explains the high magnitude of transcription difference (20-fold) reported in [291].

17.6   Coupling Constraints

Biological hierarchy   A cell is faced with numerous constraints on its function. Constraints are upwardly applicable in the biological hierarchy shown in Figure 17.10. At the top level, it may be difficult to determine the constraints that govern cellular functions on a mechanistic basis, but they can be identified phenomenologically. Conversely, at the bottom level, physico-chemical principles apply.

One can conceptualize growth and metabolic genotype–phenotype relationship in microbes using a five-layered hierarchical description, ranging from the molecular level to the whole cell (Figure 17.10). This multi-scale relationship requires a complex regulatory architecture even in the simplest organisms to coordinate the balancing act between adjacent layers. For a single cell, a number of layers of complexity and molecular functions can be delineated. Each can be described with a network model. When layers of complexity are not coupled together stoichiometrically through chemical equations, coupling constraints can be formulated to build integrated descriptions between layers.

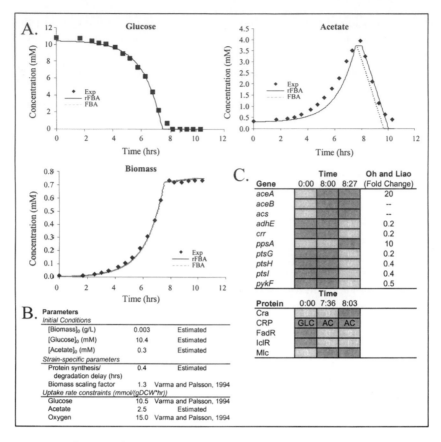

Figure 17.9   Aerobic growth on acetate with glucose reutilization. Panel A, Experimental data [438], as well as the corresponding simulations performed using FBA and rFBA. In the acetate plot, the regulatory and metabolic model predictions differ from that of the regulatory plot alone, as shown. Panel B, a table containing the parameters required to generate the time plots. Panel C, *in silico* arrays showing the up- or down-regulation of selected genes or activity of regulatory proteins in the regulatory network (dark gray, gene transcription/protein activity; light gray, transcriptional repression/protein inactivity). Data from [291] showing the experimentally determined transcriptional fold changes for certain genes (acetate:glucose) are shown where applicable. From [85].

Coupling levels in the hierarchy   Coupling constraints bound the maximum and minimum ratios of fluxes in biochemical network models. The involvement of enzymes is implicit in metabolic models and not explicitly represented in a reaction mechanism. An example is an enzyme in a metabolic reaction (Figure 17.11A). However, in a network (see the E-matrix in Chapter 7) describing macro-molecular synthesis, proteins are included explicitly as molecules in the reactions they catalyze (Figure 17.11B). The four explicit reactions (v_1 through v_4) are equivalent to the reaction (v_0) in the implicit formulation. It follows that the synthesis of the recycled reactant E is not essential to permit steady-state flux through v_1 through v_4, as it

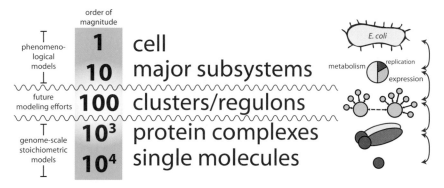

Figure 17.10 Cells are hierarchical. A five-layer conceptualization of the phenotype–genotype relationship is illustrated. Phenomenological equations can predict the relative levels of proteins belonging to major subsystems within a cell accurately (e.g., metabolism, macromolecular synthesis, and others). On the other hand, genome-scale stoichiometric models make predictions taking all single molecules and protein complexes into account. Order of magnitude estimates are given for each component in the hierarchy. The right-most arrows illustrate the fact that the layers impose describable biological constraints on each other. Prepared by Joshua Lerman.

is recycled by the last reaction (v_4). Thus, the conversion of $A + B \rightarrow C$ will occur regardless of whether the model is synthesizing E.

Consequently, additional constraints are needed to enforce the synthesis of E if its set of explicit reactions is active in a particular steady state. We require the condition

$$\text{if } v_4 > 0 \text{ then } v_{\text{synthesis E}} > 0 \tag{17.13}$$

where $v_{\text{synthesis E}}$ is the synthesis reaction rate of E. Furthermore, it would be desirable to relate the flux through reaction v_4 and the synthesis of E with some proportionality,

$$v_4 \propto v_{\text{synthesis E}} \tag{17.14}$$

even though the exact proportion factor can only be approximated (see below). Note that both v_4 and $v_{\text{synthesis E}}$ are non-negative quantities. The relationships expressed in Equations (17.13) and (17.14) can be represented by an inequality

$$v_4 \leq c_{\text{max}} \cdot v_{\text{synthesis E}} \tag{17.15}$$

where c_{max} ($0 < c_{\text{max}}$) is the upper bound on the proportion factor, termed a *coupling coefficient*.

Formulation of coupling constraints   Metabolic models can be coupled to network models of protein expression to form an integrated network. To describe the function of the integrated model, one needs to deploy coupling constraints (Figure 17.12):

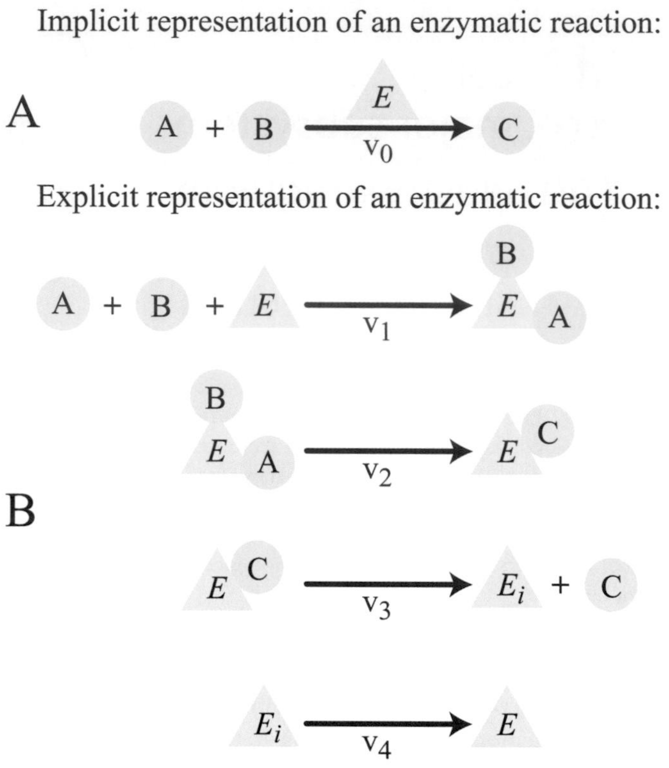

Implicit representation of an enzymatic reaction:

Explicit representation of an enzymatic reaction:

Figure 17.11 Schematic representation of the implicit and explicit participation of enzymes in network reactions. Traditionally, in metabolic network formulations, enzyme (E) participation in a reaction is implied but not modeled explicitly (Panel A). However, a network of reactions associated with transcription and translation produces enzymes; hence, the explicit incorporation of enzymes in the reactions they catalyze is needed in an integrated network description (Panel B). The state E_i conceptualized as a conformational state of the enzyme post product release that then is relaxed to the conformation that can bind a substrate again and repeat the catalytic cycle. The same approach can be applied if the reactant E is a tRNA molecule or a protein. From [421].

Panel (A):   genome-scale metabolic models (M-models) provide for a metabolic description of genotype–phenotype relationship without accounting explicitly for synthesis of enzymes. M-models employ Boolean logic statements relating genes, proteins, and reactions, or the Gene–Protein–Reaction associations, or GPRs. A reaction can only carry a non-zero flux if its GPR statement evaluates to 'True.'

Panel (B):   integrated models of metabolism and expression (ME-Models) account explicitly for the genotype–phenotype relationship. Macromolecular expression is directly integrated with cellular metabolism.

To approximate dilution of macromolecules to daughter cells and limits on translation efficiency, three constraints are employed that provide limits on transcription,

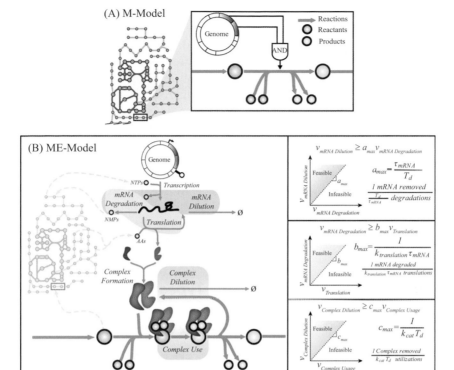

Figure 17.12   Genome-scale modeling of metabolism and expression. Panel (A) metabolism with GPRs (and M-model); panel (B) metabolism and protein expression fully integrated (an ME-model). Abbreviations: T_d, cell doubling time; k_{cat}, enzyme turnover number; τ_{mRNA}, mean mRNA lifetime; $k_{translation}$, peptide production rate. The coupling parameters a_{max}, b_{max}, and c_{max} are described in [229]. Prepared by Joshua Lerman and Nathan Lewis.

translation, and dilution rates. To accomplish the integration of the two networks (metabolism and protein expression), three coupling constraints are required:

constraint (a)   approximates mRNA dilution by requiring the cell to lose a particular mRNA molecule to a daughter cell after $1/a_{max}$ mRNA degradation events; thus preventing infinite cycling between nucleotide monophosphates (NMPs) and nucleotide triphosphates (NTPs);

constraint (b)   poses an upper limit on the number of peptides that may be translated from a specific mRNA molecule by requiring degradation of the mRNA molecule after $1/b_{max}$ translation events; and

constraint (c)   approximates dilution of catalysts to a daughter cell by requiring the model to transmit the catalyst after it has been used $1/c_{max}$ times.

Coupling constraints are relatively new in the COBRA field and are likely to find increased use as the scope of network reconstructions and models grow. Various omics data types will be useful to determine their numerical values.

17.7   Simultaneous Satisfaction of All Constraints

Myriad constraints   All the constraints discussed above need to be met simultaneously. In addition, measured omics data can be used as constraints. The imposition of polyomic data sets is described in more detail in Chapter 24. The myriad constraints that a cell has to satisfy are illustrated in Figure 17.13. These constraints can be represented mathematically. When combined with the network equations based on the topology of the reconstructed network, one gets a set of equations that confine all the allowable states to a solution space.

But still not enough   In spite of their high number, all these constraints do not suffice to shrink the solution space to a single point. Thus, we will always have a range of solutions and many candidate functional states of the network to consider and analyze. Such analysis can be performed through the deployment of optimization methods, that in turn implicitly imply that the evolutionary process has an objective that is being optimized over time.

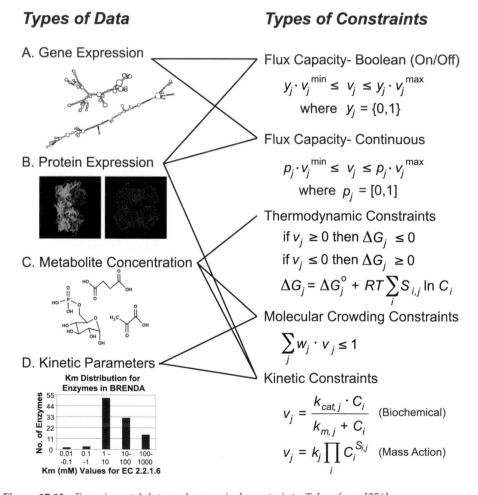

Figure 17.13   Experimental data and numerical constraints. Taken from [356].

17.8   Summary

- Cellular functions are constrained by myriad different factors.
- Mathematically, constraints are described with equations (balances) or inequalities (that describe physical and other limitations).
- Constraints can be dominant or redundant. Good information about dominant constraints in a given environment is important to achieving good estimates of phenotypic functions.
- Physical and chemical constraints can be derived from first principles.
- Biological constraints that reflect regulation are either data-derived or based on regulatory logic that has been determined experimentally. These constraints are subject to change and reflect results from distal causation.
- The imposition of the myriad definable and imposable constraints does not lead to a single solution to estimate the functional state of a network.

18 Optimization

It is all about the fluxes – Jens Nielsen

Even under myriad constraints, a reconstructed network can have many possible functional states. Organisms have an evolutionary past that led to the determination of which functional state is expressed, which in turn implies that the resulting state is optimal in some sense for the conditions under which the organism evolved. Thus, biologically relevant functional states of networks can, in principle, be predictable using optimization under the governing constraints. The application of constraint-based optimization methods introduces quantitative ways of estimating, computing, and understanding functional states of networks that correspond to phenotypic functions.

18.1   Overview of Constraint-based Methods

Constraint-based reconstruction and analysis (COBRA)   The procedures for analyzing the allowable phenotypic states on a genome-scale have developed since the late 1990s [344, 358]. The COBRA approach consists of three fundamental steps:

- first, a *genome-scale network reconstruction* is carried out,
- second, the appropriate constraints are applied to form the corresponding *genome-scale model* (GEM) *in silico*, and
- third, COBRA methods are applied to evaluate the properties of the GEM.

Optimization methods   The COBRA methods rely on the use of various optimization methods (Figure 18.1) including:

Linear programming (LP).   This method is used when the problem to be solved involves a linear set of constraints (equalities and inequalities) and a linear objective function. LP is the basis for flux balance analysis (FBA).

Quadratic programming (QP).   This method is used when the problem to be solved involves a linear set of constraints (equalities and inequalities) and a quadratic

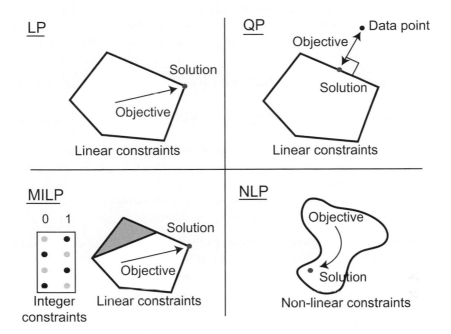

Figure 18.1 Depiction of major classes of optimization methods used in COBRA. Prepared by Daniel Zielinski.

objective function. A quadratic objective arises when one uses a Euclidean distance as an objective function. When computing the Euclidean distance, the elements of a vector are raised to a second power. The constrained 'least squares problem' is solved using QP.

Mixed integer linear programming (MILP).   The formulation of LP problems often leads to the use of discontinuous variables. Often, logical variables are introduced that only take on a value of zero or one, such as when reaction presence/absence calls are introduced as variables. MILP is used to solve this class of problems.

Non-linear programming (NLP).   The most complicated optimization problems involve the use of non-linear constraints and/or a non-linear objective function. In general, such problems are hard to solve. One fundamental issue that arises is that the solution space being searched is non-convex. In such a circumstance, one cannot guarantee finding the global optimum for the objective function in the space.

These optimization methods have been deployed in the various COBRA methods that have been developed. None of these methods are described in mathematical or algorithmic detail in this text. For such information, one should consult established textbooks in the field, e.g., [31,76].

The COBRA methods   Over 100 *in silico* methods have been developed under the COBRA framework [236]. This growing number of methods can be broadly classified into two main categories: *unbiased* methods, discussed in Part II, that are used to globally characterize a solution space; *biased* methods are the larger family of methods that uses objective functions for a variety of purposes. They are further subcategorized as follows.

1   Flux balance-derived methods that are used to assess various optimality properties of networks.
2   Methods to assess the consequences of deleting genes or inactivating gene products.
3   Strain design methods for metabolic engineering of production strains.
4   Model refinement methods that are deployed for network reconstruction and model formulations.
5   Methods that incorporate thermodynamic information.
6   Methods that incorporate regulatory mechanisms.

The second main branch describes methods to characterize the solution space globally and in an unbiased fashion, i.e., they do not rely on the use of optimization. There are two main sub-branches:

1   randomized sampling of the space for global characterization (see Chapter 14); and
2   Computation of basis vectors to characterize the entire contents of a space (see Chapters 12 and 13).

The COBRA toolbox   The widespread use of COBRA methods has spurred the development of accessible software. The COBRA toolbox [34, 375] implements the COBRA methods in MATLAB, which is an extensively used, commercially available software package. Thus, any user can implement the growing number of COBRA methods for a reconstructed network that they wish to characterize. The currently available methods in the COBRA toolbox 2.0 are summarized in Figure 18.2.

18.2   Finding Functional States

Statement of constraints   For typical biological networks, the number of reactions (n) is greater than the number of compounds (m) resulting in a plurality of feasible steady-state flux distributions. Although infinite in number, the steady-state solutions lie in a restricted region called the null space (Chapter 12), and additional constraints govern the expression of a phenotype (Chapter 17).

However, a certain set of phenotypes are expressed under particular conditions. Optimization can be used to find particular solutions of interest. Linear optimization can be used to find solutions of interest within the bounded null space (see Figure 18.3). The bounded null space is defined by:

$$\mathbf{S}_{exch} \begin{pmatrix} \mathbf{v} \\ \mathbf{b} \end{pmatrix} = \mathbf{0} \text{ where } 0 \leq v_i \leq v_{i,\max} \text{ and } b_{i,\min} \leq b_i \leq b_{i,\max} \tag{18.1}$$

where v_i are the internal fluxes and b_i are the exchange fluxes.

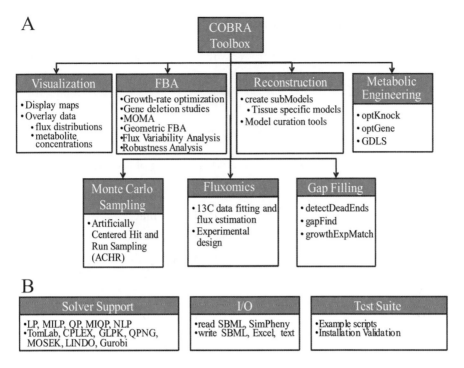

Figure 18.2 The flow chart from the COBRA 2.0 toolbox. From [375].

Statement of objective   To pick out particular solutions within this space, one has to define the desired properties of such solutions. Mathematically, the definition of the solutions sought is stated in the form of an *objective function*. A general linear objective function is defined as:

$$Z = \langle \mathbf{w}, \begin{pmatrix} \mathbf{v} \\ \mathbf{b} \end{pmatrix} \rangle = \sum_i w_i v_i + \sum_j w_j b_j \tag{18.2}$$

where the vector $\mathbf{w}$ is a vector of weights (w_i) on the internal and exchange fluxes, v_i and b_j, respectively. The weights are used to define the properties of the particular solutions sought. Z is then optimized, i.e., minimized or maximized as appropriate. The solutions to these equations give the best use of the defined network to meet the stated objective function in a steady-state.

Linear programming   With a linear objective function (Z), this constrained optimization procedure is known as *linear programming* (LP). The objective function is a user-specified statement that describes the solution to be computed. The general representation of Z in Equation (18.2) enables the formulation of a range of functionalities and network states of interest. Z can be used to represent exploration of the metabolic capabilities of a network, physiologically meaningful objectives (such as maximum cellular growth rate), or design objectives for a microbial production strain. More detailed discussion on the objective function is found in Chapter 21.

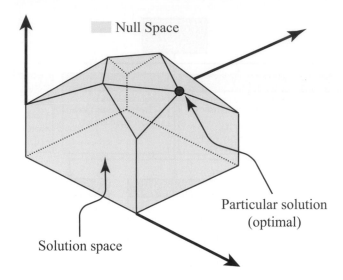

Figure 18.3 A schematic representation of the finite null space and a particular solution located within that space. Redrawn from [441].

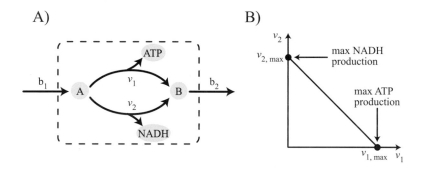

Figure 18.4 A simple LP problem. Modified from [43].

18.3   Linear Programming: the basics

Linear optimization is accomplished through LP. This optimization procedure is routinely used for the solution of a variety of different problems. The basics of LP are described in Box 18.1 and throughout this section.

How LP works An easy to understand example of an LP problem is shown in Figure 18.4A. Panel A shows a reaction network where a compound, A, is picked up by a cell and is metabolized to B via two different routes and then secreted. One route, v_1, produces high-energy phosphate bonds in the form of ATP. The other route, v_2, produces redox potential in the form of NADH. The flux balance for this system is

$$v_1 + v_2 = b_1 (= b_2) \tag{18.3}$$

Box 18.1 Linear Programming/Linear Optimization (LP)

Concept: Identifies the optimal outcome for a stated objective subject to a list of requirements (constraints) that are represented as linear relationships

Objective $Z = max\langle \mathbf{c}^T.\mathbf{x}\rangle = \max \sum_{i=1}^{n} c_i x_i$

Constraints   Problem constraints (convex polytope): $\mathbf{A} \cdot \mathbf{x} \leq \mathbf{b}$

Non-negative variables: $\mathbf{x} \geq 0$; $\mathbf{b} \geq 0$

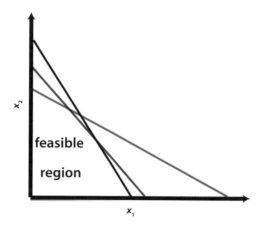

Because $v_1 \geq 0$ and $v_2 \geq 0$, these constraints define the solution space to be a line segment that is the intersection of the positive quadrant (Figure 18.4B). Once b_1 is measured and has a known numerical value, this intersection forms a closed line in the (v_1, v_2)-plane. Given a stated objective, this line segment is searched for the best solution. The optimal solutions for the maximization of ATP production or maximization of NADH production are shown in Figure 18.4B and they lie at the ends of the line segment that forms the solution space. This example shows a one-dimensional solution space and the optimal solutions for two single-valued objective functions.

Extreme points as optimal solutions   The fact that solutions lie at the edge of the allowable solution space is particularly easy to see from the example in Figure 18.4. If one maximizes ATP production, it is clear that v_2 should go to zero, and v_1 to the maximum value equal to the uptake rate. This optimal solution thus lies at the right extreme point of the solution space. Conversely, if one maximizes the redox production from this metabolite in the form of NADH, the optimal solution is v_2 equal to the uptake rate b_1, and v_1 goes to zero. That optimal solution is at the opposite end of the solution space.

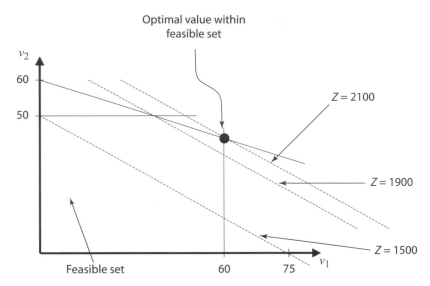

Figure 18.5   A depiction of a bounded two-dimensional solution space and an objective function that is being maximized. In this illustrative case, there are no equality constraints, but there are three inequality constraints that form the solution space: (i) $0 \leq v_1 \leq 60$, (ii) $0 \leq v_2 \leq 50$, and (iii) $v_1 + 2v_2 \leq 120$. The objective function is $Z = 20v_1 + 30v_2$.

Location of the optimal solutions   Next, we consider a slightly more complex example where a two-dimensional solution space is formed by three inequalities (Figure 18.5). We can also consider an objective function that is a combination of the two variables. For a fixed value of Z, the objective function forms a straight line in the two-dimensional plane. If the value of Z is changed, the line moves and intersects the two-dimensional polytope at a different location. As we increase the value of Z, the intersecting line moves closer and closer to the periphery of the solution space. The maximum value for the objective function or 2100 is found when it intersects the solution space at a single point, which is an extreme point in the space.

The types of solutions found   There are three types of feasible solutions encountered in solving LP problems. They are illustrated in Figure 18.6.

1   *Unique solutions.* For small networks, the optimal solution typically lies at an extreme point of the feasible set, as is the case in Figure 18.5.
2   *Degenerate solutions.* In some instances, the line formed by a constant value of an objective function is parallel to a constraint. In this case, the entire edge of the feasible set has the same value as the objective function, and all the points along the edge represent an optimal solution. This edge represents an infinite number of solutions, mathematically called *degenerate solutions*. These solutions are *alternate*, or *equivalent*, optimal solutions because they correspond to the same value of the objective function. The occurrence of alternate optimal solutions is frequent in genome-scale networks [359], and thus genome-scale networks are typically able to achieve the same overall functional network state in many different ways (see Chapter 20).

Unique solution   Degenerate solution   Unbounded solution

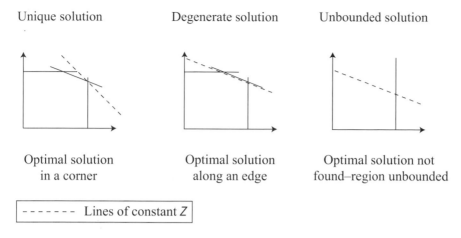

Optimal solution   Optimal solution   Optimal solution not
in a corner   along an edge   found–region unbounded

- - - - - - - Lines of constant Z

Figure 18.6 A graphical representation of the types of feasible solutions found by LP.

3 *Unbounded solutions.* Sometimes the feasible set is unbounded and the objective function increases without limit in the open direction. In this case, no solution is found. Biologically, such situations are unrealistic, and if detected, typically result from an incomplete network formulation. Clearly, the statement of global upper and lower constraints on the flux variables prevents an unbounded problem from ever occurring.

If the constraints are inconsistent, then the set of feasible solutions is empty and no solution can satisfy the stated constraints. In such cases, the constraints are formulated incorrectly.

Assessment of the sensitivity of the optimum solution The sensitivity of the optimal solution is measured by *shadow prices* and *reduced costs*.

- *Shadow prices.* The shadow prices (π_i) are the derivatives of the objective function at the boundary with respect to an exchange flux:

$$\pi_i = -\frac{\partial Z}{\partial b_i} \tag{18.4}$$

  The shadow prices define the incremental change in the objective function if a constraining exchange flux is incrementally changed. Shadow prices may change discontinuously as b_i is varied (see Chapter 22). The shadow prices can be used to determine whether an optimal functional state of a network is limited by the availability of a particular compound (Figure 18.7). The shadow prices thus essentially define the intrinsic value of the metabolites toward attaining a stated objective. This feature has proven useful for interpreting optimum solutions and for metabolic decision making [437]. We note that in some literature, the definition of a shadow price is the negative of what is stated in Equation (18.4).

- *The reduced costs.* The reduced costs (ρ_i) can be defined as the amount by which the objective function will change with the flux level through an

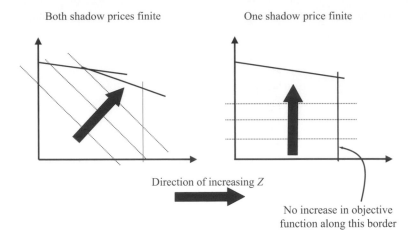

Figure 18.7   A graphical representation of a zero and non-zero shadow price at the edge of a boundary. The thick lines denote constraints, while the dashed lines represent lines of constant value of the objective function as in Figure 18.5.

internal flux that is not in the basis solution (i.e., fluxes that have a zero net flux):

$$Z = Z_0 + \langle \rho, \mathbf{v} \rangle, \quad \rho_i = -\frac{\partial Z}{\partial v_i} \tag{18.5}$$

where Z_0 is the optimal solution.

The reduced costs can be used to analyze the presence of alternate optimal flux distributions. If a reduced cost is zero, that means that the flux level through the corresponding reaction does not change the objective function. Thus, the reduced costs can be useful for examining the effect of gene deletions on the overall function of a network.

Although we introduce the shadow price and reduced cost as sensitivity measures, they are fundamentally quantities associated with the so-called dual problem in the optimization literature [364]. A discussion of the details of LP solutions and how they are interpreted is found in Chapter 19.

18.4   Genome-scale Models

When applying COBRA methods on a genome-scale, there are a few general and significant issues that arise.

Building quality models   It is important to note that only a curated and quality-controlled reconstruction can lead to organism-specific GEMs. Many automated procedures generate reaction maps that cannot be used as a basis for computation. The reasons for this vary, but often, basic rules of chemistry may be violated, pathways may have gaps, and so forth.

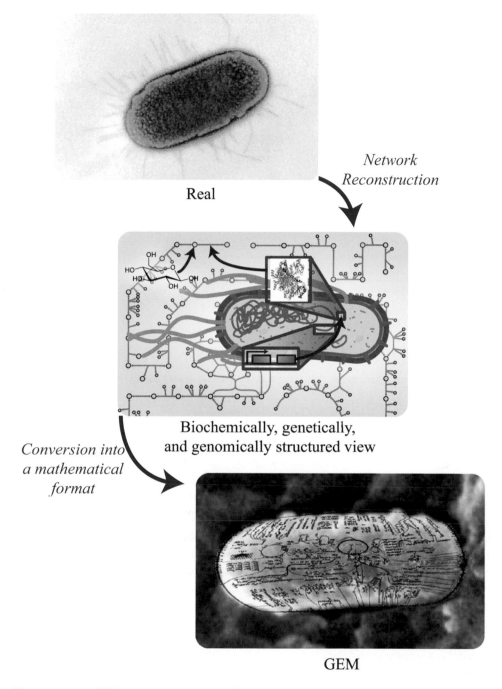

Figure 18.8 A GEM is a computer model of a real cell. GEMs are based on a reconstruction that contains an accurate chemical representation of all the known components in the corresponding real cell. Top panel from [3], middle panel courtesy of Nathan Lewis, bottom panel from [128].

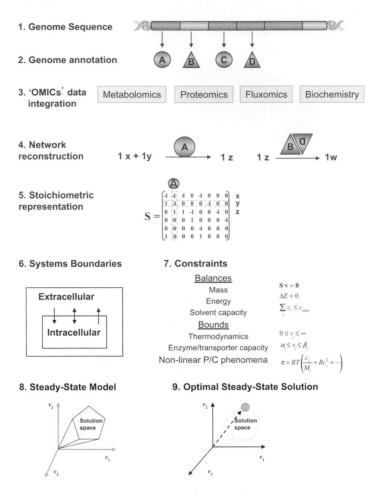

Figure 18.9 The process of forming a GEM and computing its functional states. From [426].

A GEM is a computer model of its real counterpart mediated by a BiGG knowledge base (Figure 18.8). It should accurately account for all the function of the gene products that are found on a genome. Once they are all put in context of one another, the properties of the *in silico* cell can be computed. The quality of the representation of function of the real cell depends on the coverage of the gene products found on the genome and the accuracy of the representation of their biochemical functions.

From reconstructions to models The conversion of a genome-scale reconstruction to a model is a laborious and detailed process, but it has been accomplished for a number of organisms (Chapter 3). The procedure to form an *in silico* model of a cell has many steps that are outlined in Figure 18.9. Steps 1 through 5 in this figure are detailed in Parts I and II of the text. To form a computational model, one has to define the systems boundary and the fluxes that cross it (step 6). This leads to an input/output (I/O) model formulation. The imposition of constraints follows (step 7), leading to the definition of the confined null space (step 8). Then, the objective function is formulated and the functional states are computed (step 9). Various COBRA methods can then be deployed.

Table 18.1   There are numerous methods that have been developed to analyze constraint-based reconstructions of metabolic networks using experimental data to answer biological questions. Below is a list of some of these methods and questions they can help to answer. Taken from [235].

Method	Question
Equivalent Optima	How many flux states can be attained by maximizing or minimizing an objective function (e.g., maximum growth or ATP production)?
Energy Balance Analysis	How can one evaluate the thermodynamic feasibility of FBA simulation results?
ExPa/ElMo	How does one define a biochemically feasible, unique set of reactions that span the steady-state solution space?
FBA	What is the maximum (or minimum) of a specified cellular objective function?
Flux Confidence Interval	What are the confidence intervals of flux values when fluxomic data are mapped to a constraint-based model?
Flux Coupling	What are the sets of network reactions that are fully coupled, partially coupled, or directionally coupled?
Flux Variability Analysis	What is the maximum and minimum flux for every reaction under a given set of constraints (i.e., what is the bounding box of the solution space)?
Gap-Fill/Gap-Find	What are the candidate reactions that can fill network gaps, thus helping improve the model and provide hypotheses for unknown pathways that can be validated experimentally?
Gene Annotation	Which reactions are likely missing from the network,
Refinement Algorithm	given a set of phenotypic observations? What are the candidate gene products with which corresponding reactions could fill the gap?
Gene Deletion Analysis	Which are the lethal gene deletions in an organism?
K-cone analysis	Given a set of fluxes and concentrations for a particular steady state, what is the range of allowable kinetic constants?
Metabolite Essentiality	How does metabolite essentiality contribute to cellular robustness?
Minimization of Metabolic Adjustment (MOMA)	Can suboptimal growth predictions be more consistent with experimental data in wild-type and knock-out strains?

Table 18.1   (*cont.*)

Method	Question
Net Analysis	Given metabolomic data, what are the allowable metabolite concentration ranges for other metabolites, and what are likely regulated steps in the pathway based on non-equilibrium thermodynamics?
Objective function finder/ObjFind	Which are different possible cellular objectives?
Optimal Metabolic Network Identification	Given experimentally measured flux data, what is the most likely set of active reactions in the network under the given condition that will reconcile data with model predictions?
OptKnock/OptGene	How can one design a knock-out strain that is optimized for by-product secretion coupled to biomass production?
OptReg	What are the optimal reaction activations/inhibitions and eliminations to improve biochemical production?
OptStrain	Which reactions (not encoded by the genome) need to be added in order to enable a strain to produce a foreign compound?
PhPP	How does an objective function change as a function of two metabolite exchanges?
rFBA	How do transcriptional regulatory rules affect the range of feasible *in silico* phenotypes?
Regulatory On/Off Minimization	After a gene knock-out, what is the most probable flux distribution that requires a minimal change in transcriptional regulation?
Robustness analysis	How does an objective function change as a function of another network flux?
SR-FBA	To what extent do different levels of metabolic and transcriptional regulatory constraints determine metabolic behavior?
Stable Isotope Tracers	How can intracellular flux predictions be validated experimentally, and which pathways are active under the different conditions?
Thermodynamics-based Metabolic Flux Analysis	How can one use thermodynamic data to generate thermodynamically feasible flux profiles?
Uniform Random Sampling	What are the distributions of network states that have not been excluded based on physico-chemical constraints and/or experimental measurements? What are the completely or partially correlated reaction sets?

A well-curated and validated model can be considered an *in silico* counterpart to a real cell (Figure 18.8). Such a cell should be able to recapitulate many phenotypic states of the real cell that it represents.

Global view of model properties One optimal solution is rarely of interest in isolation. The constraint-based optimization methods are scalable and can be applied repeatedly for varying environmental and genetic parameters. This scalability has spurred a growing number of analysis methods that have been developed under the constraint-based approach [236, 344].

In subsequent chapters, we describe some of the methods that are used to characterize a changing environment and genetic makeup. To date, the focus has been on the steady-state flux distributions, but now this approach is being used to study all allowable concentrations [32] and kinetic states [110], and to algorithmatize iterative model-building procedures [83, 163]. Many biologically relevant questions can be addressed with a high-quality GEM (Table 18.1).

Deployment of COBRA methods Quality GEMs and the suite of COBRA tools have led to a spectrum of applications. The COBRA methods are used to address a variety of questions (Table 18.1). A series of well-curated GEMs for model organisms have been widely used, and COBRA methods have been deployed to study their various attributes and uses (see [53, 118, 254, 263, 290]) and Part IV of text).

18.5 Summary

- The null space of **S** is bounded with the application of v_{max} values, and other governing constraints.
- Specific points within these bounded solution spaces can be determined through constraint-based optimization procedures.
- The optimization is carried out based on a stated objective.
- Objectives can be used to probe network capabilities, to represent likely physiological objectives, and to represent candidate biological designs.
- If the objective function and constraints are linear, then linear programming can be used to find the optimal solution.
- Unique optimal solutions are found in the corners of the bounded solution space.
- Frequently, for large biological systems, the solutions are found on an edge of a surface of the solution space leading to redundant solutions. In such cases, many different solutions lead to the same optimal objective value.
- Genome-scale reconstructions are mathematically represented and the governing constraints are imposed. This procedure leads to an *in silico* organism that contains all the known components of the real organism that it represents, and allows the simulation of allowable states given a set of governing constraints.
- High-quality GEMs and the COBRA methods can be used to address a variety of important biological questions.

19 Determining Capabilities

There is nothing more practical than a good theory – Kurt Lewin

The linear programming methods that were outlined and illustrated in the previous chapter are highly scalable and can be applied to complex networks. Before moving towards complicated situations, we first look at the use of objective functions to explore network properties and to determine their capabilities. We examine the methods and approaches that are used to interpret the solutions to gain better understanding of the function of the network under consideration. The corresponding objective functions are single-valued; i.e., they focus on a single flux or property. In this chapter, we will perform illustrative computations using the core *E. coli* network. We will focus on the ability to produce key co-factors and metabolic precursors to growth. These production capabilities are dependent on the substrate used as well as the electron acceptor available. This chapter is influenced by an early study of the capabilities of core metabolic pathways in *E. coli* [439], and represents progression in this text from the conceptual toward the more detailed and specific applications.

19.1 Optimal Network Performance

The core *E. coli* metabolic network can be interrogated computationally for various network capabilities, including the maximal yields of co-factors and biosynthetic precursors from any of the substrates that it can consume. In this chapter we will study the yield of key co-factors and biosynthetic precursors by the core *E. coli* metabolic network and learn how to interpret the optimal solutions.

19.1.1 Co-factors

There are three key co-factor molecules in the core metabolic network. ATP is the principal carrier of high-energy phosphate bonds. It can be made by substrate-level or oxidative phosphorylation. The two primary carriers of redox potential are NADH and NADPH. The computation of yields are implemented by putting a drain reaction on the precursor and then optimizing the flux through this reaction for a given set of inputs. The computations can be carried out for different environmental conditions,

Table 19.1 Maximum stoichiometric yields of key co-factors (ATP, NADH, and NADPH) in the core *E. coli* model from glucose. PPP designates the use of the pentose pathway. Prepared by Jeff Orth.

Co-factor	Yield	PPP (%)	ATP shadow price	Governing constraint
Aerobic				
ATP	17.50	0	0.00	Internal proton balancing
NADH	10.00	0	0.57	Energy and stoichiometry
NADPH	8.78	300	0.44	Energy and stoichiometry
Anaerobic				
ATP	2.75	0	0.00	Internal proton balancing
NADH	6.00	150	1.00	Energy and stoichiometry
NADPH	4.00	0	1.33	Energy and stoichiometry

such as the presence or absence of oxygen. Optimal co-factor yields from glucose are summarized in Table 19.1.

19.1.2   Biosynthetic Precursors

A carbon source used for growth is degraded to a set of 12 key biosynthetic precursors (Table 19.2). The biosynthetic reactions then convert this set of 12 precursors into the monomers from which all cellular macromolecules are synthesized. As for the co-factors above, the ability to produce the 12 key biosynthetic precursors can be studied (see Table 19.2). This capability is determined by introducing a drain on a precursor of interest and maximizing the drain. The computation is then repeated for all 12 of them.

The optimal solutions The yields of these 12 precursors can be calculated from any carbon source, once the LP problem has been set up. The results from maximal yield computations of the biosynthetic precursors using one unit of glucose as an input are shown in Table 19.2 for both aerobic and anaerobic conditions.

Under aerobic conditions, three glycolytic intermediates (3PG, PEP, Pyr) can be produced with 100% carbon conversion. Their maximum yield has no energy constraints, the ATP shadow price is zero, and there is a surplus production of ATP that is dissipated through a futile cycle. The production of two OA corresponds to a carbon conversion efficiency of 8/6 or 133%. This higher than 100% conversion rate corresponds to CO_2 fixation through the PPC reaction.

Constraints on the optimal solutions Not all of the co-factors can be made with a 100% conversion efficiency because of constraints on their production. The governing constraints fall into two categories: energy and stoichiometry.

Energy.   The monophosphate sugars (G6P, F6P, R5P, E4P, G3P) cannot be produced at a 100% carbon conversion. Energy is a constraint, as some of the glucose has to be metabolized to generate the required energy for their formation.

Table 19.2 Maximum stoichiometric yields of biosynthetic precursors from glucose under aerobic and anaerobic conditions. Note that the production of 2 OA requires 8 carbon molecules, 2 of which come from fixing CO_2. The core network was allowed to fix CO_2 during these computations. Table courtesy of Jeff Orth.

Metabolite	Yield	Carbon conversion (%)	ATP shadow price	Constraint
Aerobic				
3PG	2	100	0	–
PEP	2	100	0	–
Pyr	2	100	0	–
OA	2	133.33	0	–
G6P	0.8916	89.16	0.0482	Energy
F6P	0.8916	89.16	0.0482	Energy
R5P	1.0571	88.10	0.0571	Energy
E4P	1.2982	86.55	0.0702	Energy
G3P	1.6818	84.09	0.0909	Energy
AcCoA	2	66.67	0	Stoichiometry
αKG	1	83.33	0	Stoichiometry
SuccCoA	1.64	109.33	0	–
Anaerobic				
3PG	1	50	0	Stoichiometry
PEP	1	50	0	Stoichiometry
Pyr	1	50	0	Stoichiometry
OA	1	66.67	0	Stoichiometry
G6P	0.625	62.50	0.1667	Energy
F6P	0.625	62.50	0.1667	Energy
R5P	0.72	60.00	0.192	Energy
E4P	0.8491	56.60	0.2264	Energy
G3P	1.0345	51.72	0.2759	Energy
AcCoA	1	33.33	0	Stoichiometry
αKG	0.4	33.33	0	Stoichiometry
SuccCoA	1.434	95.60	0	Stoichiometry

Stoichiometry. AcCoA can only be produced with a 66.7% carbon conversion. The optimal solution is not constrained by energy. The only route to AcCoA in the core network is through a decarboxylase and thus a CO_2 is, by necessity, lost during the formation of AcCoA. Therefore, maximally, a 2/3 conversion efficiency is possible, simply due to stoichiometric constraints. The conversion to αKG is similarly limited by stoichiometric constraints.

The optimal solutions can be computed readily using linear programming. The hard part is interpreting the solutions. We will now detail the characteristics of key optimal solutions.

19.2   Production of ATP

Consumption of ATP is represented through a demand reaction:

$$\text{ATP} + \text{H}_2\text{O} \xrightarrow{v_{\text{ATPM}}} \text{ADP} + \text{P}_i + \text{H}_I^+ \tag{19.1}$$

The objective function for linear optimization is:

$$Z = v_{\text{ATPM}} \tag{19.2}$$

which becomes maximized for given network inputs. In the cases studied below, we use an input value of 1 for glucose. The calculations can be done in a dimensionless setting because we are computing yields, which are relative numbers. Later on, we have to assign units to the exchange fluxes.

19.2.1   Producing ATP aerobically from glucose

If the core *E. coli* metabolic network is provided one unit of glucose and unlimited oxygen, it will maximally produce 17.5 units of ATP. The full solution to this LP problem is shown in Table 19.3. The P/O ratio for the core *E. coli* is 5/4, as 5 protons are pumped out for each pair of electrons that goes down the ETS from the dehydrogenase and there are 4 protons entering the cytoplasm through ATPase for every ATP molecule that is produced. All the carbon in glucose is released as CO_2, representing full oxidation of the substrate.

The 17.5 ATP produced per glucose number is different than the number of 38 reported in standard biochemistry textbooks. The difference is due to the fact that the ETS in *E. coli* does not have a P/O ratio of 3, as is assumed to be the case for mitochondria in animal cells.

The optimal solution   The flux map that leads to maximal ATP production is shown in Figure 19.1A. The production and use of ATP and H_i^+ during maximal production is shown separately in Figure 19.1B using a node map:

ATP balance: five ATP molecules are produced by substrate-level phosphorylation; three in glycolysis, PYK (one ATP) and PGK (two ATP) and two in the TCA, by SUCOAS (two ATP). PYK produces only one ATP as the first phosphorylation step in glycolysis to G6P is performed by the PTS by transferring the high-energy phosphate group from PEP as PEP is converted to Pyr. Oxidative phosphorylation, the ATP synthase (ATPS4r), produces 13.5 ATP molecules, for a total production of $5 + 13.5 = 18.5$ ATP molecules. PFK uses one ATP. The ATP load, ATPM, balances the node with a flux of 17.5.

Table 19.3 The shadow prices and reduced costs for the optimal aerobic ATP yield from glucose. Prepared by Jeff Orth.

#	Reaction name	Flux value	Reduced cost
1	ACALD	0	0
2	ACALDt	0	0
3	ACKr	0	0
4	ACONTa	2	0
5	ACONTb	2	0
6	ACt2r	0	0
7	ADK1	0	0
8	AKGDH	2	0
9	AKGt2r	0	0
10	ALCD2x	0	0
11	ATPM	17.5	0
12	ATPS4r	13.5	0
13	Biomass	0	185.1603
14	CO2t	-6	0
15	CS	2	0
16	CYTBD	12	0
17	D_LACt2	0	0
18	ENO	2	0
19	ETOHt2r	0	0
20	EX_ac(e)	0	4.25
21	EX_acald(e)	0	6.5
22	EX_akg(e)	0	10.75
23	EX_co2(e)	6	0
24	EX_etoh(e)	0	7.5
25	EX_for(e)	0	0
26	EX_fru(e)	0	17.5
27	EX_fum(e)	0	7.75
28	EX_glc(e)	-1	17.5
29	EX_gln_L(e)	0	11.25
30	EX_glu_L(e)	0	12
31	EX_h(e)	0	0
32	EX_h2o(e)	6	0
33	EX_lac_D(e)	0	7.75
34	EX_mal_L(e)	0	7.75
35	EX_nh4(e)	0	0
36	EX_o2(e)	-6	0
37	EX_pi(e)	-2	0
38	EX_pyr(e)	0	6.5
39	EX_succ(e)	0	8.25
40	FBA	1	0
41	FBP	0	1
42	FORt2	0	0.25
43	FORti	0	0
44	FRD	0	0
45	FRUpts2	0	0
46	FUM	2	0
47	FUMt2_2	0	0
48	G6PDH2r	0	0
49	GAPD	2	0
50	GLCpts	1	0
51	GLNS	0	0
52	GLNabc	0	0
53	GLUDy	0	0
54	GLUN	0	0
55	GLUSy	0	0
56	GLUt2r	0	0
57	GND	0	0.4167
58	H2Ot	-6	0
59	ICDHyr	2	0
60	ICL	0	0
61	LDH_D	0	0
62	MALS	0	0
63	MALt2_2	0	0
64	MDH	2	0
65	ME1	0	0
66	ME2	0	0
67	NADH16	1	0
68	NADTRHD	2	0
69	NH4t	0	0
70	O2t	6	0
71	PDH	2	0
72	PFK	1	0
73	PFL	0	1.5
74	PGI	1	0
75	PGK	-2	0
76	PGL	0	0
77	PGM	-2	0
78	PIt2r	0	0
79	PPC	0	0
80	PPCK	0	1
81	PPS	0	1
82	PTAr	0	0
83	PYK	1	0
84	PYRt2r	0	0
85	RPE	0	0
86	RPI	0	0
87	SUCCt2_2	0	0
88	SUCCt3	0	0
89	SUCDi	2	0
90	SUCOAS	-2	0
91	TALA	0	0
92	THD2	0	0
93	TKT1	0	0
94	TKT2	0	0
95	TPI	1	0

#	Metabolite name	Shadow price
1	13dpg[c]	9
2	2pg[c]	8
3	3pg[c]	8
4	6pgc[c]	17.75
5	6pgl[c]	17.5
6	ac[c]	4.5
7	ac[e]	4.25
8	acald[c]	6.5
9	acald[e]	6.5
10	accoa[c]	5.25
11	acon-C[c]	12.5
12	actp[c]	5.5
13	adp[c]	-1
14	akg[c]	11
15	akg[e]	10.75
16	amp[c]	-2
17	atp[c]	0
18	cit[c]	12.5
19	co2[c]	0
20	co2[e]	0
21	coa[c]	0
22	dhap[c]	10
23	e4p[c]	12.9167
24	etoh[c]	7.75
25	etoh[e]	7.5
26	f6p[c]	18.75
27	fdp[c]	20
28	for[c]	0
29	for[e]	0
30	fru[e]	17.5
31	fum[c]	8.25
32	fum[e]	7.75
33	g3p[c]	10
34	g6p[c]	18.75
35	glc-D[e]	17.5
36	gln-L[c]	12.25
37	gln-L[e]	11.25
38	glu-L[c]	12.25
39	glu-L[e]	12
40	glx[c]	3.75
41	h2o[c]	0
42	h2o[e]	0
43	h[c]	-0.25
44	h[e]	0
45	icit[c]	12.5
46	lac-D[c]	8
47	lac-D[e]	7.75
48	mal-L[c]	8.25
49	mal-L[e]	7.75
50	nad[c]	-1.5
51	nadh[c]	0
52	nadp[c]	-1.5
53	nadph[c]	0
54	nh4[c]	0
55	nh4[e]	0
56	o2[c]	0
57	o2[e]	7
58	oaa[c]	8
59	pep[c]	0.25
60	pi[c]	0
61	pi[e]	0
62	pyr[c]	6.75
63	pyr[e]	6.5
64	q8[c]	0
65	q8h2[c]	0.5
66	r5p[c]	15.8333
67	ru5p-D[c]	15.8333
68	s7p[c]	21.6667
69	succ[c]	8.75
70	succ[e]	8.25
71	succoa[c]	9.5
72	xu5p-D[c]	15.8333

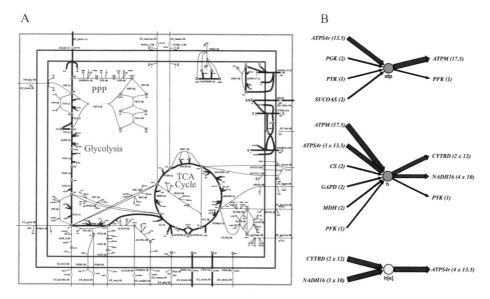

Figure 19.1 Maximal aerobic ATP production in the *E. coli* core metabolic network. Maximum ATP yield is 17.5 mol/mol glucose. Panel A: the flux map. Panel B: node maps showing the rates of production and use of ATP and H_i^+ under maximal ATP yield conditions. Prepared by Jeff Orth.

H_i^+ *balance:* the production and consumption of the proton is complicated. It is produced by 6 reactions and consumed by 3. Most of the protons are produced by two processes: the ATP load (ATPM) produces 17.5 protons, and the ATPase (ATPS4r) produces 40.5. In addition, two glycolytic steps produce protons (two by GAPD and one by PFK) and two TCA reactions produce protons (two by CS and 2 by MDH). The total production on internal protons is thus $17.5 + 40.5 + 2 + 1 + 2 + 2 = 65$. This production is balanced by the ETS by pumping out protons (24 by CYTBD and 40 by NADH16) and one is consumed by PYK in glycolysis, for a total of $24 + 40 + 1 = 65$.

Interpretation of the optimal solution The shadow prices and reduced costs can be used to interpret the optimal solutions. The shadow prices and reduced costs for the optimal aerobic production of ATP from glucose are in Table 19.3. There is much information conveyed in this table, and we discuss a few items here.

- Four protons are needed to make one ATP. Thus, every proton that comes across the membrane as a part of the steady-state flux map reduces the potential to make an ATP molecule by 0.25. Thus, the shadow price for the internal proton is –0.25. The consequence is that every intracellular process that produces internal protons reduces the ability to make ATP via oxidative phosphorylation.
- Two kinases in glycolysis involve protons: PFK (ATP + F6P → ADP + FDP + H^+) and PYK (ADP + H^+ + PEP → ATP + PYR). The shadow price for F6P is 18.75 and 20 for FDP, a difference of 1.25. The cost of this transformation is one ATP and one intracellular proton, making FDP 1.25 higher in value than

Table 19.4 Computed ATP yields (mol/mol) on the 13 substrates that the core *E. coli* model can consume. Prepared by Jeff Orth.

Metabolite	Abbrev.	Maximum aerobic yield	Maximum anaerobic yield
Glycolysis			
Glucose	glc	17.50	2.750
Fructose	fru	17.50	2.750
Pyruvate	pyr	6.50	0.750
Lactate	lac-D	7.75	0.375
Fermentation			
Ethanol	etoh	7.50	0
Acetaldehyde	acald	6.50	0.625
Acetate	ac	4.25	0
Amino acids			
Glutamine	gln-L	11.25	0
Glutamate	glu-L	12.00	0
TCA			
α-Ketoglutarate	akg	10.75	0
Succinate	succ	8.25	0
Fumarate	fum	7.75	0.375
Malate	mal -L	7.75	0.375

F6P. The shadow price for PEP is 8 and 6.75 for Pyr, also a difference of 1.25. This drop of 1.25 is due to the use of one ATP and the production of one intracellular proton. Compare this to the PGK reaction $13DPG + ADP \rightleftharpoons 3PG + ATP$, where no proton is made or consumed. The shadow price for 13DPG is 9 and 8 for 3PG, a difference of 1.

The solution also contains information about the ATP yield on all other substrates that the core model can use to make ATP aerobically.

- Of the 14 possible substrates, 13 can make ATP (formate cannot). Note that the shadow price for them is the same as their yield. For instance, the aerobic ATP yield on succinate 8.25 (see Table 19.4) is the same as the shadow price for external succinate. This equivalency is simply due to the fact that an addition of succinate to the optimal ATP production from glucose would add the corresponding amount of ATP production to the objective function, as can be obtained from succinate.

- Many of the transport reactions involve a symport with a proton. Two examples are: $Succ[e] + 2H^+[e] \rightarrow Succ[c] + 2H^+[c]$ and $Ac[e] + H^+[e] \rightarrow Ac[c] + H^+[c]$. The shadow price for Succ[e] is 8.25 and 8.75 for Succ[c]. The

A

B

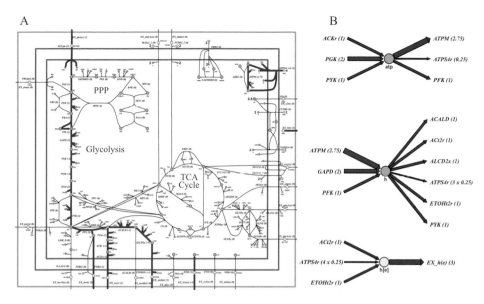

Figure 19.2 Maximal ATP anaerobic production in the *E. coli* core metabolic network. Maximum ATP yield is 2.75 mol/mol glucose. Panel A: the flux map. Panel B: node maps showing the rates of production and use of ATP and H_i^+ under maximal ATP yield conditions. Prepared by Jeff Orth.

difference is due to the $2 \times (-0.25)$ cost associated with the translocation of two protons that reduce the ability to make ATP via oxidative phosphorylation. Similarly, the shadow price for Ac[e] is 4.25 and 4.5 for Ac[c], a difference of 0.25, which corresponds to the cost of translocating the proton.

19.2.2   Producing ATP anaerobically from glucose

In the absence of oxygen as an electron acceptor, the core *E. coli* model can produce 2.75 ATPs from one glucose molecule imported. The production of ATP is solely by substrate-level phosphorylation. The ratio between the aerobic and anaerobic yield is $17.5/2.75 = 6.36$. Having an electron acceptor available makes a substantial difference. The optimal solution is found in Figure 19.2. Not all the carbon can be oxidized to CO_2, but fermentation products are secreted; one unit of acetate and one unit of ethanol. Two carbon atoms are released as CO_2. Note that there is a net production of 3 protons. This production rate will lead to acidification of the medium.

The optimal yield The flux map that leads to maximal ATP production is shown on the reaction map in Figure 19.2A. The production and use of ATP and H_i^+ during maximal production is shown separately in Figure 19.2B:

- *ATP balance:* three ATP are produced in glycolysis (two by PGK and one by PYK) and one ATP is produced in the fermentative pathway that leads to acetate secretion. PFK in glycolysis consumes one ATP and 0.25 ATP are consumed by ATPS4r to pump out one proton that is needed to balance the production of protons, see below. This leaves a balance of 2.75 ATP that are consumed by the demand reaction.

- H_i^+ *balance:* as with the aerobic case, three protons are produced in glycolysis (one by PFK and two by GAPD). In addition, the hydrolysis by the demand reaction produces 2.75 protons for a grand total of 5.75. The secretion of ethanol consumes three protons (ACALD, ALCD2x and ETOHt2r) and the secretion of acetate one (ACt2r). In addition, PYK consumes one and the ATPS4r consumes 0.75 (3:1 ratio to ATP).

19.2.3   Optimal ATP production from other substrates

The core *E. coli* model can consume 14 substrate molecules in addition to glucose, 13 of which can make ATP. No ATP can be produced from formate. We can compute the ATP yield on all of these substrates under anaerobic and aerobic conditions. Glucose and fructose have the highest and identical yields. Lactate has a higher aerobic yield than pyruvate as one NADH is produced when lactate is converted to pyruvate. The aerobic ATP yields on the TCA intermediates are in the order of their position in the cycle and their ability to generate NADH that then produces ATP through the ETS.

The anaerobic ATP yields are quite different. The network cannot make ATP from many substrates without oxygen. There are no possible flux maps with a net production rate of ATP. The substrates that can make ATP anaerobically have much lower yields than under aerobic conditions. The primary reason is that oxidative phosphorylation cannot take place and ATP production is by substrate-level phosphorylation only.

We comment further on ATP production from lactate and malate. The ATP yields are the same for these two substrates for both aerobic and anaerobic conditions (Table 19.4). Inspection of the flux maps (Figure 19.3) shows that the reason for this equivalence is that while the routes to pyruvate differ, once pyruvate is formed the ATP from then on is the same. The routes to pyruvate from the two substrates generate the same number of protons and redox equivalents. The overall yields are thus the same.

19.3   Production of Redox Potential

Consumption of redox potential (NADH or NADPH) is through a demand reaction:

$$NAD(P)H \xrightarrow{v_{NAD(P)H}} NAD(P)^+ + H^+ + 2e^- \tag{19.3}$$

The objective function then becomes:

$$Z = v_{NAD(P)H} \tag{19.4}$$

representing a demand (DM) function on redox potential.

19.3.1   Aerobic production of NADH from glucose

The optimal solution   The maximal yield of NADH from glucose is 10 (Table 19.1), and the corresponding flux map is shown in Figure 19.4A. The optimal solution uses

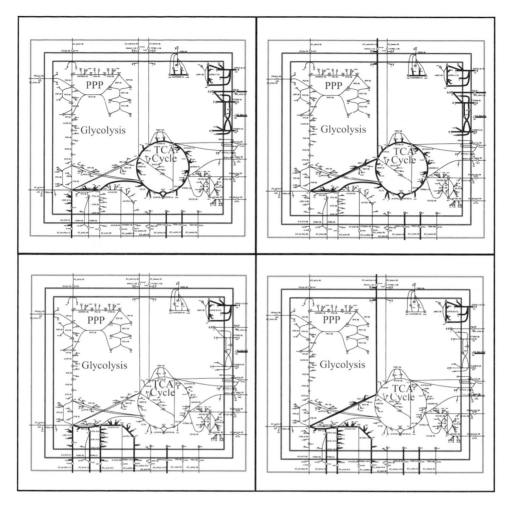

Figure 19.3 Maximal ATP anaerobic production in the *E. coli* core metabolic network from lactate (first column) and from malate (second column). Both the aerobic (top row) and anaerobic (bottom row) maps are shown. Prepared by Jeff Orth.

glycolysis, the TCA cycle, and CYTBD cytochrome system where Q8 is reduced by SUCDi. The node maps of NADH, ATP, and the protons illustrate the key features of the optimal solution.

- *NADH:* the production of NADH takes place in glycolysis (two by GAPD), two by PDH, and six in the TCA cycle (two by each of ICDHyr, AKGDH, MDH), for a total of 10 NADH that are then consumed by the demand reaction.
- *ATP:* a net production of ATP occurs in in glycolysis (–1 PFK, +2 PGK, +1 PYK) and two are produced in the TCA cycle (SUCOAS) for a total of four. The 4 ATP produced are consumed by the ATP synthase in reverse (ATPS4r).
- *External protons:* the ATP synthase (ATPS4r) run in reverse pumps out $4 \times 4 = 16$ protons. Note that 4 protons are produced externally but 3 protons are consumed internally for every flux unit through ATPS4r, the

A

B

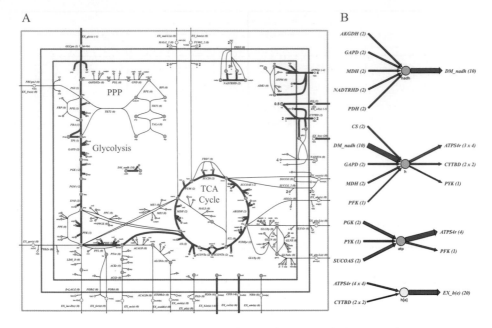

Figure 19.4 Maximal NADH production in the *E. coli* core metabolic network. Maximum NADH yield is 10 mol/mol glucose. Panel A: the flux map. Panel B: node maps showing the rates of production and use of NADH, ATP, H$^+$[e] and H$^+$ under maximal yield conditions. Prepared by Jeff Orth.

difference due to one proton being released by every ATP hydrolyzed and that gets secreted. In addition, $2 \times 2 = 4$ protons are pumped out by the CYTBD cytochrome system. Thus, a total of 20 protons are produced and secreted into the surrounding medium. Aerobic NADH production, therefore, acidifies the medium.

Interpretation of the optimal solution   The solution can be interpreted in detail from the full list of shadow prices as done for ATP production from glucose above. Here we only analyze the ATP constraint on NADH production. The complete oxidation of glucose requires a sink for 24 electrons. Thus, the metabolism of glucose could result in the reduction of 12 redox carrier molecules such as NADH. However, the maximum aerobic NADH yield is computed to be 10.

The constraints leading to this reduced NADH yield can be determined by looking at the shadow prices. There are three interconnected constraints that relate to energy (ATP), the internal proton production (H$^+$), and the stoichiometry of redox carrier coupling.

- The ATP shadow price for maximal aerobic NADH production from glucose is 0.5714 (Table 19.1); thus to get to the full 12 NADH, we need additional ATP:

$$\frac{\Delta\text{NADH}}{\Delta\text{ATP}} = 0.5714 \;\Rightarrow\; \Delta\text{ATP} = \frac{\Delta\text{NADH}}{0.5714} = \frac{12-10}{0.5714} = 3.5 \qquad (19.5)$$

Thus 3.5 ATP are needed to balance the entire map.

- In a similar fashion, we can analyze the effect of internal proton balancing. The shadow price for the proton is negative: $-0.1429 = -0.5714/4$, or negative fourth of the ATP shadow price. This ratio corresponds to the stoichiometry of ATP synthase. As above, we can compute:

$$\frac{\Delta NADH}{\Delta H^+} = -0.1429 \Rightarrow \Delta H^+ = \frac{\Delta NADH}{-0.1429} = \frac{12 - 10}{-0.1429} = -14 \qquad (19.6)$$

In other words, for every additional internal proton that would be added, the potential to make NADH is reduced by $1/7^{th}$.

- Co-factor coupling prevents the production of all 12 NADH possible from glucose. In the TCA cycle, the SUCDi reaction uses an electron pair to reduce q8 via $FADH_2$, thus preventing the production of NADH that can be directly consumed by the NADH demand function (the objective).

Alleviating the ATP constraint We can compute the effect of removing the ATPM constraint by artificially supplying intracellular ATP from another source. With ATP constraints removed and allowing unlimited ATP production, the optimal flux map looks quite different (Figure 19.5).

- The optimal solution now generates all 12 NADH by producing 12 NADPH in the pentose phosphate pathway (G6PDH2r, GND), then uses the transhydrogenase (NADTRHD). It cycles each glucose through 6 times, losing one carbon atom each time as CO_2. This cycle produces no ATP, but

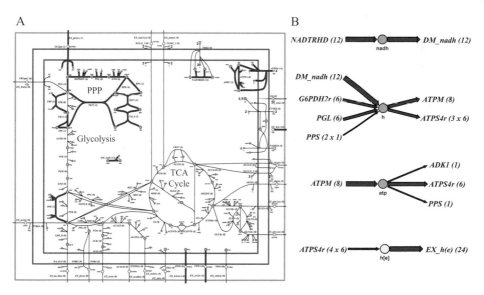

Figure 19.5 Maximal NADH production in the *E. coli* core metabolic network with auxiliary supply of ATP to release the energy constraints. Maximum NADH yield is 12 mol/mol glucose. Panel A: the flux map. Panel B: node maps showing the rates of production and use of NADH, ATP, $H^+[e]$ and H^+ under maximal yield conditions. Prepared by Jeff Orth.

leads to a net hydrolysis of a phosphate group with FBP that originates from PEP during import of glucose.

- The PEP has to be regenerated by PPS which produces an AMP and P_i. Another P_i is produced by the FBPase. These two P_i are consumed by the ATPM that is running in reverse. ATPS4r uses 6 ATP to pump out $6 \times 4 = 24$ protons. Twelve of these protons originate from glucose (that is completely oxidized to CO_2) and 12 from the 6 water molecules that are imported during the process.

Thus, relaxing the ATP constraint leads to a totally different way to stoichiometrically make NADH optimally. This example also shows how important it is to realize the diversity of functional states that a network can take on.

19.3.2 Anaerobic production of NADH

The stoichiometrically optimal anaerobic NADH production yields 6 NADH molecules from one glucose molecule (see Figure 19.6).

- This flux map generates 3 NADPH in the pentose phosphate pathway (G6PDH2r, GND) that then generates 3 NADH from the transhydrogenase. It generates 1.5 NADH in glycolysis (GAPD) and it generates 1.5 NADH with PDH for a grand total of 6 NADH produced.
- Three CO_2 are made in the pentose pathway and 3 carbons leave in the form of 1.5 acetate produced for a grand total of 6 carbon atoms that are in the consumed glucose molecule.

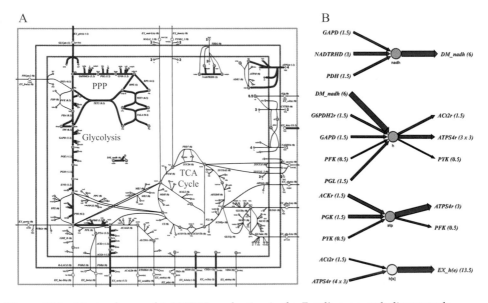

Figure 19.6 Maximal anaerobic NADH production in the *E. coli* core metabolic network. Maximum NADH yield is 6 mol/mol glucose. Panel A: the flux map. Panel B: node maps showing the rates of production and use of NADH, ATP, $H^+[e]$ and H^+ under maximal yield conditions. Prepared by Jeff Orth.

- ATP is made in this process. 1.5 ATP in glycolysis (−0.5 PFK, +1.5 PGK, + 0.5 PYK) and 1.5 ATP is generated while producing acetate. Thus, 3 ATP are consumed by ATP synthase (ATPS4r) and a total of 13.5 external protons are generated.

Relaxing the ATP constraint leads to a totally different way to stoichiometrically optimally make NADH anaerobically.

Recap   By studying these simplest of optimal solutions of the core *E. coli* network, we discover that the network appears deceptively simple. It can produce many different flux maps to satisfy different demands. It is often hard to interpret the solutions. Notice how much the proton balance complicates this task. All components of a cell have to balance while it performs a particular function. The constraints that the components put on each other are substantial, and as discussed in Chapter 17, it is difficult to satisfy them all simultaneously.

19.4   Capabilities of Genome-scale Models

The concepts and approaches discussed in this chapter can be applied to many other questions and are scale-independent. We include just two examples here. Many more examples are found throughout the text.

Functional testing of Recon 1   The generation of genome-scale models includes a validation phase (Chapter 3). The known metabolic capabilities of an organism are described with objective functions. The flux through these objective functions is then maximized for a given input. If the flux is positive, then the reconstruction passes the test; if there is no flux, then the reconstruction fails the test. In the latter case, gap-filling methods are used to determine what is missing (see Chapter 25). During the development of the RECON 1 genome-scale reconstruction of human metabolism, 288 functional tests were defined to diagnose the functionality of the reconstruction [100].

Simultaneous demands   The cases studied above are focused on only one function of a network. In physiological states, networks are performing many functions simultaneously. The detailed analysis of maximal co-factor and biosynthetic precursor production from glucose above illustrates the difference between component and systems thinking in biology. Nothing happens in isolation in the cell. All events are coupled to myriad other cellular processes and more complex objective functions are covered later (see Chapter 21).

19.5   Summary

- The capabilities of a network can be determined by simple optimization methods. These capabilities are dependent on the inputs and outputs, which, in this chapter, are the available substrates and electron acceptors.

- Optimal states can be computed and interpreted to achieve an understanding of such network functions. The governing constraints are often counterintuitive if one focuses only on a particular network function, and does not consider the effects that a particular process has on the system as a whole.
- One network can have many states and many functions. These functions have governing constraints, such as energy, proton trafficking, or stoichiometry.
- Optimal solutions can be obtained for single functions in isolation, or many simultaneous functions can be ascertained. The former can be used to diagnose a reconstruction, while the latter can be used to study physiological states.

20 Equivalent States

Living machines are not intelligently designed and will often be redundant and overly complex – Sir Paul Nurse

A reconstructed network can have many functional states even in the same environment. Constraint-based optimization problems can have multiple equivalent solutions that yield the same numerical value of the objective that is being optimized. This issue is an important one in characterizing networks and is a common occurrence in large-scale networks. In this chapter, we will cover the methods that have been developed to study what are called *alternative optimal solutions* (AOS). We demonstrate them by applying them to the core *E. coli* model, and discuss illustrative studies and issues that arise with finding AOS at the genome-scale.

20.1   Equivalent Ways to Reach a Network Objective

An historical note   Even for explorations of network properties, such as those illustrated in the last chapter, alternative optimal solutions with the same optimal value of the objective function are found. This issue arose in the early days of the development of FBA approaches when two solutions for the formation of E4P were discovered (see Figure 20.1). The optimization that was being performed was to compute the maximal yield of the biosynthetic precursor E4P from glucose as a substrate. Two different solutions were computed that gave a yield of 1.33 molecules E4P per molecule glucose, or an 88.7% carbon conversion. The two solutions differ in the utilization of the TCA cycle or the glyoxalate shunt to achieve the same value of the objective.

The existence of AOS is a fundamental issue in COBRA methods and understanding their biological implications and relevance is important. Thus, a fair amount of effort has been devoted towards computing and studying AOS.

Characterizing alternative optimal states   As outlined in the last chapter, LP can be used to find single optimal solutions. The solution to this optimization problem may not be unique, as illustrated in Figure 20.1, and AOS may exist. Various methods have been developed to characterize AOS. These methods are illustrated in Figure 20.2 and are described in this chapter.

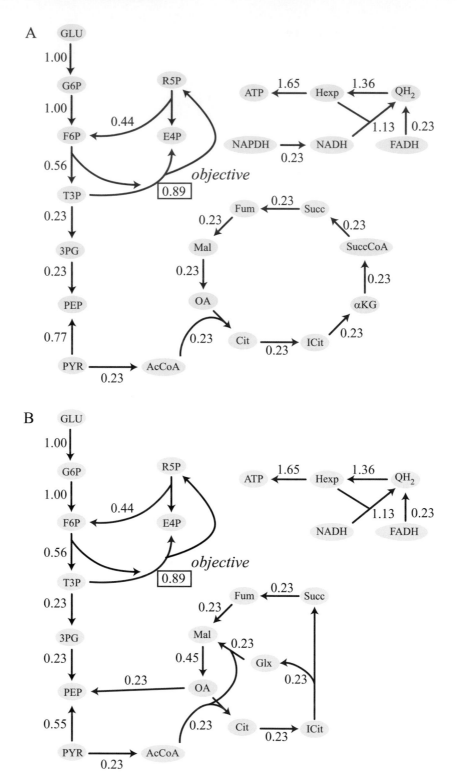

Figure 20.1 Two flux distributions for an historical core metabolic network for *E. coli* when the production of the biosynthetic precursor E4P is maximized from glucose. Taken from [439].

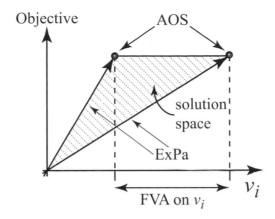

Figure 20.2 Illustration of alternative optimal solutions (AOS). An objective Z is formulated to define a functional state of a network of interest. The Z-value is put on the y-axis. In this illustration, v_i is a flux variable between the two AOS shown and has a range of values, called the flux span. There are two extreme pathways that will have the same value of the objective function. The two vertices are two AOS that can be enumerated using MILP (red dots). Note that LP will find one particular optimal solution.

Flux variability analysis (FVA):   this approach is used to find the allowable ranges of values for all the fluxes in a solution while the optimal performance as defined by the stated objective function is met. The space of AOS can be characterized in this way. This approach is computationally manageable and is highly scalable.

Enumeration of extreme pathways: extreme pathways are the vertices in the closed null space. They not only find optimal solutions, but all the extreme points. Thus, more states are found through the computation of extreme pathways than just those that are optimal. Once such a calculation is done, one can rank-order the extreme pathways by yield (i.e., the objective function) to find all the extreme pathways that have the highest numerical value of the objective function.

Direct enumeration of AOS: a MILP approach can be used to find the AOS. This approach is computationally intensive and, in principle, can enumerate all the AOS. In practice, when the number of AOS is large, the computation needs to be truncated at a certain point.

The outcome of applying these three methods to study the variability in the optimal production of the biosynthetic precursors is shown in Table 20.1. Note that the space of AOS can also be characterized by randomized sampling (Chapter 14).

The existence of AOS is a network property and represents a distinguishing feature of the *in silico* modeling of phenotypes. In comparison, solutions sought in the physico-chemical sciences are typically unique (recall Table 16.1). One can think of the biological counterpart to the AOS as being *silent phenotypes*; that is, there is actually a difference in the functional states but is hard or even impossible to determine experimentally (see Figure 15.6).

Table 20.1   Flux variability analysis for the core *E. coli* model for optimal production of the biosynthetic precursors. Prepared by Jeff Orth.

Metabolite	Yield	Carbon conversion (%)	Number of variable fluxes	Number of extreme pathways	Number of AOS
3PG	2	100	23	10	10
PEP	2	100	23	10	10
Pyr	2	100	23	69	69
OA	2	133	23	10	10
G6P	0.89	89	13	2	2
F6P	0.89	89	13	2	2
R5P	1.06	88	13	2	2
E4P	1.30	87	13	2	2
G3P	1.68	84	13	2	2
AcCoA	2	67	35	69	138
αKG	1	83	35	31	>1000
SuccCoA	1.64	109	2	2	1

Alternate optimal network states are a common property of genome-scale networks. The number of AOS varies depending on the size of the network, the chosen objective, and the environmental conditions. In general, the larger and more interconnected the network, the higher the number of AOS found.

20.2   Flux Variability Analysis

20.2.1   The concept

Definition   For a given optimal state (Z_{opt}), one can find the maximum and minimum allowable flux that a particular reaction can have while still supporting that optimal network state. This range is computed for each flux v_i of interest by solving two LP problems [256]: see Box 20.1. The computations require three constraints: the flux balance, the capacity constraints, and a constraint that fixes the value of the original objective function at its optimal value. To find the minimum value for v_j, the corresponding minimization problem is solved, and conversely the maximum value of v_j is found through its maximization. This procedure gives the range of the allowable numerical values for v_j as illustrated in Figure 20.2. One should note that if there are type III pathways in the network under study, then a flux can take on any value. Thus, such pathways have to be removed from consideration.

Characterization of results   FVA thus requires $1 + 2n$ LP problems to be solved. First, to find the optimal value of Z, and then two optimizations (min and max) on each of the n fluxes in the network. The results can be presented in a bar chart-type form

---

Box 20.1 Flux variability analysis (FVA) [256]

Concept: At a fixed objective function value (e.g., growth rate), determines the allowable ranges for all other fluxes.

Objective: max v_i and min v_i

Constraints: Flux balance $\mathbf{S} \cdot \mathbf{v} = 0$

Capacity constraints: $0 \leq v_j \leq v_{j,\text{max}}$, $j = 1, \ldots n$

Specific condition: $\langle c^T, v \rangle = Z$ (e.g., fixed growth)

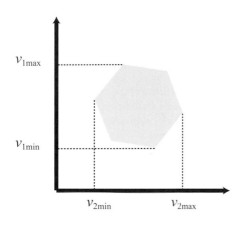

A polytope illustrating the range of fluxes with the same value of the objective function.

(Figure 20.3). All the reactions in the flux vectors are ordered on the x-axis, and on the y-axis we plot the maximum and the minimum value of each of the fluxes. This representation can be thought of as the *flux variability spectrum*. The difference between the maximum and the minimum value in each bar is sometimes called the *flux span*. The average of all flux spans in a network is sometimes used as a measure of flexibility of the network.

Dependency of the flux variables FVA results must be interpreted with care as the flux variables are treated as being independent. FVA thus draws a hyperbox around the space of AOS (Figure 20.3). One specific example of this dependency is illustrated in Figure 20.3. The flux variability spectrum is shown in the insert in the figure. The flux span for v_1 is $3 - 1 = 2$ and for v_2 it is $5 - 2 = 3$.

Note, however, that when flux v_2 is maximized to a value of 5 (open circle at the top of the pentagon), then flux v_1 takes on a single value of 2, while its flux span is 3. Conversely, when flux v_2 is minimized to a value of 2 (solid line at the bottom of the pentagon), then flux v_1 can only take on values between 1.5 and 2.5, while its flux span goes from 1 to 3.

Thus, one might get a false impression of what the flux variability actually is in a network. One way of dealing with this limitation is to do a secondary FVA fixing the value of a particular flux of interest and then determine the FVA for the remaining

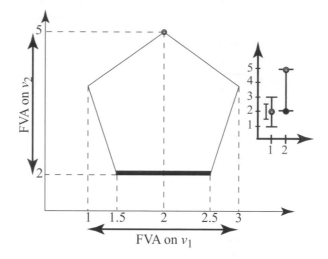

Figure 20.3 Illustration of the interaction of the flux spans of two flux variables. The pentagon represents a solution space and the insert represents the flux variability spectrum. The open and closed circles and the small side bar on the flux span for v_1 represent shrunken flux ranges of secondary fluxes once the primary one is fixed at an extreme.

fluxes. This is illustrated for the minimal value of v_2 in Figure 20.3 and the solid dot and shorter line in the flux variability spectrum.

The dependencies between flux variables in the confined null space can be quite intricate. A number of more sophisticated methods are available to characterize such dependencies, such as the flux coupling finder (FCF) [56], and randomized sampling of allowable solutions followed by the computation of correlation coefficients (see Chapter 14).

20.2.2   Flux variability in the core *E. coli* model

We can demonstrate the FVA procedure by choosing the production of a biosynthetic precursor from Table 20.1. We choose 3PG as an example. It has an optimal yield of 2 from glucose. Fixing Z at the optimal yield and performing the procedure in Box 20.1 generates the range of fluxes allowable under optimal conditions (Figure 20.4).

There are 23 variable fluxes of the 95 total (not including the variable fluxes through SUCDi and FRD7, which form a type III pathway), the rest are fixed. The variable fluxes fall into two categories. First there are fluxes that are flexible, but are always non-zero. They are needed for the production of 3PG. The second category are fluxes that can be zero or finite. They are not needed to produce 3PG, but can be utilized.

An easy way to interpret the flux variability spectrum is to show it on a reaction map (Figure 20.5). This map, for instance, shows us that the glycolysis–pentose split is fixed; there is no flux into the pentose pathway from the imported glucose and all of it flows down glycolysis through the isomerization reaction catalyzed by *pgi*. The pyruvate formate lyase (PFL) reaction is always used, but its flux can vary between 1 and 1.5.

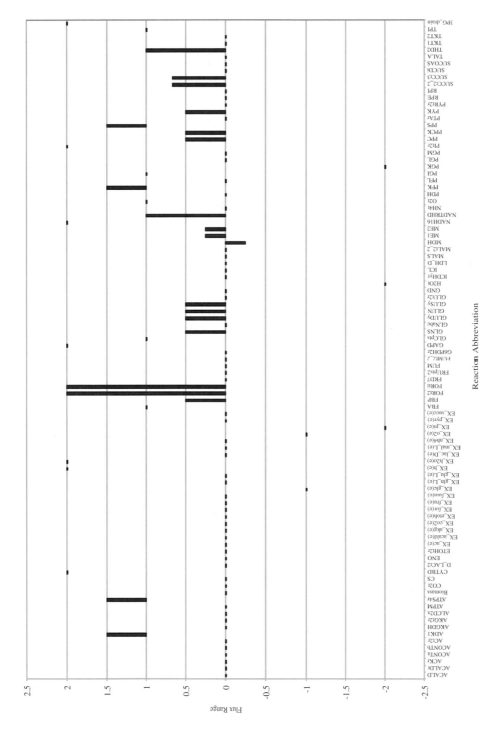

Figure 20.4 Flux spans for the optimal production of the biosynthetic precursor 3PG from glucose in the core metabolic *E. coli* model. Prepared by Jeff Orth.

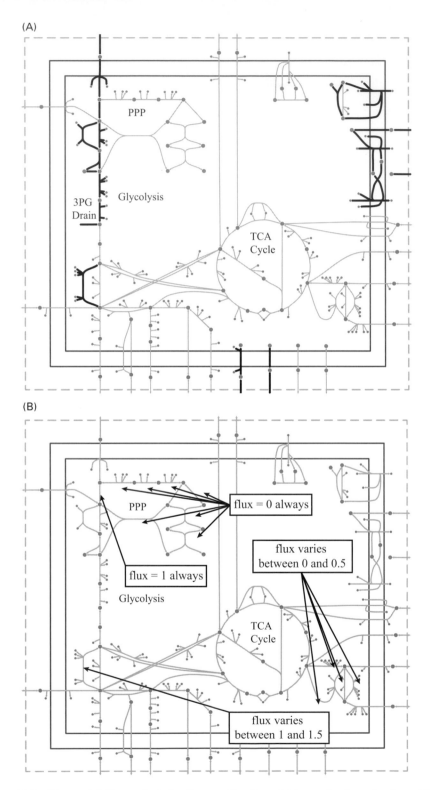

Figure 20.5 Flux variability analysis for the core *E. coli* model for optimal production of 3PG shown on a reaction map. Panel A: the optimal solution that has the fewest active fluxes. The 3PG drain arrow designates an added 3PG output for the yield calculations. Panel B shows some of the variable fluxes. This data representation puts the flux variability in an easy to interpret context.

The flux variations are associated with alternative ways to burn the excess ATP made. Basically, the glycolytic production of two 3PG comes with the net production of one ATP and one NADH and one pyruvate. To balance the map the pyruvate has to be recycled to make PEP that costs 2 ATP.[1] One ATP comes from the glycolytic 3PG production; to make the other ATP, the electrons on NADH go down the ETS to produce 1.25 ATP, resulting in a net excess of 0.25 ATP. This excess ATP can be 'burned' by several alternative futile cycles giving arise to the flux variability and alternative optimal solutions.

The core *E. coli* metabolic network provides a simple system where FVA computations are easily performed and the results interpreted. The reader can pick another biosynthetic precursor and repeat these computations, data visualization, and interpretation, see [299].

20.2.3   Genome-scale results

FVA has been used in a number of genome-scale studies. We give some examples below.

Equivalent reaction sets: the first FVA study [256] identified the redundancies in a network, called *equivalent reaction sets*, that lead to AOS. These are sets of reactions that vary between the AOS. The equivalent reaction sets in the *E. coli* iJR904 model are shown graphically in Figure 20.6. The study also demonstrated the extent of flux variability for *E. coli* growth caused by these reaction sets to be highly dependent on the nutrients that the cell grows on.

Input/output analysis: FVA can be used in the context of an input–output analysis such as for a fermentation process. Metabolic by-product secretion profiles for yeast have been calculated as a function of increasing oxygen uptake rates [101]. Because AOS exist, the flux span of secretion rates can be found among them for a fixed growth rate. Remarkably, there was less than 1% difference between the maximum and minimum allowable secretion rates for a fixed maximal growth rate. This lack of variability increases the focus of the predicted secretion rates.

Metabolic demands: diverse data sets available for human mitochondria have led to the reconstruction of their metabolic networks [446]. The capabilities of these metabolic network to fulfill three key metabolic functions in mitochondria (ATP production, heme synthesis, and mixed phospholipid synthesis) have been studied using FVA. The results show that the network has high flexibility for the biosynthesis of heme and phospholipids but modest flexibility for maximal ATP production. A subset of all of the AOS, computed with respect to the three metabolic objectives individually, was found to be highly correlated, suggesting that this set may contain physiological meaningful fluxes.

Energy needs for growth: the flexibility of a genome-scale metabolic network in *Lactobacillus plantarum* has been studied using FVA, with and without the coupling of ATP production to cellular growth in a chemostat [417]. When the ATP production

---

[1] Recall that glucose is imported in enterobacteria via the PTS system that converts PEP to Pyr as glucose is phosphorylated to G6P on entry to the cell.

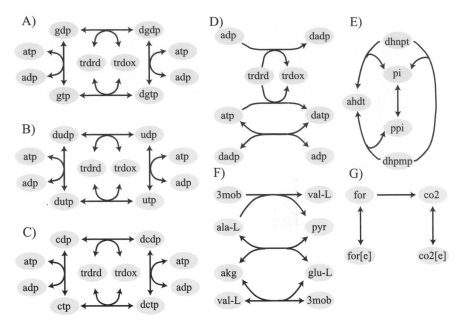

Figure 20.6 The equivalent reaction sets in the *iJR904* genome-scale *in silico* model for *E. coli* for growth on glucose. (A–D) The equivalent reaction sets in the nucleotide metabolism. (E) Diphosphate phosphohydrolase. (F) Transamination reactions among pyruvate, α-ketoglutarate, valine and oxoisovalerate. (G) Formate dehydrogenase reactions (note that this equivalent reaction set does not represent redundancy and that the protons and water produced/consumed are not incorporated in the model reactions. This set implies that formate and CO_2 are equivalent sinks). Metabolites: trdrd: reduced thioredoxin; trdox: oxidized thioredoxin; ahdt: 2-amino-4-hydroxy-6-(erythro-1-2-3-trihydroxypropyl) dihydropteridine; dhnpt: dihydroneopterin, dhpmp: dihydroneopterin phosphate; 3mob: oxoisovalerate; ala: alanine, glu: glutamate; akg: α ketoglutarate. Taken from [256].

is made equal to that for ATP demand for growth, the flux variability spectrum is much narrower than without the ATP growth demand (Figure 20.7). This result again illustrates the constraints that cellular processes put on each other when considered simultaneously.

20.3   Extreme Pathways and Optimal States

20.3.1   The concept

A subspace of AOSs is spanned by a set of extreme pathways. As described in Chapter 12, extreme pathways are vertices in the flux solution space and thus can represent optimal solutions. Therefore, one can compute all the extreme pathways, and rank-order them based on the objective of interest, such as yield of a metabolite of interest from a given substrate. Thus, the extreme pathways that have the same yield represent AOS. In fact, any non-negative linear combination of them:

$$Z_{opt} = \Sigma \alpha_i \mathbf{p}_i \quad \text{where} \quad \alpha_i \geq 0 \quad \text{and} \quad \Sigma \alpha_i = 1 \tag{20.1}$$

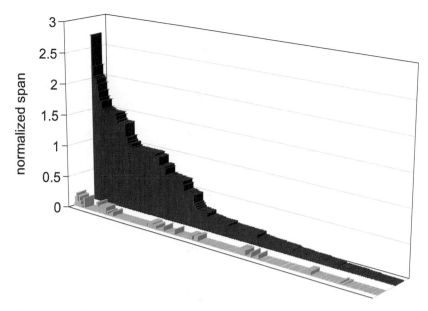

Figure 20.7 Flux variability in *Lactobacillus plantarum* with (green) and without (blue) coupling of ATP to the growth demand. The normalized flux spans were rank-ordered based on the computations without the coupling. Taken from [417].

will also be an optimal solution. Thus, there is a space of AOS that is spanned by these convex basis vectors.

Extreme pathways can be computed using a variety of approaches [38, 453, 465]. However, in general, it is difficult to compute extreme pathways at the genome-scale due to a combinatorial explosion, that is, in part, due to the occurrence of equivalent reaction sets.

20.3.2   Extreme pathways in the core *E. coli* metabolic network

The extreme pathways from glucose as an input to 3PG as an output can be computed. In these computations, all outputs can be active but 3PG must be active. Thus, we compute only extreme pathways that lead to 3PG production. These extreme pathways can be rank-ordered based on 3PG yield from glucose and displayed as a bar chart (Figure 20.8). There are 10 extreme pathways that have the same maximum yield of 2 mole 3PG per mole of glucose. These 10 vectors span the space of all AOS.

These extreme pathways can be graphically displayed on the reaction map (Figure 20.9). We can visualize the pathways. Inspecting them shows us where the main variability is. All the extreme pathways have the same basic structure but variations in the way the excess ATP is burned. This extreme pathway structure is consistent with the FVA of this same situation in the previous section.

20.3.3   Genome-scale results

Extreme pathways are difficult to compute on a genome-scale, but a few studies have appeared. We describe two examples.

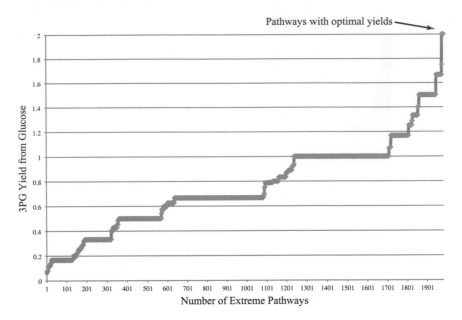

Figure 20.8   The generation of 3PG from glucose in the core *E. coli* model. The molar yield of 3PG on glucose (mol/mol) of all 1984 3PG producing extreme pathways are shown. There are 10 extreme pathways with a yield of 2. Prepared by Jeff Orth.

Understanding multiple simultaneous network functions.   The extreme pathways in the human red blood cell metabolic network were calculated and interpreted in a biochemical and physiological context [453]. The extreme pathways were divided into groups based on such criteria as their co-factor and by-product production, and carbon inputs including those that (1) convert glucose to pyruvate; (2) interchange pyruvate and lactate; (3) produce 2,3-diphosphoglycerate that binds to hemoglobin; (4) convert inosine to pyruvate; (5) induce a change in the total adenosine pool; and (6) dissipate ATP. This set of extreme pathways helped to develop an understanding of how the red cell meets its various metabolic demands.

Comparing systems properties of two networks.   The genome-scale metabolic networks reconstructed from two genomes of similar size have been compared. *Helicobacter pylori* and *Haemophilus influenzae* have genome sizes of 1.667 Mb and 1.830 Mb, respectively, and their metabolic networks have been reconstructed [105, 381]. All the extreme pathways that have the same metabolic inputs and outputs were computed for both organisms [343]. The outputs included in the comparison were acetate, succinate, and the individual nonessential amino acids common to both organisms. Even with similarly sized genomes, there was a vastly different degree of input/output redundancy between these organisms. There was an average of two extreme pathways with equivalent external states in *H. pylori*. The same outputs for *H. influenzae* resulted in an average of 46 extreme pathways per external state. Although the differences in the minimal medium requirements for each of the organisms make direct comparisons difficult, this difference

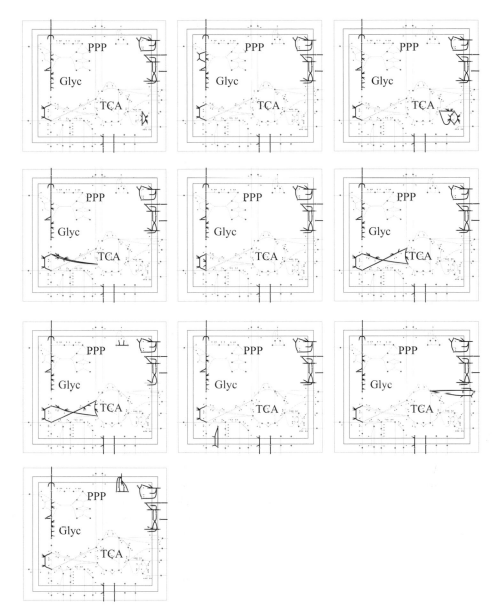

Figure 20.9 A panel of reaction map representations of the 10 ExPas that have an optimal yield of two 3PG molecules per molecule of glucose. Prepared by Jeff Orth.

in pathway redundancy does indicate a much more rigid metabolic network architecture for *H. pylori* than for *H. influenzae.*

20.4   Enumerating Alternative Optima

The concept   The basic LP problem gives a single optimal solution. Methods used to compute alternate optima involve MILP [226, 333]. A MILP-based approach can

---

Box 20.2 Enumeration of alternate optimal solutions (AOS) [359]

Concept: MILP algorithm to identify all the alternate optimal solutions for a cellular objective, using metabolic networks

Objective: $Z = \max < c^T, v >$

Constraints: Flux balance $S \cdot v = 0$

Capacity constraints: $\alpha \leq v_i \leq \beta$

Change of basis: $\sum_{i \in NZ^{K-1}} y_i \geq 1$

Generate all alternate bases:

$\sum_{i \in NZ^k} w_i \leq |NZ^k|, k = 1, 2, \dots, K-1$

$y_i + w_i \leq 1$ for all i

$\alpha w_i \leq v_i \leq \beta w_i$ for all i

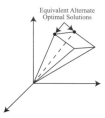

Equivalent Alternate
Optimal Solutions

---

be implemented iteratively by adding additional constraints; so-called *integer cut constraints*. This algorithm is illustrated in Box 20.2. The outcome of such calculations is the identification of all extreme points where the value of the objective function is identical (see Figure 20.2). Any point on the boundary in between such extreme points is also an equivalent optimal solution.

Enumerating AOS in the core *E. coli* network   The number of AOS are the same as the number of extreme pathways (Table 20.1), except in two cases: AcCoA and αKG. In those two cases there are intermediate vertices that are not extreme pathways but convex combinations of them.

Genome-scale studies   The algorithm in Box 20.2 was used to calculate and study a subset of the AOS for the *i*JR904 metabolic model of *E. coli* under a wide variety of environmental conditions [359]. Analysis of the calculated sets of AOS found that: (1) only a small subset of metabolic reactions have variable fluxes across the AOS, (2) sets of reactions that are always used together in AOS (i.e., the correlated reaction sets, Chapter 13) showed moderate agreement with the transcriptional regulatory network structure in *E. coli* and mRNA expression data, and (3) reactions that are used under certain environmental conditions can provide clues about network regulatory needs. In addition, calculation of suboptimal flux distributions, using FVA, identified reactions which are used under significantly more environmental conditions suboptimally than optimally. Together these results demonstrated the use of FVA and enumeration of alternate AOS to determine the utilization of metabolic reactions under a variety of different growth conditions. Extreme pathways cannot be computed for this network.

20.5 Summary

- Constraint-based optimization problems can have non-unique solutions, or so-called degenerate solutions. This feature leads to equivalent network states that are called alternative optimal solutions (AOSs).
- The number of AOSs tends to grow with network size.
- AOSs have a biological relevance and are related to the concept of silent phenotypes.
- Three methods used to characterize AOSs are described in this chapter: (1) FVA that gives the allowed variation in the fluxes under optimal conditions, (2) extreme pathways computations, and (3) direct enumeration based on MILP procedures.
- Each one of these methods has been found to have utility for a variety of basic and applied problems as illustrated with the examples given.

21 Distal Causation

Causality in biology is a far cry from causality in classical mechanics
– Ernst Mayr

COBRA methods require the statement of an objective. Objective functions are used to perform a search over the space of possible solutions to find particular functional states. There are many types of objective functions that can be used to find different types of functional states that reconstructed networks can attain. Perhaps the most interesting group are the objective functions that are those used to represent expected physiological functions. From the early days of development of COBRA methods, the objective function has been the center of attention, and attempts were made to determine appropriate objective functions in a data-driven fashion [372]. Importantly, objective functions can be used to describe selection pressures and thus distal causation.

21.1 The Objective Function

Dual causality can be represented conceptually in a plane with two axes (Figure 21.1). Proximal causation is represented on the y-axis. It represents the response of an organism to environmental stimuli against a constant genetic background. For example, the diauxic shift in substrate uptake from glucose to acetate upon depletion of glucose when *E. coli* is grown anaerobically. Given a constant genetic background, this response is dominated by physico-chemical factors.

Distal causation is represented on the x-axis. These changes take place over multiple generations where the genotype is changing, such as in a population of cells optimizing growth rate through acquisition of advantaged mutations. Applying the same environmental stimulus at each generation may lead to different proximal responses.

The processes of generation of diversity and selection drive phenotypic changes over generations. The selection process is hard to describe based on fundamental principles. However, an *objective function* can be used to describe the driving force for this change. Because we do not always know the driving force, we formulate the objective function based on intuition or guesses about the evolutionary history of the organism. Adaptive laboratory evolution (see Chapter 26) has risen over the

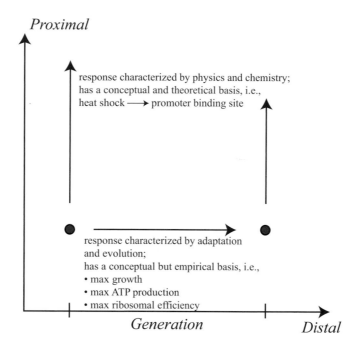

Proximal

response characterized by physics and chemistry;
has a conceptual and theoretical basis, i.e.,
heat shock $\longrightarrow$ promoter binding site

response characterized by adaptation
and evolution;
has a conceptual but empirical basis, i.e.,
• max growth
• max ATP production
• max ribosomal efficiency

Generation

Distal

Figure 21.1   A schematic of dual causation. In this conceptual representation, motion in the vertical direction is driven by physical phenomena, while biological selection forces drive motion in the horizontal direction. The proximal response can change with subsequent generations and is indicated with the change in the vertical arrow.

past decades that allows us to begin to study the process of evolution experimentally [228, 316].

21.2   Types of Objective Functions

Mathematical representation   The mathematical form of the objective function is:

$$Z = \langle \mathbf{w}, \mathbf{v} \rangle \tag{21.1}$$

where $\mathbf{w}$ is a vector of weights, and $\mathbf{v}$ is the vector of network fluxes. The vector $\mathbf{w}$ is specified by the user and represents the network function that one wishes to optimize. This network function can be an internal biological reaction of interest or an intricate equation estimating a biological objective.

Types of objective functions Z allows the representation of a wide variety of biological functionalities. The objective functions that have been used in COBRA studies fall roughly into three categories:

1   exploration of network properties (see Chapter 19),
2   physiological functions, and
3   bioengineering objectives (see Chapter 27).

Examples of objective functions   A number of different objective functions have thus been used to analyze metabolic networks. Specific examples include the following.

Minimize ATP production: this objective is stated to determine conditions of optimal metabolic energy efficiency, and has been used to study the properties of the mitochondrion [353, 446].

Minimize nutrient uptake: this objective has been used to study conditions under which the cell performs a particular metabolic function while consuming the minimum amount of available nutrients. This objective has been used to study yeast cultures grown in chemostats [109].

Minimize the Manhattan (absolute) norm of the flux vector: this objective can be applied to satisfy the strategy of a cell to minimize the sum of the flux values, or, in other words, to channel metabolites through the network using the lowest overall flux [170, 234]. Using the Euclidean norm of the flux vector would lead to *quadratic programming* (QP) [386].

Maximize metabolite production: this objective function has been used to determine the biochemical production capabilities of a particular cell, such as the maximal production rate of a chosen metabolite (i.e., lysine or phenylalanine). It also has been used to study the biochemical production capabilities of many organisms (e.g., [436]).

Maximize biomass formation: this objective has been widely used to determine the maximal growth rate of a cell in a given environment [104, 177]. We will discuss this objective function in detail in Section 21.3.

Maximize biomass and metabolite production: by weighing conflicting objectives appropriately, one can explore the tradeoff between cell growth and forced metabolite production in a production strain [58, 334, 335].

Physico-chemical properties: objective functions that take thermodynamic and kinetic properties into consideration have been formulated [170].

A set of detailed examples is given in Table 21.1. The table shows the variety of functions that can be represented and their mathematical forms.

21.3   Producing Biomass

From experimentally determined cellular composition to objective functions   Growth can be represented through experimental determination of the requirements to synthesize cellular biomass. Such requirements are determined based on measured values of biomass composition. Thus, biomass generation can be a linked set of reaction fluxes that drain metabolites in appropriate ratios. The summation of such requirements can be represented as an objective function. This concept is illustrated in Figure 21.2, which shows a schematic of a cell where a variety of substrates can enter the cell to produce all the compounds that comprise cellular macromolecular structures.

Table 21.1 Objective functions implemented in COBRA. From [386].

Objective function	Mathematical definition	Explanation	Rationale
Max biomass	$\max\dfrac{v_{\text{biomass}}}{v_{\text{glucose}}}$	Maximization of biomass yield	Evolution drives selection for maximal biomass yield ($Y_{X/S}$)
Max ATP	$\max\dfrac{v_{\text{ATP}}}{v_{\text{glucose}}}$	Maximization of ATP yield	Evolution drives maximal energetic efficiency ($Y_{ATP/S}$)
Min$\sum v^2$	$\min\sum\limits_{i=1}^{n} v_i^2$	Minimization of the overall intracellular flux	Postulates maximal enzymatic efficiency for cellular growth (analogous to minimization of the Euclidean norm)
Max ATP per flux unit	$\max\dfrac{v_{\text{ATP}}}{\sum\limits_{i=1}^{n} v_i^2}$	Maximization of ATP yield per flux unit	Cells operate to maximize ATP yield while minimizing enzyme usage
Max biomass per flux unit	$\max\dfrac{v_{\text{biomass}}}{\sum\limits_{i=1}^{n} v_i^2}$	Maximization of biomass yield per flux unit	Cells operate to maximize biomass yield while minimizing enzyme usage
Min glucose	$\min\dfrac{v_{\text{glucose}}}{v_{\text{biomass}}}$	Minimization of glucose consumption	Evolution drives selection for most efficient usage of substrate
Min reaction steps	$\min\sum\limits_{i=1}^{n} y_i^2, y_i \in \{0,1\}$	Minimization of reaction steps	Cells minimize number of reaction steps to produce biomass
Max ATP per reaction step	$\min\dfrac{v_{\text{ATP}}}{\sum\limits_{i=1}^{n} y_i^2, y_i \in \{0,1\}}$	Maximization of ATP yield per reaction step	Cells operate to maximize ATP yield per reaction step
Min redox potential	$\min\dfrac{\sum\limits_{n} v_{\text{NADH}}}{v_{\text{glucose}}}$	Minimization of redox potential	Cells decrease number of oxidizing reactions, thus conserving their energy or using their energy in the most efficient way possible
Min ATP production	$\min\dfrac{\sum\limits_{n} v_{\text{ATP}}}{v_{\text{glucose}}}$	Minimization of ATP producing fluxes	Cells grow while using the minimal amount of energy, thus conserving energy
Max ATP production	$\max\dfrac{\sum\limits_{n} v_{\text{ATP}}}{v_{\text{glucose}}}$	Maximization of ATP producing fluxes	Cells produce as much ATP as possible

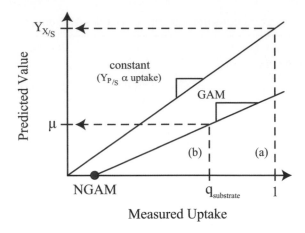

Figure 21.2   The growth rate and product yield as a function of absolute or relative glucose uptake rate, respectively. The two cases shown correspond to: (a) biomass synthesis, where $Y_{X/S}$ is the biomass yield; and (b) growth rate, i.e., biomass synthesis rate with simultaneous fulfillment of the maintenance requirements (GAM and NGAM), where μ is the growth rate and $q_{substrate}$ is the uptake rate. Prepared by Adam Feist.

The biomass objective functions are specific for a given cell type. For example, a Gram-negative cell will contain less murein and more lipids (including lipopolysaccharide) than will a Gram-positive cell. As such, simulations using a similar metabolic core model to estimate biomass production in a Gram-positive cell will draw less acetyl-CoA for lipid production and more fructose-6-phosphate for murein production.

In animals, the composition of different cell types can vary drastically. As an example, stem cells tend to be small and have a high nuclear to cytoplasmic ratio, while adipocytes are large and have a high lipid content (Chapter 6).

Biomass synthesis from the core metabolic network   The requirements for making one gram of *E. coli* biomass from key co-factors and biosynthetic precursors have been estimated [282]. Thus, for *E. coli* to grow, all these components must be provided in the appropriate relative amounts. Their relative requirements to make one gram of *E. coli* biomass are given in Figure 21.3. These numbers are put into the **w** vector in the equation for the objective function:

$$Z_{precursors} = \langle \mathbf{w}, \mathbf{v} \rangle \tag{21.2}$$

The experimentally determined biomass composition of a cell thus serves to define the weight vector **w** in this objective function. The optimization problem then maximizes the value of $Z_{precursors}$ so that all these precursors are drained in the appropriate ratios from a given substrate that enters the network.

Cost of driving biosynthesis   In addition to the material that is needed to form the biomass of the cell, charged co-factors are needed to drive the biosynthesis process.

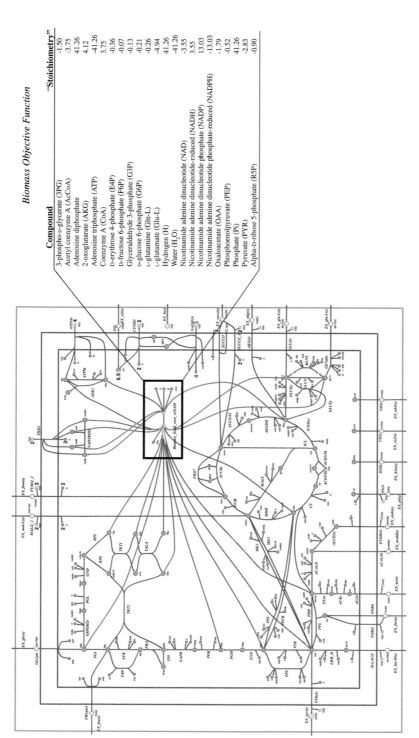

Biomass Objective Function

Compound	"Stoichiometry"
3-phospho-D-glycerate (3PG)	-1.50
Acetyl coenzyme A (AcCoA)	-3.75
Adenosine diphosphate	41.26
2-oxoglutarate (AKG)	4.12
Adenosine triphosphate (ATP)	-41.26
Coenzyme A (CoA)	3.75
D-erythrose 4-phosphate (E4P)	-0.36
D-fructose 6-phosphate (F6P)	-0.07
Glyceraldehyde 3-phosphate (G3P)	-0.13
D-glucose 6-phosphate (G6P)	-0.21
L-glutamine (Gln-L)	-0.26
L-glutamate (Glu-L)	-4.94
Hydrogen (H)	41.26
Water (H₂O)	-41.26
Nicotinamide adenine dinucleotide (NAD)	-3.55
Nicotinamide adenine dinucleotide-reduced (NADH)	3.55
Nicotinamide adenine dinucleotide phosphate (NADP)	13.03
Nicotinamide adenine dinucleotide phosphate-reduced (NADPH)	-13.03
Oxaloacetate (OAA)	-1.79
Phosphoenolpyruvate (PEP)	-0.52
Phosphate (Pi)	41.26
Pyruvate (PYR)	-2.83
Alpha-D-ribose 5-phosphate (R5P)	-0.90

Figure 21.3 Biomass synthesis as an input–output problem subject to constraint-based optimization. The figure shows the core metabolic network of *E. coli* and the production of the 12 biosynthetic precursors. As described in the text, there are additional co-factor and maintenance requirements that need to be met. The relative amounts of the precursors given in the insert are estimated from experimentally determined biomass composition [296]. These precursors need to be produced in the ratios given in Equation (21.4).

The energy-carrying co-factor requirement to synthesize the major monomers of the cell from the precursors (amino acids, fatty acids, nucleic acids) and to polymerize them into macromolecules is estimated to be:

$$Z_{cofactors} = 41.26\,v_{\text{ATP}} + 3.547\,v_{\text{NAD}} + 13.03\,v_{\text{NADPH}} \tag{21.3}$$

To balance the equations the corresponding amount of ADP, NADP, and NADH must be produced. Protons and inorganic phosphate are also produced as ATP is hydrolyzed with the consumption of water. Note that the biosynthetic reactions generate net redox potential in the form of NADH and thus the co-factor NAD has to be consumed in the formation of biomass.

Computing optimal biomass yields   The precursor and co-factor requirements to generate *E. coli* biomass are given by:

$$Z_{\text{biomass}} = Z_{\text{precursors}} + Z_{\text{co-factors}} \tag{21.4}$$

These requirements are numerically summarized in the insert in Figure 21.3. If Z_{biomass} is maximized from an input of 1 mmol glucose per g DW (grams dry weight) per hour, we get the flux distribution that is shown in Figure 21.4. To compute the mass yield we have to compute the mass flow of glucose in using its molecular weight of 180 g/mol, which then corresponds to an uptake rate of 0.18 g Gluc/g DW/h. The mass yield then is 0.55 g DW/g Gluc. Note that the growth rate is proportional to the uptake rate, but the yield is always the same. Thus, the time ordinate is arbitrary in this simulation and is set by the uptake rate that one uses in the computations.

Maintenance energy requirements   Biomass maintenance requirements have to be accounted for to simulate realistic growth situations. Energy is used to meet both growth-associated (GAM) and non-growth-associated maintenance (NGAM) requirements. These requirements are estimated to be 7.6 mmol ATP/g DW/h and 13.0 mmol ATP/g DW, respectively, for *E. coli* [435]. The former represents use of ATP that is proportional to the biomass being produced, while the latter is a constant drain that needs to be satisfied even in the absence of growth. This distinction is important, as it leads to the introduction of time into the computations and will distinguish between yield and growth rate computations.

The growth-associated maintenance requirement is accounted for by changing the ATP requirements to 54.26 mmol/g DW ($= 41.26 + 13.0$),

$$Z_{\text{growth}} = -54.26v_{\text{ATP}} - 3.547v_{\text{NAD}} - 13.03v_{\text{NADPH}} + Z_{\text{precursors}} \tag{21.5}$$

while the non-growth-associated maintenance term is added as a constant drain and is included in the constraints during the optimization. The effects of the maintenance energies is shown in Figure 21.2 and discussed below. The computed growth rate with the maintenance energies using an uptake rate of 10 mmol glucose per g DW per hour (a representative experimental value) is 0.87/h, or a doubling time of 0.80 h ($= \ln(2)/0.87$), or 48 min.

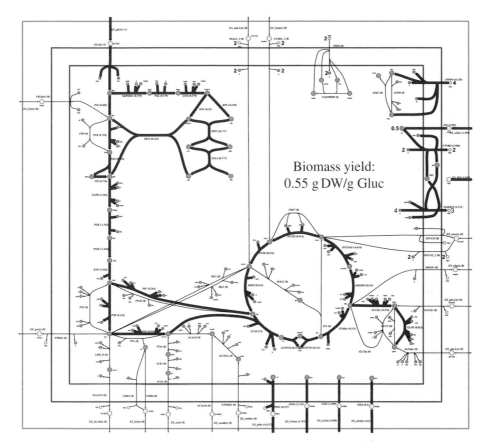

Figure 21.4   The flux distribution for the core metabolic network in *E. coli* for biomass production alone, Equation (21.4) from glucose (Gluc) as the input. The thick arrows indicate the active reactions. The computed biomass yield is 0.55 g DW/g Gluc. Prepared by Jeff Orth.

It should be noted that the GAM value of 13 mmol/g DW is subject to the value of the P/O ratio, specific growth conditions, and genome-scale issues [115, 295]. The exact numerical value can be 2 to 3 times 13 mmol/g DW. Sensitivity calculations are performed in the next chapter.

Yield versus growth   The growth rate can be computed as a function of the glucose consumed (Figure 21.2). The fulfillment of the stoichiometric biomass synthesis requirement is a straight line through the origin. Once the two maintenance requirements are added, the line intersects the x-axis and it becomes slightly affine. The glucose uptake rate at intercept with the x-axis represents the glucose that must be consumed to satisfy only the non-growth-associated maintenance requirement. The growth rate increases more slowly with the glucose uptake rate when the growth-associated maintenance requirement is imposed as not all the glucose consumed can be used directly to synthesize biomass constituents.

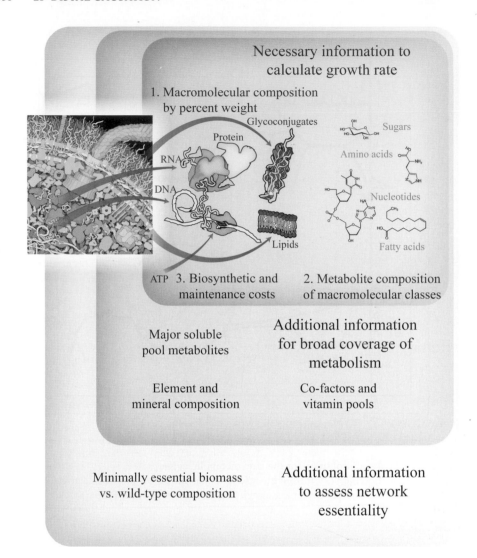

Figure 21.5 Information needed to generate a detailed biomass objective. The top box contains the necessary information needed to calculate accurately a growth rate and this content determines the bulk of metabolic activity (i.e., flux). Addition of information from the second box enables a broader coverage of metabolism and increases the accuracy of predictions of the growth rate and network essentiality. The addition of information from the bottom box allows for the generation of a 'core' biomass objective function that can be used for even greater accuracy of network essentiality prediction. Prepared by Nathan Lewis. Modified from [119].

21.4   Formulating The Biomass Objective Function

A common objective function used to compute physiological states is the biomass objective function. The formulation of a detailed biomass objective function is dependent on knowing the chemical composition of the cell and its energetic requirements, detailed in Figure 21.5. The formulation of the detailed biomass objective function can take place at a different level of detail.

Basic level One can start with the definition of the macromolecular content on the cell (i.e., the weight fractions of protein, RNA, lipid, etc.) and then detail the metabolites needed to form each macromolecular group (e.g., amino acids, nucleotide triphosphates, etc.). With this information, it is possible to detail the required amount of metabolites that are needed for growth along with associated reaction pathways.

Intermediate level One can increase the level of resolution of the biomass objective function and calculate the necessary biosynthetic energy that is needed to synthesize the macromolecules. For example, it is known that it takes approximately two ATP molecules and two GTP molecules to drive the polymerization of each amino acid into a protein molecule [282]. Additional energy is required when considering processes such as RNA error checking and corrections in transcription.

These energetic requirements can be included in the biomass objective function. This addition details the necessary energy that the cell has to make to drive macromolecular synthesis processes that at a lower resolution is treated as a part of the maintenance energies. This energy requirement is over and above the energy that is necessary to synthesize the macromolecular building blocks themselves. An additional detail is the inclusion of the products from macromolecular biosynthesis, such as water from protein synthesis and diphosphate from RNA or DNA synthesis. These polymerization products are directly available to the metabolic network and reduce the amounts of nutrients the cell needs to take up from the media. The addition of these chemical details increases the appropriateness of the biomass objective function.

Advanced level Further detailing of biomass objective functions can be performed by detailing the necessary vitamins, elements, and co-factors required for growth as well as determining core components necessary for cellular viability. Inclusion of vitamins, elements, and co-factors allow for the analysis of a broader coverage of network functionality and required network activity.

An advanced approach is to not only define the wild-type biomass content of the cell, but to generate a separate biomass objective function that contains the minimally functional biomass content of the cell (Figure 21.5). This objective function (referred to as the 'core' biomass objective function [115]) can result in increased accuracy when predicting gene, reaction, and metabolite essentiality and is formulated using experimental data from genetic mutants and knock-out strains. Workflows for how a biomass objective function is formulated have appeared [117, 427]. The detailed delineation of actual data used for formulating both a wild-type and core biomass objective function is available for *E. coli* [115] and can be used as a template for other organisms.

Additional details The scope of network reconstructions continues to grow (Chapter 7). With the reconstructions of the entire protein synthesis machinery [420] we can expect the level of detail in biomass objective functions will further increase and make the mathematical representation of the growth requirements more and more realistic. This increase in scope will likely lead to a simplification of biomass objective functions as network processes (e.g., protein synthesis) will be explicitly built on the basis of mechanism and approximation of these processes will be necessary.

Table 21.2 Studies examining objective functions (baker's yeast). From [119].

Ref	Objective function(s) examined	Modeling approach	Metabolic reconstruction and model used	Source of experimental data	Simple statement
[420, 427]	(1) Max. of growth rate, (2) Min. of ATP production, (3) Minimizing total nutrient uptake, and (4) Minimize redox metabolism through minimizing NADH production.	Linear programming	Hybridoma cell line central metabolism (83 reactions, 42 metabolites) [427]	(1) Growth, uptake, secretion, and protein production rates [140]	Optimization of biomass production can be used to examine growth characteristics and explain observed phenomena.
[372]	ObjFind Algorithm – Optimization-based framework to infer best objective function	Linear programming	E. coli core central metabolism model (62 reactions, 48 metabolites) (see [372])	(1) Aerobic and (2) anaerobic growth isotopomer-based flux distributions [167]	Optimization of biomass production (growth) was identified as the most significant driving force in both cases examined.
[373]	(1) Max. of growth rate, (2) Min. of the production rate of redox potential, (3) Min. of ATP production rate, (4) Max. of ATP production rate, and (5) Min. of nutrient uptake rate	Linear programming and Bayesian discrimination technique	E. coli genome-scale metabolic network iJR904 (1320 reactions, 625 metabolites) [79]	(1) Aerobic growth, substrate, production rates [58]	Min. of the production rate of redox potential to be the most probable objective function.
[386]	(1) Max. of biomass yield (production), (2) Max. of ATP yield (energy expenditure), (3) Min. of the overall intracellular yield, (4) Max. of ATP yield per unit flux, (5) Max. of biomass yield per unit flux, (6) Min. of glucose production, (7) Min. of reaction steps, (8) Max. of ATP yield per reaction step, (9) Min. of redox potential, (10) Min. of ATP producing reactions, (11) Max. of ATP producing fluxes	Linear programming and non-linear programming	E. coli genome-scale metabolic network iJE660 (720 reactions, 436 metabolites) [106] and the E. coli genome-scale metabolic network iJR904 (1320 reactions, 625 metabolites) [360]	(1) Aerobic, (2) anaerobic, (3) anaerobic with nitrate growth in batch, and (4) carbon- and (5) nitrogen-limited growth in chemostat; Isotopomer-based flux distributions. [116, 330, 335]	No single objective describes the flux states under all conditions. Unlimited growth on glucose in oxygen or nitrate respiring batch cultures is best described by non-linear Max. of the ATP yield per flux unit. Under nutrient scarcity in continuous cultures, in contrast, linear Max. of the overall ATP or biomass yields achieved the highest predictive accuracy.
[57]	Biological Objective Solution Search (BOSS) Algorithm – Optimization-based framework to infer best objective function	Linear programming	S. cerevisiae core central metabolism model (62 reactions, 60 metabolites) [126]	(1) Aerobic growth isotopomer-based flux distribution [126]	Growth is the best-fit objective function for the examined network and conditions.
[213]	GrowMatch Algorithm – Minimizes modifications (addition of reactions or activation of secretion of metabolites) in the metabolic model to match growth phenotype data	Linear programming (bi-level optimization)	E. coli genome-scale metabolic network iAF1260 (2077 reactions, 1039 metabolites) [282]	(1) Growth phenotype data for wild-type and mutant E. coli; (2) pathway content data; MetaCyc/KEGG [360,385,449]	GrowMatch is a useful model-refinement tool for curating/refining metabolic reconstructions and can be used to increase predictivity of phenotype data.
[306]	(1) Maximizing production of biomass (growth rate), (2) Max. of plasmid production rate (Max plasmid), and (3) maximizing maintenance energy expenditure (Max ATPm).	Linear programming	E. coli genome-scale metabolic network iJR904 (1320 reactions, 625 metabolites) with plasmid/protein product reactions [79]	Aerobic (1) wild-type and (2) plasmid-bearing substrate and product rates and isotopomer-based flux distribution [331]	Wild-type can best be determined with the objective function of maximizing growth rate, and maximizing expenditure of ATP best predicts overall metabolism and phenotype of plasmid-bearing E. coli.

21.5   Studying the Objective Function

Over the past two decades, a number of studies have been carried out to examine the use of objective functions [119] (see Table 21.2). These studies can roughly be divided into two categories: (1) studies examining assumed objective functions and comparison with experimental data, and (2) studies examining optimization methods to discover or predict algorithmically biological objective functions from experimental data. The former is a user-biased approach, while the latter is unbiased.

Biased search for cellular objectives One can state hypotheses about cellular objectives, perform optimization, and then compare the outcome with experimentally obtained data.

- An early study utilized extensive metabolic and growth data available for a hybridoma cell line to examine a series of candidate objective functions. It was found that for this cell line, that computation using the minimization of redox production fit the experimental data the best [372].
- A subsequent study performed a computational analysis of various different objective functions and compared the results to experimentally determined growth characteristics of *E. coli* [386]. This study found that growth under batch (unlimited) and chemostat (limited) conditions was best described by different cellular objectives.
- Cellular composition changes at very fast growth rates. Thus, growth-rate-dependent biomass objective functions have been studied [339].
- The metabolic burden of plasmid-based expression in a host cell has been studied [306]. This additional metabolic demand is helpful for strain design.

Unbiased search for cellular objectives Computational algorithms have been developed to determine best-fit cellular objective functions [55, 140, 217]. The details of each algorithm will not be discussed here, but these optimization-based frameworks each approach the determination of a predictive objective function in a different manner, and can also be utilized as tools to improve reconstructed network content [217].

In contrast to the studies where objective functions are first identified and then tested, one can examine the entire range of all objective functions to determine which one fits the data best. This approach is considered unbiased as no objective function is initially assumed. Two such separate studies where an objective function was not initially assumed concluded that optimization of biomass production or growth is the best fit for predicting growth data for *E. coli* [55] and for *S. cerevisiae* [140].

21.6   Objective Functions in Practice

Although, 'everything in biology should be viewed through the eyes of evolution' implies some optimal performance based on the organism's past history, we are only beginning to decipher what the cellular objectives might be that drive evolution of phenotypic states. The objective functions for an organism are likely condition-dependent, as suggested based on the studies discussed above, and to change with adaptation. The former relates to proximal causation while the latter to distal causation. Both can now be studied computationally with realistic genome-scale models and experimentally in the laboratory. One can therefore anticipate that many studies of the objective function are to appear.

Some lessons learned   Several important lessons have been learned from the many studies of the objective function carried out to date.

Natural environments.   Nutritionally rich environments, as those typically used in the laboratory, are the exception rather than the norm in natural environments. In general, we might begin to conceptualize cellular survival strategies in order to formulate useful objective functions. Consider three different environments: (1) nutritionally rich, as above; (2) nutritionally scarce; and (3) elementally limited. From a natural habitat standpoint, and the experiences of microorganisms, these are perhaps listed from the least likely to the likeliest; however, no computational studies of the third case have appeared. For the first and second cases, data from batch growth (nutritionally rich, case one) and chemostat growth experiments (nutritionally scarce, case two) suggest that optimal biomass yield or growth rates are meaningful objectives [55, 140, 306, 372, 373, 386].

Prospectively studying the objective function in the laboratory.   The rise of adaptive laboratory evolution (ALE) allows us to begin to study the objective function experimentally and its biochemical and genetic constraints. Cumulated data suggest that strains, such as the widely studied *E. coli* strains, that have been grown over long periods of time in laboratory settings have acquired an optimal growth phenotype on commonly used substrates in growth media [104]. When confronted with an unfamiliar substrate, optimal growth phenotypes can be generated using ALE from initial suboptimal growth states [127, 177, 417, 418]. Evolved strains can now be re-sequenced to find all mutations generated, thus illuminating the underlying genetic and molecular biological basis for optimal growth phenotypes [79, 167]. The use of ALE is detailed in Chapter 26.

Evolution for bioengineering purposes.   Predictable phenomena become the basis for design. Thus, if the outcome of ALE experiments can be predicted, then the possibility of designing the endpoints arises by defining the genetic composition of the starting strain. For example, growth coupling of a bioengineering production objective has emerged as a strain design strategy [58, 116, 330, 335], with ALE being a tool to produce such designs [126]. The use of ALE for strain design is discussed in Chapter 26.

Philosophical aspects.   After reflecting on distal causation and thinking about the fundamental role of the objective function, the reader might develop thoughts about its relationship with concepts like 'the will to live', 'the purpose in life,' and 'what keeps us going.'

21.7   Summary

- Constraint-based methods require the statement of an objective function. A variety of objective functions can be formulated from relatively simple mathematical equations.
- Objective functions are formulated to explore properties of networks, compute likely physiological states, and to represent bioengineering objectives.
- The formulation of the biomass growth function is well established for metabolic networks. It is multilayered in terms of the detail and resolution sought. The more detail is included, the better various cellular functions can be represented.
- A well-formulated growth optimization problem has many applications.
- The objective function is perhaps the most 'interesting' part of COBRA methods as it can be used to describe the selection process and analyze the evolutionary history of an organism.

PART IV
Basic and Applied Uses

Ask not what you can do for your reconstruction, but what your reconstruction can do for you – with apologies to JFK

Part I described the network reconstruction process, Part II detailed its conversion into a mathematical format, and Part III discussed the characterization of network properties using constraint-based methods. What is all this effort good for? A well-curated genome-scale model, in principle, can be used to study all the phenotypic functions that can be produced from the genome of the target organism. Thus, the possible range of applications of GEMs is broad.

In June 2013, 645 papers had appeared that used COBRA tools to explain existing data or predict biological functions [52]. Analysis of these publications showed that the history of GEMs and their uses can be divided into three phases. Shortly after the appearance of the first GEMs in 1999 and 2000, there was a period of creativity around algorithms and analysis method development. In the mid-2000s, experimental validation studies began to accumulate resulting in the availability of well-curated and validated GEMs. Around 2010, a series of studies begun to appear that demonstrated that a variety of predictions of biological functions could be made using GEMs.

In this part of the book we describe the use of genome-scale networks. Chapters 22 and 23 describe how environmental and genetic parameters are represented and studied with genome-scale models. GEMs have proven to be remarkably useful for studying these two types of parameters. Media composition and growth requirements can be studied productively using GEMs. The prediction of the outcome of phenotypic screens of KO strain collections using GEMs with tens of thousands of outcomes represent perhaps the largest-scale effort for predicting biological functions.

Then we move onto specific application areas in a series of four chapters. First, we discuss how GEMs give a useful context for the analysis of omics data sets. Mapping such data against a known background proves to increase the resolution and use of the data. Second, we show how gaps, or missing parts, of a reconstruction can be filled using computational methods. The outcome of this process is the generation of a hypothesis that can then be addressed experimentally. Third, we describe the use of genome-scale models to understand and predict complex biological phenomena. We will focus on the outcomes of adaptive laboratory evolution. Fourth, we describe a significant application: metabolic engineering. This application of genome-scale models is likely to have a significant socio-economic consequence as we move towards more sustainable lifestyles.

22 Environmental Parameters

Prediction is very difficult, especially if it is about the future
– *Niels Bohr*

In Part III we discussed the properties of a single optimization and the shadow prices and reduced costs that give the sensitivity of the solution at the optimal state to changes in variables. However, we are often interested in a much broader view of the optimal properties of a reconstructed network than simply studying the changes in optimality properties close to a single point. In this chapter we show how computations are performed by varying one or two parameters over a finite range of numerical values and how network properties and states can be studied in this fashion.

22.1 Varying a Single Parameter

The changes in the optimal state of a network can be assessed by changing a parameter over a numerical range of interest by repeatedly computing the optimal state over the defined range of parameter values. Both environmental and internal parameters can be considered in this fashion. Varying a single parameter over a finite range of values has been termed *robustness analysis* [107].

22.1.1 Robustness analysis

A parameter can be varied in a step-wise fashion and the LP problem solved for every incremental value. If we are interested in varying v_j between two values, i.e., $v_{j,\min}$ and $v_{j,\max}$, we can solve l optimization problems, as described in Box 22.1. The results will generate a series of l values for Z ($Z_k, k = 1, \ldots, l$), and the associated shadow prices and reduced costs. Once the computations have been performed, these quantities can be graphed as a function of the parameter varied and the results studied.

22.1.2 The effects of oxygen on ATP production

The relative oxygen to glucose uptake determines the ATP yield The ability of the core *E. coli* model to produce ATP from substrate-level phosphorylation to oxidative phosphorylation can be computed by varying the oxygen input rate. At no oxygen

Box 22.1 Robustness analysis [107]

Concept: Determines the sensitivity of the optimal state of a network to a particular parameter. Usually applied to assess the effect of an environmental parameter (exchange rate) or an internal reaction on the optimal growth state or any other objective.

Objective   max $Z_k = <\mathbf{c}^T, \mathbf{v}>$ for $k = 1, ..., l$

Constraints   Flux balance $\mathbf{S} \cdot \mathbf{v} = 0$

Fixing the parameter

$v_j = v_{j,\min} + \frac{(k-1)}{(l-1)} * (v_{j,\max} - v_{j,\min})$

Capacity constraints: $v_{i,\min} \leq v_i \leq v_{i,\max}, i = 1, ..., n, i \neq j$

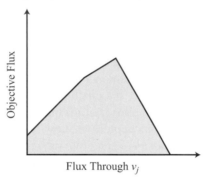

input, the anaerobic ATP production of 2.75 (see Figure 22.1A) is all by substrate-level phosphorylation. Conversely, at a ratio of 1:6 for glucose to oxygen uptake, glucose is fully oxidized and maximal oxidative phosphorylation takes place to generate 17.5 ATP (see Figure 22.1A).

The oxygen uptake rate can be varied between these two extremes and the ATP production computed (Figure 22.1A). A continuous curve is generated between the extreme conditions of 2.75 and 17.5 ATP/glucose. For oxygen uptake rates above 6, there is no solution as the equations cannot be balanced.

Three types of optimal states or 'phenotypic phases' The computed ATP per glucose yield curve formed is piecewise linear. The kinks in the curve correspond to changes in secretion rate (see Figure 22.1B). At no uptake rate of oxygen there is no CO_2 produced, but two formate, one ethanol, and one acetate molecules are produced. Three protons are also produced. At the other extreme where 6 oxygen molecules are taken up, all the glucose is converted to CO_2. In between the two extremes there are three distinct metabolic states:

Dropping oxygen uptake from 6 to 2: as the oxygen uptake drops from 6, acetate and protons are secreted due to a shift away from the TCA cycle where pyruvate is fully oxidized to CO_2. As oxygen update drops, pyruvate begins to be transformed into acetate by phosphotransacetylase (PTAr) and acetate kinase (ACKr) due

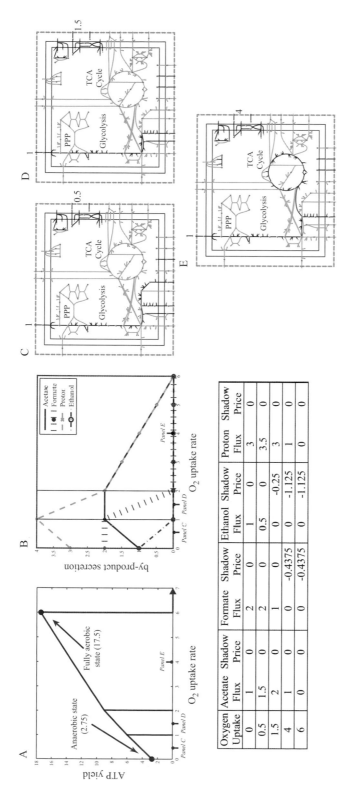

Figure 22.1 The effects of varying the oxygen uptake rate on the ability of the core *E. coli* network to generate ATP. Oxygen uptake rate increases from 0 to 6 while the glucose input is kept at unity. These are in arbitrary units because yields are being computed. Panel A: the ATP yield, panel B: the generation of by-products. The table shows the shadow prices and secretion rates at select values of the oxygen uptake rate. The optimal flux distributions that correspond to the three states of ATP production from glucose, depending on the oxygen availability. Panel C: oxygen uptake is 0.5; D: oxygen uptake is 1.5; and E: oxygen uptake is 4. Prepared by Jonathan Monk.

Oxygen Uptake	Acetate Flux	Acetate Shadow Price	Formate Flux	Formate Shadow Price	Ethanol Flux	Ethanol Shadow Price	Proton Flux	Proton Shadow Price
0	1	0	2	0	1	0	3	0
0.5	1.5	0	2	0	0.5	0	3.5	0
1.5	2	0	1	0	0	-0.25	3	0
4	1	0	0	-0.4375	0	-1.125	1	0
6	0	0	0	-0.4375	0	-1.125	0	0

361

to the lack of oxygen as an electron acceptor. When the oxygen uptake rate reaches 2, there are two acetate molecules and two protons made, and the ATP yield has dropped from 17.5 to 9.0. The corresponding flux map is shown in Figure 22.1E.

Dropping oxygen uptake from 2 to 1: as oxygen uptake drops below 2, pyruvate begins to be converted into formate by pyruvate formate lyase (PFL). The proton production rate increases further, but the acetate production rate stays at 2. Once an oxygen uptake rate of 1 is reached, then 2 moles of acetate, two moles of formate, and four protons are produced. The ATP yield is now 6.0. The corresponding flux map is shown in Figure 22.1D.

Dropping oxygen uptake from 1 to 0: as oxygen uptake drops below 1, ethanol fermentation begins where acetyl-CoA is transformed into acetaldehyde by acetaldehyde dehydrogenase, into ethanol by alcohol dehydrogenase, and then secreted. Acetate, formate, and protons are also secreted. Finally, the ATP yield drops to 2.75. The corresponding flux map is shown in Figure 22.1C.

These three phases represent distinct optimal phenotypes that are computed by linear optimization. The optimal ATP yield relies on different pathways through the metabolic network.

The use of shadow prices to interpret phenotypic phases Shadow prices give insight into the constraints being imposed on the cell. In Figure 22.1 we see the shadow price of ethanol decrease as more O_2 is added. The magnitude of the shadow price indicates the amount that the objective function of ATP generation will be increased by adding more ethanol. When there are low levels of O_2 in the cell, ethanol is secreted as a by-product of substrate-level phosphorylation. Ethanol is a fermentation product and still contains chemical energy because it has not been fully oxidized. However, it cannot be further metabolized until there is available oxygen. Therefore, we see that ethanol goes from being a by-product (shadow price of 0) to offering itself as a carbon source (negative shadow price) for further ATP generation as the O_2 uptake rate increases.

22.1.3   The effects of oxygen uptake rate on growth rate

Single vs. multiple variable objectives The generation of ATP from glucose represents a single variable objective: maximization of the flux that utilizes ATP forcing its maximal rate of generation. Most realistic physiological states rely on coordinating multiple processes simultaneously. Growth is such a process, as all the biosynthetic requirements must be met simultaneously. We now examine the effect of oxygen on growth rate.

The relative oxygen to glucose uptake determines growth rate The effects of varying the oxygen uptake rate on the core *E. coli* metabolic network to support optimal growth can be computed. The results from varying the oxygen uptake rate from zero (completely anaerobic growth) to all the oxygen required for fully oxidizing the substrate on the growth rate is shown in Figure 22.2.

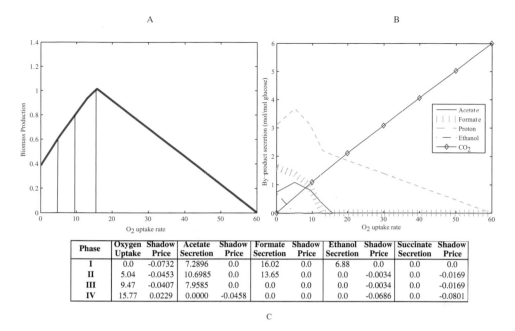

Phase	Oxygen Uptake	Shadow Price	Acetate Secretion	Shadow Price	Formate Secretion	Shadow Price	Ethanol Secretion	Shadow Price	Succinate Secretion	Shadow Price
I	0.0	-0.0732	7.2896	0.0	16.02	0.0	6.88	0.0	0.0	0.0
II	5.04	-0.0453	10.6985	0.0	13.65	0.0	0.0	-0.0034	0.0	-0.0169
III	9.47	-0.0407	7.9585	0.0	0.0	0.0	0.0	-0.0034	0.0	-0.0169
IV	15.77	0.0229	0.0000	-0.0458	0.0	0.0	0.0	-0.0686	0.0	-0.0801

C

Figure 22.2   The effects of varying oxygen uptake rate on the ability of the core *E. coli* network to support growth. The uptake rate of glucose is fixed at 10 mmol/g DW/h. The optimal point corresponds to the perfect conversion of glucose into biomass with no by-product formation. Panel A: the biomass production value (g DW/h) as a function of O_2 uptake. Panel B: the by-product secretion value as a function of O_2 uptake. Panel C: the shadow prices. These calculations use no growth-associated maintenance and the ATP demand is 41.257 mmol/g DW/h. Prepared by Jonathan Monk.

As the oxygen uptake rate increases from zero, the growth rate increases (Figure 22.2). There are three linear segments in the rising part of the curve; the shadow price structure changes at each (Figure 22.2C).

- In the first segment, the shadow prices for acetate, formate, and ethanol are zero. These three metabolites do not increase the growth rate (the objective function) and are thus secreted.
- In the second segment, the shadow prices for acetate and formate are zero and are secreted, while ethanol has value to the growth process as indicated by the negative shadow price.
- In the third segment, only acetate has a zero shadow price and is secreted.
- At the peak of the curve, the *carbon to oxygen uptake* (C/O) ratio is perfect for biomass formation and no by-products are secreted.
- Past the peak in the curve, too much oxygen is taken up relative to glucose and the growth rate drops due to forced dissipation of the excess oxygen that leads to the forced formation of CO_2. Eventually all the glucose is converted to CO_2. The last segment represents an unrealistic physiological situation as the cell would simply take up less oxygen to increase its growth rate. This part of the curve could be described as being *phenotypically unstable*.

These two examples give the essence of performing robustness analysis. The analysis can then go deeper. For instance, flux variability analysis (FVA) can be performed to see if the optimal solutions are unique. If they are not, one can randomly sample the space of optimal solutions to see what the global characteristics of the optimal solutions are. One can compute the extreme pathways (only useful for small models) to understand what the extreme optimal states look like. Various additional analyses can be performed and the curious model-builder will undoubtedly try several to obtain a deep understanding into the optimal states of the network that has been reconstructed.

22.1.4   Sensitivity with respect to key processes

In the previous subsections we showed how we can vary a single parameter to compute a series of phenotypic states that are optimal given a particular set of environmental parameters. In a fully formulated model, robustness analysis can be used to perform sensitivity analysis over a finite range of the values of key parameters. We now show how the optimal aerobic growth state can be examined with respect to uncertainty in certain key parameters and processes.

1   *Pathway usage*: the flux splits of the optimal solution between the major pathways can be examined and compared to measurements.
2   *Maintenance requirements*: cells have maintenance energy requirements that are separate from the stoichiometric requirements for growth used above.
3   *Metabolic stoichiometry*: while the stoichiometry of metabolic reactions are fixed, the same is not true for reactions in transducing membranes that can vary in a condition-specific manner.

Pathway usage   The effect of restricting flux through specific pathways can be examined on the optimal biomass yield. Such a flux can correspond to the first reaction in a pathway. A good example of this type of inquiry is performed by restricting the flux into the pentose phosphate shunt (PPS). By restricting the flux through the PPS all the way to zero, we see that maximal biomass yield is only decreased by $\approx$ 1.5% (Figure 22.3A). This result indicates that the cell has significant flexibility in determining metabolic distribution around the G6P node without experiencing detrimental effects on the biomass yield.

An additional question can be addressed: how does the optimal glycolysis PPS split depend on the ATP maintenance cost? The insert in Figure 22.3A shows the optimal percent PPS flux plotted as a function of the ATP maintenance requirements. This graph shows that the optimal PPS flux is very sensitive to the maintenance requirements. Maintenance requirements of 4–6 ATP result in physiologically consistent optimal PPS fluxes of 20–40%. This and other questions can be answered easily through similar computations.

Maintenance requirements   Cells use energy for metabolic functions other than growth. These include cellular motility, maintenance of cellular osmolarity, macromolecular turnover, error correction, and many more functions. Because it

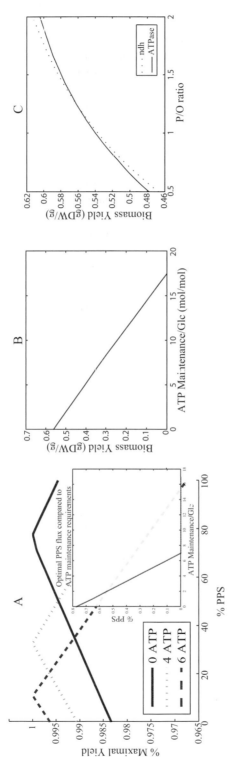

Figure 22.3 Robustness analysis used for parameter sensitivity computations. panel A: the glycolysis–pentose phosphate shunt (PPS); panel B: ATP maintenance requirements; and panel C: the P/O ratio. Prepared by Jonathan Monk.

is difficult to quantify each of these functions individually, a measure known as the maintenance coefficient is used to represent all these functions. Because most maintenance requirements are of energy, they are termed ATP maintenance require- ments. These requirements impose a significant drain on metabolic resources at low growth rates. The biomass yield is very sensitive to ATP maintenance requirements as indicated in Figure 22.3B. If 17.5 ATP are used for maintenance per glucose consumed, then no biomass is produced, consistent with the aerobic ATP/glucose yield calculations above (Figure 22.1).

Energy-transducing membranes: the P/O ratio   The stoichiometry of metabolic reac- tions is well known. However, the stoichiometry of energy-transducing membranes is not fixed. The P/O ratio is the number of high-energy phosphate bonds formed during the transfer of a pair of electrons to oxygen from NADH via the electron transfer sys- tem (ETS). In *E. coli*, this ratio is primarily determined by two independent factors: the number of protons translocated by the NADH dehydrogenases and the stoichiometry of ATPase.

NADH dehydrogenase transfers hydrogen from NADH to one of several quinones. Both energy-linked and unlinked activities have been observed. Use of the non-energy-linked activity results in a P/O ratio of 0.667, whereas use of the energy-linked dehydrogenase results in a P/O ratio of 1.33 assuming an ATPase stoichiometry of 3 H^+/ATP. The relative expression of the genes involved would allow the cell to control the P/O ratio. The effect of varying this P/O ratio on biomass yield is shown in Figure 22.3C.

The energy of the transmembrane proton gradient is used to produce ATP using ATPase. The exact stoichiometry of this conversion is not known. A value of 3 H^+/ATP is often assumed ($4H^+$/ATP were used earlier in the book); however, non-integer values may be more accurate to the long-term behavior of the process. The effects of varying the ATPase stoichiometry are also shown in Figure 22.3C. We can see that varying the P/O ratio has a strong effect on biomass yield.

22.1.5   Uses of robustness analysis

Given the ease of robustness analysis and the wealth of information and insight it provides, it has been deployed in a number of investigations. Below, we give a few simple examples.

Over- and under-expression of key enzymes   The initial robustness analysis was focused on internal fluxes. The effects of forced activity of key enzymes over and under the optimal level was computed (Figure 22.4A). The optimal solution here, growth on glucose in minimal medium, is used as a reference. Then the flux through key steps in metabolic pathways is fixed at lower and higher fluxes than the optimal.

Four reactions that correspond to essential genes were examined. These reactions are in glycolysis, the pentose pathway, and the TCA cycle. Lowering the flux below the optimal solution reveals that the growth rate does not change much until a critical level is reached, and there are compensating metabolic routes. For instance, the TCA cycle flux has to be dropped below 20% of optimal before the growth rate starts dropping noticeably (Figure 22.4A). Further reduction leads to a straight line into the origin,

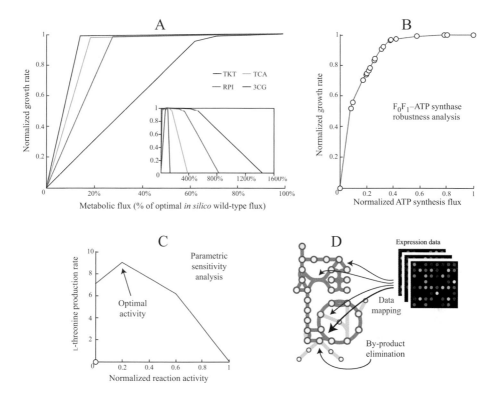

Figure 22.4 Using robustness analysis. Panel A: original robustness diagrams. The effect of altered metabolic flux in the essential metabolic reactions on the normalized growth flux is illustrated. The optimal flux is used for normalization. Enzymes examined are the transketolase (TKT), ribose-5-phosphate isomerase (RPI), enzymes in the 3-carbon stage of glycolysis (3CG), and the first three enzymes of the TCA cycle. From [107]. Panel B: F_0F_1–ATP synthase robustness analysis. The normalized growth rate corresponding to F_0F_1–ATP synthase flux is plotted. From [68]. Panels C, D: robustness analysis in metabolic engineering. The computation of the strain engineering objective can be computed as a function of the activity level of a key enzyme and the results used to tune expression levels. From [225]. Prepared by Nathan Lewis.

indicating that the TCA cycle flux has become the growth-limiting process. The insert in Figure 22.4A shows a similar result for forcing a high flux, that would correspond to a forced over-expression of a gene. Eventually the growth rate drops to zero.

ATP F_0F_1 synthase in a pathogen The effects of the activity of the ATP synthase in *Leishmania major* on the growth rate of the organism has been studied in a genome-scale model (Figure 22.4B) [68]. This synthase is the target of metabolic intervention with drugs like oligomycin. This robustness analysis gives information about how much inhibition is needed *in vivo* to achieve a given growth rate reduction. Based on these computations, the activity of the synthase has to be reduced by 90% to reduce the growth rate by one-half.

Metabolic design Metabolic engineering is focused on altering metabolic fluxes to achieve a production phenotype of interest. The design of such phenotypes typically revolves around tuning metabolic fluxes so that the substrate consumed is directed towards a secreted compound of interest. The design of a threonine production strain relied on such a procedure [225]. The activity of a key enzyme in threonine production was controlled by manipulating gene expression. A graph that provides the computed relationship between L-threonine production and the activity of particular reaction is shown in Figure 22.4C. This *in silico* parametric sensitivity analysis guided the level of expression necessary for increased production of the amino acid in the strain. Figure 22.4D is a map of central metabolism representing the metabolic reconstruction of *E. coli*. In the analysis, expression data were mapped onto the network to guide the elimination of negative regulation and the network was used to over-express a reaction that diverted flux away from a by-product (by-product elimination) towards the desired product. Chapter 27 describes metabolic engineering in more detail.

22.2   Varying Two Parameters

Robustness analysis represents the effects of varying a single parameter over a finite numerical range on the optimal performance of a network. In a similar fashion, two parameters can be varied simultaneously over a finite numerical range [38, 103]. Performing simultaneous analysis of two parameters is computationally more difficult than performing robustness analysis, but the interpretation becomes more interesting as we are now looking at the optimality properties of a network from a broader perspective.

22.2.1   Phenotypic phase planes

The concept   The plane formed by changing two parameters is called the *phenotypic phase plane* (PhPP). Although any parameters can be chosen, a set of parameters that have been of particular interest are the substrate and oxygen uptake rates for microbial growth. Optimal flux maps can be calculated for all points in a plane formed by using the substrate uptake rate on the x-axis and the oxygen uptake rate on the y-axis. A 3D surface can be graphed above this (x, y)-plane (Figure 22.5). The floor in this 3D representation is the PhPP.

The PhPP in some ways resembles the PVT (pressure, volume, temperature) phase planes used in physical chemistry, which define the different states (i.e., liquid, gas, or solid) of a chemical system depending on the external conditions. To compute the state of the compounds of interest, the Gibbs free energy is minimized. In contrast, in the PhPP, the extrinsic, or external, variables are inputs to and outputs from the network and the optimal state is computed based on an objective function of interest that characterizes the network property of interest.

Finite number of phenotypic phases   Although an infinite number of optimal solutions in the PhPP can be computed, it turns out that there are a finite number of fundamentally different optimal functional states present in the PhPP. The demarcations between the regions of different functional states can be determined from the shadow prices of the variables that are represented on the axes of the PhPP. As the robustness analysis shows, the shadow prices can be used to interpret shifts from one optimal

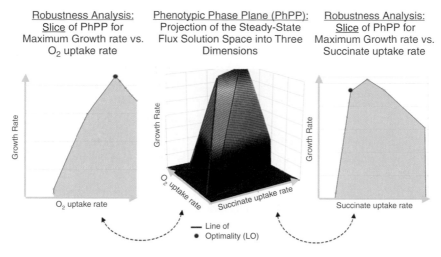

Figure 22.5   Phenotypic phase planes. The center figure shows a three dimensional rendering of a maximal growth rate plotted as a function of two variables: the O_2 and succinate uptake rate. A phase plane is a projection of this surface into two dimensions (the floor of the 3D figure). The line of optimality (LO) corresponds to the conditions where the objective function is optimal (in this case, growth rate). Robustness analysis of the two uptake rates individually is shown in the two side panels. The graph on the left shows the effect on growth rate of varying O_2 uptake at a fixed succinate uptake rate (as in Figure 22.2) and represents a slice through the 3D surface at a specific oxygen uptake rate. Conversely, the graph on the right shows the effect on biomass generation of varying the succinate uptake rate at a fixed oxygen rate. The red dots lie on the LO. From [344].

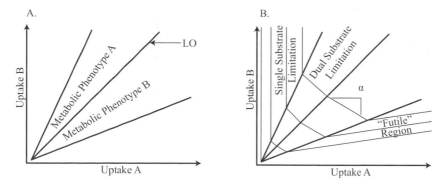

Figure 22.6   Characteristics of the PhPP. (A) Two phases (A and B) in a PhPP with and LO separating them, as well as infeasible regions; and (B) isoclines and their use to classify different phases.

flux distribution to another. This procedure leads to the definition of distinct regions, or phases, in the PhPP in which the optimal use of the network is fundamentally different, corresponding to a different functional state (Figure 22.6A). One can denote each phase as: $Pn_{x,y}$, where P represents phenotype, n is the number of the demarcated region for this phenotype, and x, y are the two uptake rates on the axes of the plane.

Isoclines   The regions in the PhPP can be defined based on the contributions of the two parameters represented on the x and y axes to the objective function. To facilitate such an interpretation, we define the ratio of the relative shadow prices for the two variables on the axes of the PhPP:

$$\alpha = -\frac{\pi_x}{\pi_y} \tag{22.1}$$

where π is the shadow price (see Equation (18.4)) and x and y refer to the variables on the x and y axes. The negative sign on α is introduced in anticipation of its interpretation.

The ratio α is the relative change in the objective function for changes in the two key exchange fluxes. In order for the objective function to remain constant, an increase in one of the exchange fluxes will be accompanied by a decrease in the other, and thus we introduce the negative sign on the definition of α. The parameter α is thus the slope of a line in the PhPP along which the value of the objective function is a constant. This line is called an *isocline*. Note that the isocline is equivalent to an isotherm or an isobaric line in a PVT phase plane. When the objective function is graphed above the (x, y) plane (see below) an isocline corresponds to a constant height.

Characteristics of phases in the PhPP   The slope of the isoclines within each phase of the PhPP is calculated from the shadow prices. Thus, the slope of the isoclines will be different in each region of the PhPP. Based on these considerations we identify four types of regions on the PhPP (Figure 22.6B):

1   In phases where the α value is negative, there is dual limitation of the substrates. Based on the absolute value of α, the substrate with a greater contribution toward obtaining the objective can be identified. If the absolute value of α is greater than unity, the substrate along the x-axis is more valuable toward obtaining the objective, whereas if the absolute value of α is less than unity, the substrate along the y-axis is more valuable to the objective.

2   The phases where the isoclines are either horizontal or vertical are phases of single substrate limitation; the α value in these phases will be zero or infinite, respectively. These phases arise when the shadow price for one of the substrates goes to zero, and thus has no value to the cell.

3   Phases in the PhPP can also have a positive α value, termed 'futile' phases. In these phases, one of the substrates is inhibitory toward obtaining the objective function, and this substrate will have a positive shadow price. The metabolic operation in this phase is wasteful, in that it consumes substrate that is not needed to improve the objective, i.e., the post-peak segment in Figure 22.2. Phases with positive α values are expected to be *phenotypically unstable*, i.e., a cell would not be expected to choose a state in such a region. Under selection pressure, cells would move their phenotype state out of the phase.

4   Finally, due to stoichiometric limitations, there are infeasible steady-state phases in the PhPP. If the substrates are taken up at the rates represented by

these points, the metabolic network is not able to obey the mass, energy, and redox constraints while generating biomass.

22.2.2 Using the PhPP at a small scale

PhPP for ATP production   In Figure 22.1 we computed the ATP yield as a function of oxygen uptake rate at a fixed glucose input rate of unity. We can form a PhPP by considering all possible glucose uptake rates (Figure 22.7B). The results are quite similar to Figure 22.1, where we see that the maximal ATP yield is 17.5 mol ATP/mol glucose. From there, the yield decreases to the minimum yield of 2.75 mol ATP corresponding to the reduction in O_2 uptake rate. The phase plane now shows the three phases in Figure 22.1 for a range of glucose and oxygen uptake rates. Here, $P1_{O_2,gluc}$ represents a phenotypic state that corresponds to Figure 22.1C, and $P2_{O_2,gluc}$ and $P3_{O_2,gluc}$ to Figure 22.1D,E, respectively.

PhPP for growth   The growth rate can be computed as a function of the glucose and oxygen uptake rates (Figure 22.8A). The PhPP (Figure 22.8B) has four phases. They are consistent with the robustness analysis in Figure 22.2.

- At low oxygen uptake rates, the phase $P4_{O_2,gluc}$ is characterized by the secretion of acetate, formate, and ethanol. $P3_{O_2,gluc}$ has acetate and formate secreted, and $P2_{O_2,gluc}$ has only acetate secretion. These three phases have isoclines with negative slopes.
- The line of optimality (LO) represents the condition where all the carbon taken up is fully oxidized or incorporated into biomass.
- $P1_{O_2,gluc}$ has isoclines with a positive slope and are thus phenotypically unstable, i.e., if the glucose rate is fixed, then lowering the oxygen uptake rate towards the LO will increase the growth rate. Thus, the maximum allowable oxygen uptake rate would not be chosen to maximize growth rate (recall Figure 22.2A).

22.2.3 Using the PhPP at the genome-scale

PhPPs have much utility in developing an understanding of particular organisms and their functions. A few examples are given below.

Analyzing carbon and nitrogen requirements of *Haemophilus influenzae*   Figure 22.9 is the first PhPP published showing the different optimal metabolic phenotypes and their characteristics that can be derived from the *H. influenzae* metabolic network depending on the substrate (fructose and glutamate) uptake rates. There are six distinct optimal metabolic phenotypes in the PhPP.

- In phase 1, the capability of the *H. influenzae* metabolic network to meet growth requirements is limited by its ability to generate the biosynthetic precursors derived from fructose.
- In phases 2, 3, and 4, cellular growth is limited by the ability of the metabolic network to produce high-energy phosphate bonds and redox potential. These limitations have different optimal pathway use in each phase.

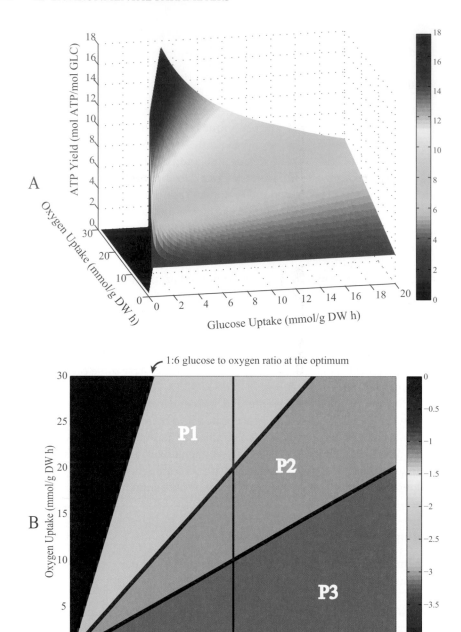

Figure 22.7   The phenotypic phase plane for the core *E. coli* model for ATP yield as a function of glucose and oxygen uptake rates. Panel A: a 3D surface of maximal ATP yield as a function of glucose fixed oxygen and glucose uptake rates. Panel B: the PhPP which is the floor of the polytope in Panel A. The lines are projections of the edges of the polytope in Panel A. The vertical line corresponds to Figure 22.1, i.e., a fixed glucose uptake rate but a varying oxygen uptake rate. Prepared by Jonathan Monk.

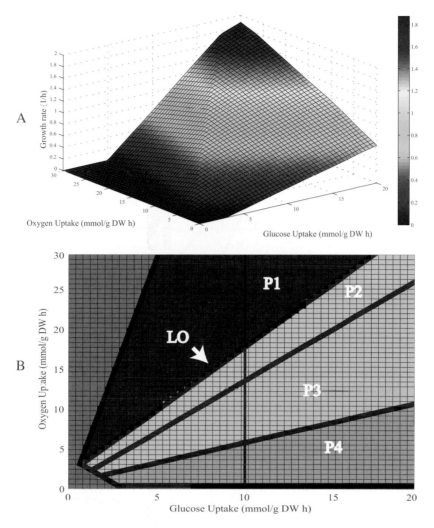

Figure 22.8 The phenotypic phase plane for the core *E. coli* model for growth on glucose. Panel A: a 3D surface of maximal biomass production as a function of glucose fixed oxygen and glucose uptake rates. Panel B: the PhPP which is the floor of the polytope in Panel A. the lines are projections of the edges of the polytope in Panel A. The vertical line corresponds to Figure 22.2. Prepared by Jonathan Monk.

- Phase 5 is characterized by the excess redox potential. The large glycolytic flux leads to a condition in which the ability to eliminate the redox potential is limiting growth. The oxidative branch of the PPP is not utilized under optimal conditions for this region, and thus the biosynthetic precursors are generated by the non-oxidative branch.
- Glutamate, the nitrogen source, is the limiting factor in phase 6. The optimal metabolic phenotype in this region is characterized by conversion of the non-limiting substrate into metabolic by-products. There is excess energy and redox potential in this region.

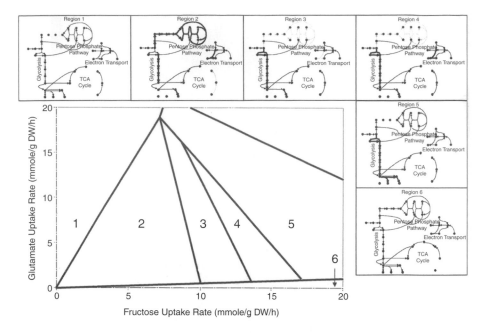

Figure 22.9   The PhPP for *Haemophilus influenzae*. From [105].

Such detailed delineation of optimal states is of great use in studying particular target organisms. Typically, the user will be knowledgeable about the target organism and can use PhPPs to understand the functions of the organisms better and to design experiments to probe its functions.

Experimental study of the optimal growth states of yeast   A *Saccharomyces cerevisiae* genome-scale metabolic network was used to compute a PhPP that displays the maximum allowable growth rate and distinct patterns of metabolic pathway utilization for all combinations of glucose and oxygen uptake rates [101]. *In silico* predictions of growth rates and secretion rates and *in vivo* data for three separate growth conditions: (1) aerobic glucose-limited (AGL), (2) oxidative-fermentative (OF), and (3) microaerobic (MA) were found to be consistent (Figure 22.10). The function and capacity of yeast's metabolic machinery are clearly interpreted using the PhPP, and it can be used to accurately predict metabolic phenotypes and to interpret experimental data in the context of a genome-scale model.

Characterization of growth and metabolism of an extremophile   A model organism *Natronomonas pharaonis*, an archaeon adapted to two extreme conditions (high salt concentration and alkaline pH) is used for the study of extremophilic life. A genome-scale, manually curated metabolic reconstruction for the microorganism is available [145].

A PhPP was used to perform theoretical analysis of aerobic growth of *N. pharaonis* on acetate (Figure 22.11). The surface in the figure represents the theoretical maximum growth of *N. pharaonis* as a function of two parameters: (1) the acetate to oxygen ratio, and (2) the maintenance energy. This twist to a PhPP formulation thus uses a substrate

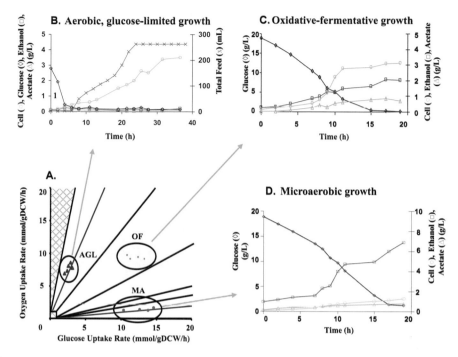

Figure 22.10   Growth experiments shown on the PhPP. Panel A: the three groups of experimental data displayed on the *S. cerevisiae* PhPP were used as an index for the time course profiles in panels B, C, and D. Panel B: aerobic glucose-limited (AGL) growth controlled by fed-batch operation. Panel C: oxidative-fermentative (OF) growth with unlimited glucose and oxygen availability. Panel D: microaerobic (MA) growth with unlimited glucose and very low oxygen availability. The AGL (B) and MA (D) data sets are located on lines of optimality and as a result are stable metabolic states with only one degree of freedom (glucose for AGL and oxygen for MA). OF (C) is an unstable metabolic state with two degrees of freedom (glucose and oxygen), making it more difficult to control this type of growth condition. By perturbing the environmental conditions, cells in OF can be shifted to either AGL or MA. From [101].

to electron acceptor ratio to combine two key uptake rates on one axis and a hard to define quantity, the maintenance energy, on the other.

The green-shaded region corresponds to experimentally observed values of the acetate:oxygen ratio, while the orange-shaded region corresponds to experimentally determined values of the maintenance energy. For different values of the maintenance energy, the computed growth is found at different acetate to oxygen ratios. This optimality relationship is represented by the red broken curve and a two-dimensional projection into the x, y-plane is shown in the inset.

Optimal community composition   Phase plane analysis has been expanded to the analysis of a two-species microbial 'community' [274]. Coupled GEMs and multi-omic data sets were used to investigate *direct interspecies electron transfer* (DIET) mechanisms in the syntrophic association of two species, *Geobacter metallireducens* and *Geobacter sulfurreducens*. Phase plane analysis revealed that a co-evolved culture locates itself on the LO in a PhPP where the DIET is maximized (Figure 22.12). A suboptimal

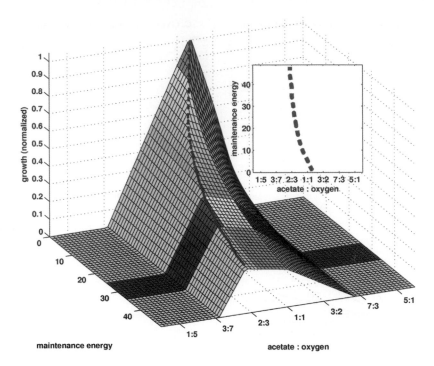

Figure 22.11   Phenotypic phase plane analysis of the extremophile *Natronomonas pharaonis*. From [145].

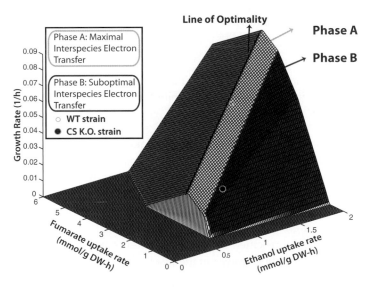

Figure 22.12   Phenotypic phase plane representation of syntrophic growth that involves direct interspecies electron transfer (DIET). PhPP analysis reveals the presence of two distinct phases with maximal and suboptimal efficiency of DIET. A suboptimal state was created by knocking out citrate synthase (CS). From [274].

state was generated by the knock-out of citrate synthase (the CS K.O. condition in Figure 22.12).

Detailed experimental and omic analysis of the optimal and suboptimal states of this two-species community shows that while *G. sulfurreducens* adapts to rapid syntrophic growth by changes in DNA sequence and at the transcriptomic level, *G. metallireducens* responds only at the transcriptomic level. The approach of using multi-omic data generation and GEMs thus can enhance understanding of adaptive responses and factors that shape the evolution of optimal syntrophic microbial communities. PhPP analysis may thus be productively used to analyze more complex biological situations than the optimal growth rates of a clone.

Distal causation   Evolutionary experiments in the laboratory under controlled conditions have been performed to allow organisms to develop optimal states. PhPPs have proven to be useful in the design and interpretation of adaptive laboratory experiments in the laboratory. This issue is detailed in Chapter 26.

22.3   Summary

- Constraint-based optimization can be repeated systematically for a finite range of parameter values of interest.
- Robustness analysis is based on varying one parameter and studying the changes in the characteristics of the system under consideration.
- Phenotypic phase plane analysis is based on varying two parameters simultaneously and studying the changes in the characteristics of the system under consideration.
- Variation in performance of a metabolic network to responses in environmental parameters can be studied at both a small and genome-scale.
- Other parameters can be varied in the same fashion, such as internal rates and maintenance coefficients.

23 Genetic Parameters

Prediction is at least two things: important and hard — Howard Stevenson

Genetic parameters can be represented explicitly in genome-scale models enabling the assessment of their effects on phenotypic functions. There are three issues associated with genetic parameters discussed in this chapter. First, the analysis of the consequences of the loss of function (LOF) or gain of gene function (GOF) will be explored. For *essential genes*, it is important to understand how the loss of a gene product may lead to cell death. Further, sophisticated methods have been developed to assess the minimal perturbation of phenotypic states resulting from non-lethal gene knock-outs. Second, the simultaneous removal of two genes allows one to address issues related to *synthetic lethality* and *epistasis* in general. Such understanding has implications for many studies, such as finding drug targets. Third, sequence variations in genes or their *gene dosage*, i.e., gene copy number, can determine their activity level. Gene dosage may be of interest in many situations, ranging from *imprinted genes* in humans (where one copy is active), to aneuploidy in cancer, to cell line engineering (where gene dosage can be manipulated). The coordinated functions of genes can be assessed through reaction co-sets, and genetic variations in a co-set may have similar consequences on network functions, and thus phenotypic states. The following contains some examples of how to describe genetic parameters within GEMs; there will undoubtedly be many more applications developed in the future.

23.1 Single Gene Knock-outs

23.1.1 Concept

The network function of a gene is traced through its gene–protein–reaction (GPR) association to the reactions with which the gene product is involved. Thus, if a gene is deleted, then flux constraints for the reaction(s) affected by the loss of a gene product are put to zero and thus no flux can take place through the corresponding reaction. Therefore, it is simple to analyze the functions of a network with respect to the loss of a single gene function by comparing the optimality properties of the wild-type network with one where a reaction flux has been restricted to zero.

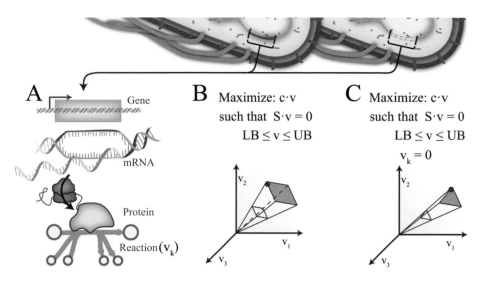

A Gene

mRNA

Protein

Reaction (v_k)

B Maximize: $c \cdot v$

such that $S \cdot v = 0$

$LB \leq v \leq UB$

v_2

v_1

v_3

C Maximize: $c \cdot v$

such that $S \cdot v = 0$

$LB \leq v \leq UB$

$v_k = 0$

v_2

v_1

v_3

Figure 23.1   A schematic showing the reduction in the size of a solution space as a consequence of gene knock-out. (A) GEMs contain gene–protein–reaction associations (GPRs), which allow one to represent genetic alterations. (B) Flux balance analysis allows one to search the solution space to find the solution with the optimal combination of flux values (red dot). (C) When using GPRs to simulate gene deletions, portions of the solution space can be removed, and the optimal combination of flux values can change (red dot). Prepared by Nathan Lewis.

Implementation   The process of eliminating a gene from a network is shown in Figure 23.1. The result of the removal of a gene is the nullification of fluxes through one or more reaction in the network. Mathematically, the loss of a flux shrinks the size of the solution space and changes its shape. The capabilities of the network to optimize a phenotypic function without these reactions can be assessed and compared to the full network. The implementation of such computations is relatively easy.

Three types of consequences   There are basically three outcomes from such a computation. Note that computations are condition-specific. (1) A gene can be non-dispensable and its removal makes it impossible to satisfy the phenotypic function of interest. When the phenotypic function sought is growth, such genes would be termed essential. (2) The removal of a gene can lead to reduction in, but not elimination of, a phenotypic function. (3) A gene removal can have no consequence on phenotypic function and be redundant; this is the case when alternative optimal solutions are found (Chapter 20). The first two situations are discussed in this section.

23.1.2   Core *E. coli* metabolic network

Gene dispensability is condition-dependent   All the genes in the core *E. coli* network can be removed one at a time and the optimal growth rate can be computed. The computed optimal growth rate of the *in silico* knock-out strain can then be normalized relative to the wild-type growth rate. The relative growth rates of all the KO strains can then be rank-ordered and presented as a histogram (Figure 23.2). Such computations can be repeated for different growth conditions. The ability of the core *E. coli* metabolic

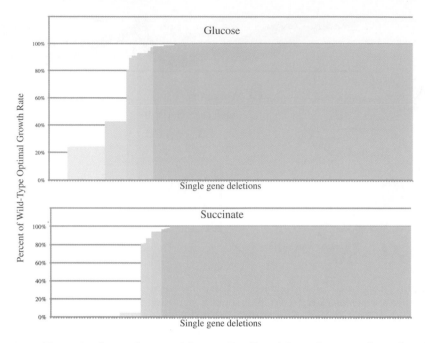

Figure 23.2   The optimal growth rates of the core *E. coli* model as a function of gene knock-out. Growth on glucose and succinate are shown. The growth rate is scaled relative to the wild-type optimal growth rate. The essential genes for glucose and succinate are different. For glucose as a substrate, the genes (given in b-numbers) are: b2415, b2416, b0720, b1136, b1779, b2779, b2926. for succinate as a substrate they are: b0721, b0722, b0723, b0724, b3528, b0720, b1136, b1779, b2779, b2926, b3919, b4025. The suboptimal portions (less than 100% WT percentage) are lighter shades than gene knock-outs achieving optimal growth rate. Prepared by Alex Thomas.

network to support growth on glucose and succinate is presented in Figure 23.2, which shows the effects of loss of all the genes individually on the growth rate.

The histograms show the three types of consequences of loss of gene function discussed above. First, there is a set of genes that, if removed, no growth solution is found. These genes are predicted to be essential. Second, there is a set of genes that reduce or retard the growth rate. These genes are said to show a phenotype; growth is possible without the gene, but at a lower rate. Third, there is a set of genes that, when removed, the exact same growth rate is computed. These are redundant, and would appear in alternate optimal solutions, or in near-optimal solutions if there is a small reduction in growth rate that is below experimental error and thus indeterminate. Predictions of gene lethality in the core *E. coli* metabolic model are largely consistent with experimental data (see Table 23.1).

Determining cause of lethality   We can analyze the cause of lethality in more depth. A failure to grow *in silico* is due to the fact that one or more components of the biomass cannot be provided without the presence of an active gene product from the deleted gene. Such predictions could be tested in the lab by providing the precursor for which the KO strain effectively becomes an auxotroph.

Table 23.1 Comparison of the predicted mutant growth characteristics from a gene deletion study to published experimental results with single KO strains . From [106].

Gene	Glucose	Glycerol	Succinate	Acetate	Reference
aceA	+/+		+/+	-/-	[280]
aceB				-/-	[280]
aceEF	-/+				[89]
ackA				+/+	[218]
acn	-/-			-/-	[280]
acs				+/+	[218]
cyd	+/+				[60]
cyo	+/+				[60]
eno	-/+	-/+	-/-	-/-	[280]
fba	-/+				[280]
fbp	+/+	-/-	-/-	-/-	[280]
frd	+/+		+/+	+/+	[89]
gap	-/-	-/-	-/-	-/-	[280]
glk	+/+				[280]
gltA	-/-			-/-	[280]
gnd	+/+				[280]
idh	-/-			-/-	[280]
mdh	+/+	+/+	+/+		[82]
ndh	+/+	+/+			[431]
nuo	+/+	+/+			[431]
pfk	-/+				[280]
pgi	+/+	+/-	+/-		[280]
pgk	-/-	-/-	-/-	-/-	[280]
pgl	+/+				[280]
pntAB	+/+	+/+	+/+		[154]
ppc	±/+	-/+	+/+		[82, 444]
pta				+/+	[218]
pts	+/+				[280]
pyk	+/+				[280]
rpi	-/-	-/-	-/-	-/-	[280]
sdhABCD	+/+		-/-	-/-	[280]
sucAB	+/+		-/+	-/+	[89]
tktAB	-/-				[280]
tpi	-/+	-/-	-/-	-/-	[280]
unc	+/+		±/+	-/-	[46, 281, 447]
zwf	+/+	+/+	+/+		[280]

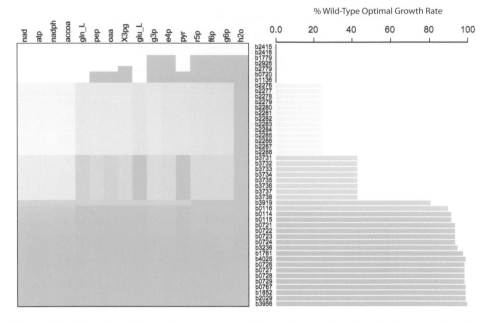

Figure 23.3   The effect of removing genes (rows) from the core *E. coli* metabolic network on its ability to produce the biosynthetic precursors (columns) from glucose. Suboptimal single deletions from Figure 23.2 are selected, and the corresponding percent of WT growth rates are transposed and shown on the right. The objective function is set to produce a precursor and the optimization is repeated for a series of cases where the genes are deleted. White indicates that the reaction deleted completely inhibits production, whereas shades of color indicate low inhibition (darker) or high inhibition (lighter). Prepared by Alex Thomas.

The analysis is performed by setting the objective function to maximize the production of one of the biosynthetic precursors. Then repeated optimization computations are performed, in each of which a flux is constrained to zero for a reaction that is rendered inactive due to gene deletion. The procedure is then repeated for all the biosynthetic precursors of interest (Figure 23.3).

- The knock-out of the GLCpts (i.e., the PTS system) prevents the network from producing any of the biosynthetic precursors. The core model has no other way to import glucose. Similarly, if the phosphate transporter (PIt) is removed, none of the precursors that contain phosphate can be made.
- GAPD, PGK, and PGM, all in lower glycolysis, prevent the production of all the precursors except those that are found in the pentose pathway and in upper glycolysis. Without these enzymes, the core network cannot get carbon to flow into the TCA cycle.
- The removal of three TCA cycle reactions (ACONT, CS, and ICDyr) and of transhydrogenase (NADTRHD) and enolase (ENO) prevents the production of αKG.
- The removal of three reactions in the pentose pathway (RPI, TALA, and TKT1) prevents the production of R5P.

Detailed studies of this type have been performed at a genome-scale for *Haemophilus influenzae* [382].

23.1.3   Genome-scale studies of essential genes

The study of gene knock-out strains and their growth properties has been performed at the genome-scale with an increasingly large set of knock-outs and growth conditions [23,138,194,214,284]. A summary of such studies at the end of the section shows that genome-scale models have a good success record of being able to predict the consequences of a loss of gene function [217].

The initial *in silico* KO study [106] *E. coli* *i*JE660 was subjected to the deletion of each individual gene product in the central metabolic pathways (glycolysis, pentose phosphate pathway (PPP), tricarboxylic acid (TCA) cycle, respiration processes), and the maximal capability of each *in silico* mutant metabolic network to support growth was assessed. The simulations were performed under an aerobic growth environment on minimal medium.

The results identified the genes required for growth in central metabolism. For growth on glucose, the essential gene products were involved in the three-carbon stage of glycolysis, three reactions of the TCA cycle, and several points within the PPP. The remainder of the central metabolic genes could be removed and *E. coli* *i*JE660 maintained the potential to support cellular growth. The *in silico* gene deletion study results were compared with growth data from known mutants. The growth characteristics of a series of *E. coli* mutants on several different carbon sources were examined and compared with the *in silico* deletion results (Table 23.1). From this analysis, 68 of 79 cases, or 86%, of the *in silico* predictions were consistent with the experimental observations.

KO strain collections made large-scale deletion analysis studies possible [130] The initial *E. coli* knock-out study looked at relatively few cases (i.e., 79 cases). An initial large-scale *in silico* evaluation of gene deletions in *Saccharomyces cerevisiae* was conducted using a GEM, and the computational results are summarized in Table 23.2.

The effect of 599 single gene deletions on cell viability was evaluated *in silico* and compared to published experimental growth data on yeast. Sixty-nine cases were removed from the analysis as these genes were associated with dead ends (see Chapter 25). Remarkably, of the 475 cases examined, 89.6% of the *in silico* results were in agreement with experimental observations when growth on synthetic complete medium was simulated. Viable phenotypes were correctly predicted in 445 out of 487 cases and lethal phenotypes were correctly predicted in 30 out of 43 of the cases considered. This *in silico* evaluation was solely based on well-established reaction stoichiometry and no interaction or regulatory information was accounted for in the *in silico* model. Such a large-scale study provides a thorough validation of the *in silico* model.

False predictions were analyzed on a case-by-case basis and divided into four categories: (1) incomplete media composition, (2) substitutable biomass components, (3) incomplete biochemical information, and (4) missing regulation. This large-scale manual and detailed analysis eliminated a number of false predictions and suggested

Table 23.2 Summary of an *in silico* KO study of metabolic genes in *S. cerevisiae*. Taken from [130]. True-positive classifications indicate that the model correctly predicts growth. True-negative predictions indicate that the model correctly assigns no growth . True-negative predictions are generally considered to be a stronger and more valuable prediction than true-positives.

Simulation	Outcome
Number of deletions	530
True-positive	445
True-negative	30
False-positive	42
False-negative	13
Overall prediction	89.6% (475/530)
Positive prediction	91.4% (445/487)
Negative prediction	69.8% (30/43)

a number of experimentally testable hypotheses. A genome-scale *in silico* model can thus be used to systematically reconcile existing data and fill in our knowledge gaps about an organism (see Chapter 25).

Phenotypic screens [83]   The *i*JR906 *E. coli* was expanded to include 104 regulatory genes whose products together with other stimuli regulate the expression of 479 of the 906 genes in the reconstructed metabolic network. This model, accounting for 1010 genes and called *i*MC1010, is able to predict the outcomes of high-throughput growth phenotyping experiments.

KO strains were grown in different media creating a data set of 13,750 experimental conditions that could be compared to computational outcomes. Comparison with the growth phenotypes showed that experimental and computational outcomes agreed in 10,828 (78.7%) of the cases examined (see Figure 23.4). In addition, 2512 (18.3%) of the cases were predicted correctly only when regulatory effects were incorporated into the metabolic model.

Classifying outcomes [194, 357]   The outcome of a large-scale comparison between computations and experiments with KO strains can be systematically analyzed (Figure 23.5). The result of the comparison can be agreement, disagreement, or an inability to make the comparison. The agreements can be considered a validation of a model, whereas the disagreements are opportunities for discovery. If a KO strain can grow in a given environment but the corresponding *in silico* strain cannot grow, then it is likely that the model is missing a component. The initial analysis of a handful of such cases where experiment and computation disagreed led to a series of hypotheses that were addressed experimentally and new gene functions were discovered [357].

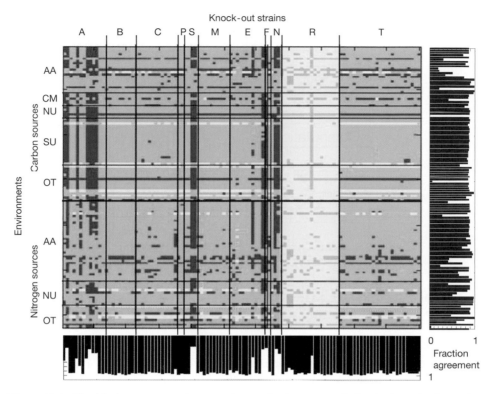

Figure 23.4 Double (genes vs. environment) perturbation screen. Individual results for each knock-out under each environmental condition are shown: green, agreement between computation and experiment; red, disagreement between computation and experiment; yellow, incorporation of regulatory rule (not discussed here). The environments involve variation of a carbon or nitrogen source and are further divided into subgroups: AA, amino acid or derivative; CM, central metabolic intermediate; NU, nucleotide or nucleoside; SU, sugar; OT, other. The knock-out strains are also divided by functional group: A, amino acid biosynthesis and metabolism; B, biosynthesis of co-factors, prosthetic groups and carriers; C, carbon compound catabolism; P, cell processes (including adaptation and protection); S, cell structure; M, central intermediary metabolism, E; energy metabolism; F, fatty acid and phospholipid metabolism; N, nucleotide biosynthesis and metabolism; R, regulatory function; T, transport and binding proteins; U, unassigned. Taken from [83].

Algorithmic formalization [217]   The success with predicting the outcomes of phenotypic screens with genome-scale models has spurred interest in automating and improving the analysis of the outcome of the experiment. The comparisons between experiment and computation are used to suggest model modifications, and identifying these modifications has often been performed manually. Such an approach is effectively the initiation of automated workflows that systematically reconcile all available data and information and represents a step towards the automation of biological discovery processes.

As an example, an automated procedure called GrowMatch has been developed to address this challenge. This algorithm improves comparisons between experiment and computation resulting from automated analysis and can be used for discovery (see Chapter 25).

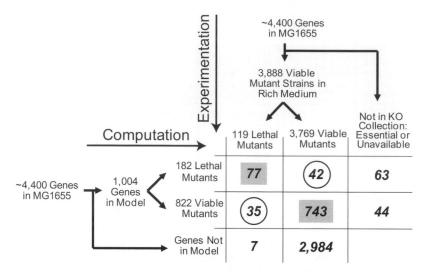

Figure 23.5 Tabulation of the outcome of comparison between computation and experiment for growth of *E. coli* KO strains on glycerol. From [194].

A decade's summary of analysis of KO data from microorganisms The extent of agreement between organism function *in vivo* and *in silico* has reached a high level for a number of organisms. Such detailed and extensive comparisons between the actual and the simulated function of a target organism represent a thorough validation of a genome-scale model. The high degree of agreement gives elevated confidence in the *in silico* model and its ability to predict the metabolic genotype–phenotype relationship. The summary of predicting the growth consequences of gene KOs over the past year is found in Figure 23.6. This effort represents the largest-scale effort of predicting phenotypes, reaching over 100,000 cases in some studies.

Mouse KO studies [397] The prediction of the consequences of gene knock-outs are being extended past microorganisms. The reconstruction of human metabolism, Recon 1, has been used to generate a model of metabolism in the mouse by homology mapping and by using metabolic information specific to the mouse. Gene lethality can be predicted using this model. Homozygous knock-out phenotypes for 17 of the predicted lethal genes was found in the literature. Of those, 14 genes had been confirmed to have lethal phenotypes. With a growing number of mouse KO strains becoming available, the potential to perform large-scale KO validation of mammalian metabolic reconstructions becomes possible.

23.1.4   Studying non-lethal gene KOs

When a gene is knocked out and results in a viable but suboptimal phenotype, it is worthwhile to predict the consequences of the gene knock-out. Minimization of metabolic adjustment (MOMA) is a mathematically based method that utilizes minimization of Euclidean distance between the flux distribution of the wild-type and KO strains to predict the flux state following a gene knock-out.

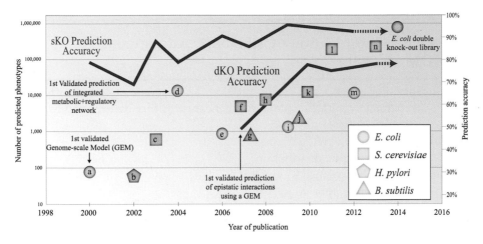

Figure 23.6 Historical overview of improving scope and prediction accuracy of GEMs of experimental outcomes of phenotypic screens. The number of phenotypic predictions from (Environment × Gene knock-out) screens has grown steadily over the past 15 years. Red line indicates accuracy of single gene knock-out (sKO) prediction accuracy. Blue line indicates prediction of double gene knock-out (dKO) accuracy. Studies shown are as follows: (a) [106], (b) [380], (c) [129], (d) [83], (e) [194], (f) [157], (g) [292], (h) [401], (i) [217], (j) [162], (k) [159], (l) [412], (m) [298], (n) [300]. Prepared by Jonathan Monk.

Concept MOMA was developed to predict the changes in the location of the flux vector within the solution space if the function of a gene product is lost [388]. A loss of gene product will reduce the size of the solution space. If an optimal functional state for the wild-type strain was in the portion of the solution space that is eliminated with the loss of the function of a gene product, the growth solution will have to be projected into the reduced solution space to represent functions of the knock-out strain. MOMA finds a new solution in the reduced solution space such that the Euclidean distance between the wild-type state and the reduced solution space is minimized. Because the Euclidean distance is not a linear function, this procedure uses quadratic programming (QP). MOMA is mathematically described in Box 23.1. Note that MOMA is based on a mathematical argument, i.e., a Euclidean distance, and not a biologically based argument. The solution to the quadratic programming problem is unique. The main limitation of MOMA is that it may not be clear which wild-type solution to choose for the mapping. This issue has been studied [256]. The absolute (i.e., Manhattan) norm of the flux vector has also been used as a distance criterion, leading to an LP problem.

MOMA analysis of core *E. coli* The MOMA projection of the optimal growth state of the core *E. coli* model following the deletion of TPI corresponds to a biomass of 0.01699 g/mol glucose. This solution is closest (via Euclidean distance) to the optimal, wild-type biomass production of 0.08739 g/mol glucose. However, FBA after TPI deletion has an optimal yield of 0.0704 g/mol glucose. These three cases are compared in Figure 23.7. The deletion of TPI cuts off flow through glycolysis during its preparatory phase, drastically diminishing the efficiency of the latter phase of glycolysis. The LP optimization after TPI deletion attempts to alleviate the inefficiency in

---

Box 23.1 Minimization of Metabolic Adjustment (MOMA) [388]

Concept: For a knock-out strain, provides an approximate solution for a suboptimal growth flux state that is nearest in flux distribution to the unperturbed state.

Objective   min $||\mathbf{v} - \mathbf{w}||^2$

   w is wild-type flux distribution

Constraints   Flux balance $\mathbf{S} \cdot \mathbf{v} = 0$

   Capacity constraints: $v_{j,min} \leq v_j \leq v_{j,max}$

   Gene deletion: $v_j = 0, j \in A$ A - set of reactions associated with deleted genes.

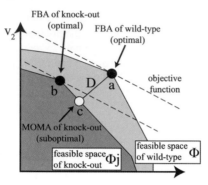

---

glucose-3-phosphate production by shuttling more flux through the PPP and bypassing most of the TCA cycle (Figure 23.8B). In addition, NADTRHD (transhydrogenase) converts NADPH to NADH and then donates electrons to the electron transport chain to improve ATP production. The MOMA solution (Figure 23.8C) also has a more active pentose pathway and utilizes NADTRHD, but also has more reactions in common with the wild-type flux state. Overall, the MOMA solution splits its allegiance between anabolic (pentose pathway) and catabolic (TCA cycle) pathways in terms of creating energy, but sacrifices efficiency to result in a lower biomass production than both WT and the FBA TPI knock-out solutions.

The MOMA flux state might be taken as a prediction of initial response to the loss of TPI. Adaptation of the KO strain would lead to a state that can be predicted by FBA. Such a study has been carried out for *pgi* [67], and adaptive laboratory evolution is discussed in Chapter 26.

Minimization of regulatory off/on modification (ROOM)   Other methods for predicting the effect of deletion of non-essential genes have been developed. ROOM has a similar goal as MOMA, except it uses biological reasoning for how cells deal with the consequences of the loss of a gene product. The criterion that ROOM uses is a

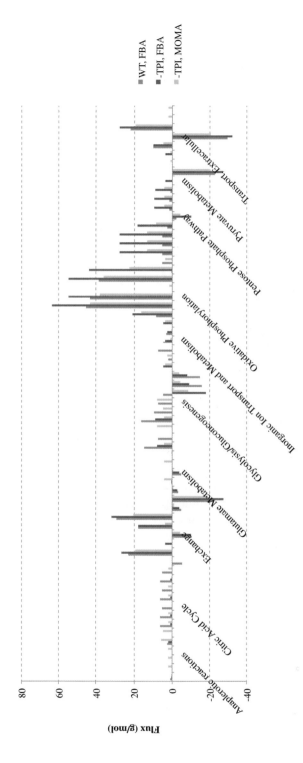

Figure 23.7 Comparison of flux distributions from single knock-out methods. Knock-out of TPI was used to illustrate the general differences between the algorithms. Zero flux reactions across the three models are not included in the figure above. Prepared by Alex Thomas.

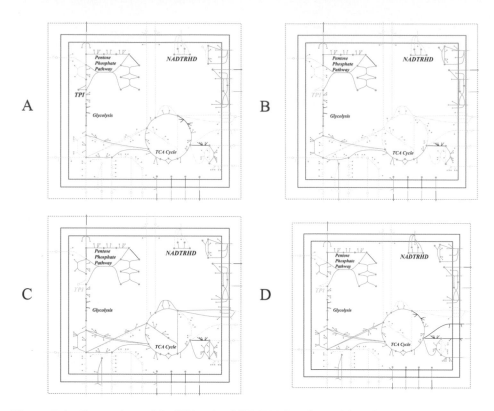

Figure 23.8 Comparison of the WT optimal (A), FBA knock-out (B), MOMA knock-out (C), and ROOM knock-out (D) growth state of the core *E. coli* model following the deletion of triose-phosphate isomerase (TPI). Prepared by Alex Thomas.

minimization in the number of genes whose expression has to change to deal with the loss of the gene product. The mathematical formulation is more complicated than a relatively simple quadratic programming problem and is shown in Box 23.2. The ROOM prediction of optimal growth state after the deletion of TPI corresponds to a biomass of 0.02581 g/mol glucose, which is between the MOMA and FBA solutions. The most striking quality is that the ROOM solution closely resembles the wild-type flux state with its conservation of the TCA cycle (Figure 23.8D). This is in contrast to MOMA's Euclidean measure which enforced flux through malate synthase and isocitrate lyase. Therefore, the ROOM solution conserved gene products with respect to the wild-type state, whereas MOMA minimized flux magnitudes with respect to the wild-type flux. Neither MOMA or ROOM have been extensively tested against experimental data and their use needs further validation.

23.2 Double Gene Knock-outs

Concept   Gene interactions can be studied *in silico* by deleting two genes simultaneously as described for single genes in the previous section. Today, it is much easier to perform a gene interaction study of this sort *in silico* than in a real organism. Gene

---

Box 23.2 Regulatory on/off minimization (ROOM) [394]

Concept: For a knock-out strain, ROOM finds a flux distribution that satisfies the same constraints as FBA while minimizing the number of significant (large enough) flux changes.

Objective   min $\sum_{i=1}^{m} y_i$ (y_i is a binary variable)

  w is wild-type flux distribution

  ϵ and δ are threshold constraints

Constraints   Flux balance $\mathbf{S} \cdot \mathbf{v} = 0$

  Capacity constraints: $v_{j,min} \leq v_j \leq v_{j,max}$

  Gene deletion: $v_j = 0, j \in A$, A - set of reactions associated with deleted genes.

$$v_i - y_i(v_{max,i} - w_i^u) \leq w_i^u$$
$$v_i - y_i(v_{min,i} - w_i^l) \geq w_i^l$$
$$y_i \in \{0,1\}$$
$$w_i^u = w_i + \delta|w_i| + \epsilon$$
$$w_i^l = w_i - \delta|w_i| - \epsilon$$

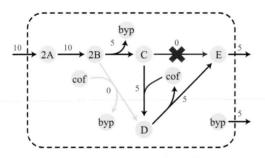

interactions are referred to as *epistatic* interactions. An epistatic interaction that is of particular interest is one where the removal of the genes singly has no or small effect on phenotype while the removal of both simultaneously leads to a strong phenotype. A lethal phenotype is of particular interest and two genes that meet this criteria are called a *synthetic lethal pair*.

23.2.1   Core *E. coli* metabolic network

Synthetic lethals in the core metabolic network in *E. coli* can be identified. One can present the data as a two-dimensional version of Figure 23.2. For each single KO strain, one can compute the growth properties of all strains where a second gene has been removed (see Figure 23.9). The strains that are lethal from a single KO obviously need no further consideration (the red region in Figure 23.9).

The viable single KO strains that grow are now evaluated. Non-lethal KO strains can be analyzed with respect to their synthetic lethal (SL) partners. SL pairs have two

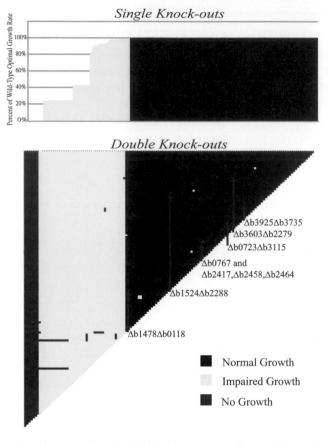

Figure 23.9 The identification of synthetic lethals in core *E. coli* metabolism. The single KOs are the same as in Figure 23.2. Prepared by Alex Thomas.

major categories: (1) they knock-out essential reaction(s) through one Boolean OR relationship in the reaction's GPR relationships, or (2) they knock-out two or more non-essential, compensatory reactions (depending on the extent of the gene's influence). Figure 23.9 shows SL combinations (in red) found among the maximal WT growth area (blue) and in the 'sick' single KO stain area (yellow).

23.2.2   Genome-scale studies

Synthetic gene interactions   Over the past decade there has been significant growth in quantitative data on genetic interactions. However, there is only a limited understanding of the molecular mechanisms through which one genetic alteration modifies the phenotypic effect of another. Although the general properties of genetic interaction networks have been explored phenomenologically, mechanistic understanding of these interactions is lacking.

GEMs offer a mechanistic basis for interpreting and understanding the outcome of gene–gene interaction screens. A number of studies have appeared that use GEMs to address this challenge [157, 159, 401, 412]. These studies show that, broadly speaking,

GEMs need to be improved in their content to have a high overall success rate with synthetic interaction screens, especially for parts of metabolism that are poorly characterized. The results of such screens thus offer the opportunity to validate and improve the reconstructions.

A comprehensive study of gene–gene interactions in yeast covering a large number of pairwise interactions shows how genome-scale reconstructions can be improved based on such data [412]. The original iFF708 yeast metabolic reconstruction had three pathways leading to the synthesis of nicotinate mononucleotide (Figure 23.10). The screen showed synthetic lethality between genes in two of these pathways, but the GEM predicted growth through the use of the third pathway. Further examination revealed that the third pathway does not exist in yeast. It had been included in the yeast model by extrapolation from the E. coli model.

Target identification [133]   Fumarate hydratase (FH) is an enzyme in the TCA cycle. FH catalyzes the hydration of fumarate to malate. Germline mutations of FH are responsible for hereditary leiomyomatosis and renal cell cancer. The absence of FH

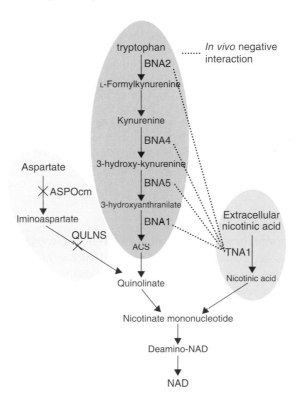

Figure 23.10   Synthetic gene interactions can be used to improve GEMs. Biosynthetic routes to nicotinate mononucleotide in the iFF708 yeast metabolic network reconstruction. Genes involved in the *de novo* pathway from tryptophan show negative genetic interactions with the nicotinic acid transporter gene *in vivo* but not *in silico* because of the presence of a two-step biosynthetic route from aspartate to quinolinate in the reconstruction (ASPOcm, aspartate oxidase; QULNS, quinolinate synthase). From [412].

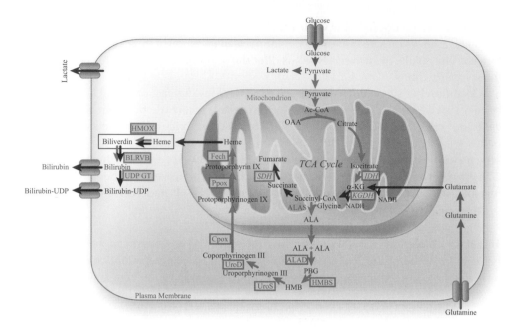

Figure 23.11   The basic metabolic pathways illustrating the normal and interrupted balancing of NADH redox equivalents. Blue arrows indicate synthetic lethal metabolic pathways predicted by the *in silico* model; red arrows indicate genes and reactions found to be unregulated in $Fh^{-/-}$ cells. The scheme also demonstrates the truncation of the TCA cycle observed in $Fh^{-/-}$ cells. Fumarate and succinate are significantly accumulated (in red). The flux through the first part of the TCA cycle is reduced in $Fh^{-/-}$ cells due to decreased pyruvate entry and the absence of recycling of metabolites through the TCA cycle. Glutamine uptake and glycolytic production of lactate (in green) are induced in $Fh^{-/-}$ cells. Modified from [133].

leads to an interruption in the flux through the TCA cycle and an accumulation of fumarate, and thus an interruption of normal balancing of NADH in the central metabolic pathways. A mechanism that explains the ability of cells to survive without a functional TCA cycle was identified using the reconstruction of human metabolism, Recon 1.

Recon 1 predicted and subsequently experimentally validated a linear metabolic pathway that balances the NADH that begins with glutamine uptake and ends with bilirubin excretion. This pathway, which involves the biosynthesis and degradation of heme, enables the use of accumulated TCA cycle metabolites and permits partial mitochondrial NADH production (see Figure 23.11).

Targeting the heme biosynthetic pathway with a metabolic intervention renders FH-deficient cells non-viable, while sparing wild-type FH-containing cells. Thus, in cells with FH deficiency, there are synthetic lethal targets in the heme biosynthetic pathway. These results demonstrate a model-driven identification of a synthetic lethal in metabolism that was subsequently validated experimentally, and give rise to the validation of a novel target for development of a cancer therapeutic.

Extending the synthetic lethal concept [410] One can extend the concept of synthetic lethality by considering gene groups of increasing size where only the simultaneous elimination of all genes is lethal, whereas individual gene deletions are not. An optimization-based procedure for the exhaustive and targeted enumeration of multi-gene (and by extension multi-reaction) lethals for genome-scale metabolic models has been developed [410]. This definition has been applied to the *i*AF1260 *E. coli* model leading to the complete identification of all double and triple gene and reaction synthetic lethals as well as the targeted identification of quadruples and even some of higher order.

Graph representations of these synthetic lethals reveal a variety of motifs ranging from hub-like to highly connected subgraphs. This provides an overall view of the possible ways in which metabolic fluxes can be redirected. Further, it gives a map of epistatic interactions. The analysis of functional classifications of the genes involved in synthetic lethals shows connections within and across clusters of orthologous groups of functional genes.

23.3   Gene Dosage and Sequence Variation

Because the genetic basis for a reconstructed reaction network is explicit, many genetic parameters can be varied. This includes gene dosage and sequence variations.

Reduction or increase in gene product activity The level of gene expression can be explicitly studied. As shown with robustness analysis in Chapter 22, the activity level of a gene product can be altered directly. The effects of increased and decreased expression can thus be assessed in a continuous manner. Similarly, if the gene copy number is amplified, then a quantal effect can be analyzed by using a multiple of the expression level of a particular gene. This latter situation is important when looking at aneuploid effects in cancer cells or in multiplying transfected protein production cell lines. It is also important in analyzing haplo-pro and-in-sufficiency.

Gene imprinting The ability of metabolic network reconstructions to calculate functional states or phenotypes enables the study of the metabolic effects of genetic and epigenetic properties such as dosage sensitivity. Gene imprinting is an epigenetic phenomenon where only one of the parental alleles is expressed and the other is silenced. *Haig's parental conflict theory* has been used to explain imprinting through an inherent competition between the mother and the father [272]. Recon 1 contains nine known or predicted imprinted genes. These can be analyzed using COBRA methods and such analysis finds that four of nine genes had a metabolic effect as predicted by the Haig's parental conflict theory [398].

Epigenotype models can be generated *in silico*. Initial analysis of the epigenotypes follows a six-step procedure (Figure 23.12).

Panel A   Two abnormal epigenotypes resulting in the expression of either no allele (epigenotype I) or the expression of both alleles (epigenotype III) compared to normal epigenotype (epigenotype II).

Panel B   The altered expression resulting from the different epigenotypes can be translated into a set of constraints for the metabolic network through altered

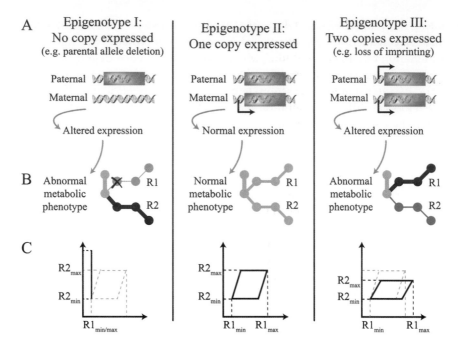

Figure 23.12 Generation of epigenotype models *in silico*. Modified from [398]. Prepared by Nathan Lewis.

expression level. The network homeostasis is then simulated allowing input of essential metabolites while maximizing biomass. Pathways indicated in red reflect increased flux capacity and those in green reflect decreased flux capacity. The abnormal metabolism is then compared to the normal metabolism to predict the phenotypes resulting from the different epigenotypes.

Panel C A diagram detailing FVA (Chapter 20) of imprinting. The comparisons of FVA are done to determine flux capacity changes. For epigenotype II, all fluxes through reactions R1 and R2 resulting in the maximum biomass are within the parallelogram. For epigenotypes I and III, the FVA is compared to the FVA for epigenotype II. In the example shown here, simulation of epigenotype I demonstrates a zero flux capacity for R1 but increased flux capacity for R2 compared to normal. Simulation of epigenotype III demonstrates an increased flux capacity for R1 but decreased flux capacity for R2.

Example: This procedure has been applied to known imprinted genes in human metabolism. The *ATP10A* gene is a maternally expressed gene located within the imprinted cluster in chromosome 15. Simulations of maternal deletion of *ATP10A* using Recon 1 indicated an anabolic metabolism consistent with the known clinical phenotypes of obesity.

Mapping causal SNPs using co-sets [181] Analysis of functional states of networks shows that the activity of biochemical reactions can be highly correlated, forming so-called co-sets (recall Chapter 13). Co-sets represent the functional modules of the network. Thus, detrimental sequence defects in any one of the gene-encoding members of a co-set can result in similar phenotypic consequences as they would alter the functional state in the same way. The relationship between genes and co-sets is shown in Figure 13.7 and three different relationships can be defined. Causal SNPs in genes encoding mitochondrial metabolic functions in human cells can be classified and correlated using co-sets [181].

23.4 Summary

- Genetic parameters can be represented explicitly in genome-scale models, and they can thus be altered to compute phenotypic consequences of changes in the genotype.
- Many genetic parameters have been studied using genome-scale models, including gene dispensability, synthetic lethals, changes in gene dosage, and changes in gene expression or activity of an expressed gene product.
- The application of these methods to address a variety of questions has demonstrated the utility of this approach.

24 Analysis of Omic Data

Network reconstructions provide context for 'content' — Mick Savage

Since the late 1990s, there has been an explosion in the development of technologies that measure cellular content on a genome scale. These methods generate large amounts of data, generally referred to as *omic data*; sometimes referred to as *content*. The quality and coverage of omic data sets has improved steadily with time. Individual data points in omic data sets are treated as independent variables. They can then be correlated statistically to find patterns in the presence of cellular components. Such statistical correlations do not confer causality. GEMs, given their mechanism-based construction, can be used as a context for the analysis of polyomic data sets [358]. Such contextualization can lead to the establishment of causation. Furthermore, GEMs can integrate mechanistically multiple omic data sets, thus aiding in the determination of how various components come together to produce cellular functions and phenotypic states. For some, this *big data to knowledge* conversion represents a grand challenge in biology. Here we will cover the basic principles of omic data analysis using GEMs. More detailed reviews have appeared describing these data-mapping methods and their use [41, 175, 209, 232, 253].

24.1 Context for Content

Over the past 10–15 years, many ingenious methods have been developed to profile the molecular content of cells and to determine the interactions between components. There are vast repositories of the resulting omic data available on various websites, and there are many sources that describe the omics data types, how they are generated, and their availability. We will not repeat these here, but instead turn our attention to the use of COBRA methods to analyze omics data sets.

Networks as a backdrop or context for omics data-mapping Two main network approaches are used to extract biological insights from omic data sets [348]: *inference-based* and *knowledge-based*. Both approaches use an interconnected network of biological components to interpret omic data sets. There are significant differences between the two approaches in how the networks are constructed, and therefore in the range of

Table 24.1   A comparison of key features of biochemical and statistical network models. From [348].

Biochemical reaction networks	Statistical influence networks
Directly mechanistic	Not generally mechanistic
Require significant knowledge of the system	Can be applied without needing prior knowledge (although can be incorporated)
Broadly applicable where biochemistry is known	Broadly applicable without knowledge of biochemistry
Laws of physics and chemistry can be directly applied	Physico-chemical laws typically not applicable or applied
Relate more closely to phenotype (i.e., fluxes)	Relate more directly to omic data
Once reconstructed from biochemical data, network not likely to change (other than additional reactions)	Additional data can lead to significant network rewiring

biological questions that they can answer. The contrast between the two approaches is given in Table 24.1.

Inference-based approaches use statistical methods to construct network models from correlation or recurring patterns in omic data sets [95]. In contrast, BiGG network reconstructions are knowledge bases of organized biochemical, genomic, and genetic data. They enable the structured mapping and analysis of multiple omic data sets simultaneously (Figure 24.1).

GEMs provide a context for integration and interpretation of omics data [358] Given their biochemical, genetic, and genomic structure, bottom-up network reconstructions provide a context for omic data-mapping (see Figure 24.1). This figure outlines the conceptual and formal basis for such omics data-mapping and interpretation procedures. There are many advantages to using GEMs for omics data analysis.

- They help tie the measurements together functionally and are not treated as individual measurements, but rather as *state variables* of a system.
- Network reconstructions capture the relationship between genes and enzyme activities explicitly. The embedded relationships between genomic location, mRNAs, proteins, protein complexes, and enzymatic activities provide explicit ways to integrate omic data with GEMs (Figure 24.1A). Model simulations of phenotypic states, such as specific growth rate (μ), afford the opportunity for comparison of polyomic data sets with phenomics data.
- It is possible to map transcriptome, proteome, and metabolome data on a network map and elucidate the active pathways under a given experimental condition (Figure 24.1B).

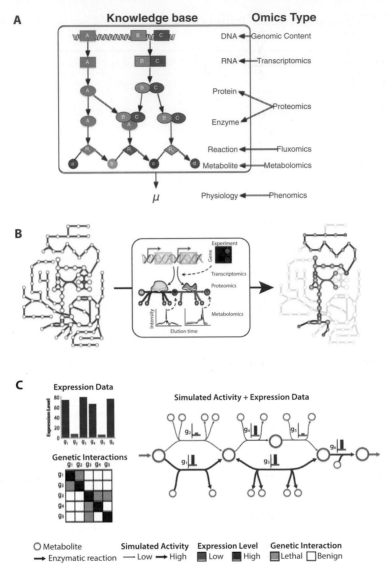

Figure 24.1 Direct comparison of omic data sets and models derived from bottom-up metabolic reconstructions. From [175].

- The examination of omic data sets in the context of GEMs can direct research and provide biological insight (Figure 24.1C). For example, when mRNA expression levels are overlaid on a model simulation we see a high expression level for gene g_4, but the predicted flux for the associated reaction is relatively low. This discrepancy could be due to a measurement error, g_4 encoding for another unknown activity, or indicate that g_4 is regulated post-transcriptionally. Examining genetic interaction data in the context of the network model reveals the underlying reason for lethalities. The double mutants $\Delta g_1 \Delta g_2$, $\Delta g_3 \Delta g_4$, and $\Delta g_3 \Delta g_5$ are synthetic lethal pairs because they

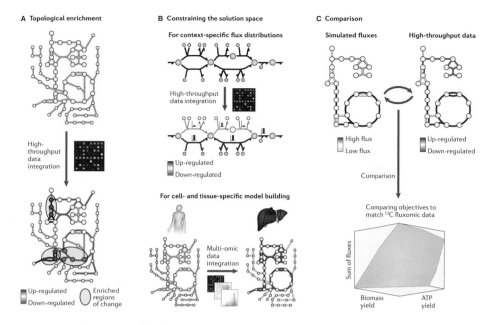

A Topological enrichment

B Constraining the solution space

For context-specific flux distributions

High-throughput
data integration

Up-regulated
Down-regulated

For cell- and tissue-specific model building

Multi-omic
data
integration

High-
throughput
data
integration

Up-regulated
Down-regulated

Enriched
regions
of change

C Comparison

Simulated fluxes   High-throughput data

High flux
Low flux

Up-regulated
Down-regulated

Comparison

Comparing objectives to
match ^{13}C fluxomic data

Sum of fluxes

Biomass
yield

ATP
yield

Figure 24.2   Three uses of high-throughput data with constraint-based models. From [52].

render the network non-functional. Synthetic lethality was discussed in
Chapter 23.

Major classes of application   COBRA methods can be used to interpret and augment
the information content in omic data sets using a biochemically validated underlying
cellular network, and thus help with the *big data to knowledge* conversion [4]. A number
of COBRA methods have been developed to enable omics data-mapping onto reconstructed networks [236]. Three types of applications are described in the following
sections.

1   Similar to pathway enrichment and interaction networks, high-throughput
    data can be integrated with the metabolic network topology to determine
    enriched regions and even significantly perturbed metabolites
    (Figure 24.2A).

2   Omic data add an additional layer of constraints for reaction fluxes
    (Figure 24.2B). Integrated expression-profiling data can determine
    context-specific flux distributions (pathway depicted in red), which increases
    the fidelity of the data as well as the accuracy of flux prediction (upper part).
    Omics data can be used to build cell-/tissue-specific models of human
    metabolism by removing unexpressed reactions (depicted as discolored
    reactions) from the global human metabolic network (lower part). Such
    applications were foreshadowed in Chapter 6.

3   Constraint-based analysis predictions can be compared and validated
    against high-throughput data sets. One can compare COBRA solutions of
    different objectives against data types to assess the prediction of the detailed
    molecular phenotype (Figure 24.2C).

24.2   Omics Data-mapping and Network Topology

Reconstructions can be displayed graphically as reaction maps. Omics data can be displayed on these maps, next to the component that they represent. This procedure organizes the omics data with the known underlying biochemistry at a large scale, possibly a genome scale, although such maps would become very large. The reaction maps allow one to look at omics data at the systems level and help decipher their relationships and the consequences of changed cellular states.

Reporter metabolites [329]   Differential expression data for genes can be mapped onto metabolic networks and used to determine which parts of the network are likely to change the most between the two conditions compared. The metabolites whose concentrations are likely to change the most can be identified, and they are referred to as *reporter metabolites*.

The procedure is outlined in Figure 24.3. The metabolic map in a cell can be represented with a bipartite undirected graph. In this graph, a metabolite node is linked to all the enzymes that catalyze reactions involving this metabolite. The enzyme node is linked to all the metabolites taking part in a particular reaction. A second unipartite undirected graph is constructed to link enzymes that share a common metabolite. Transcriptomic data are then mapped onto these graphs. The algorithm then computes the reporter metabolites as well as coordinately regulated subnetworks. These computations were found to elucidate metabolic perturbations between the conditions being compared. Reporter metabolites are used in an example below (see Figure 24.5).

Expression dynamics of cellular metabolic network [202]   Network structure can be used to correlate multiple data types and genomic features. One would expect intuitively that genes that encode for neighboring enzymes in a metabolic map are coexpressed. Detailed analysis of expression profiling in yeast shows that this is the

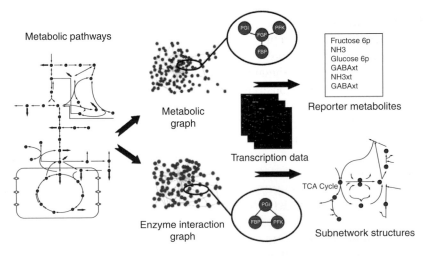

Figure 24.3   The algorithm used to identify reporter metabolites and subnetworks that call out transcriptionally regulated modules. Copyright 2005, National Academy of Sciences, U.S.A. From [329].

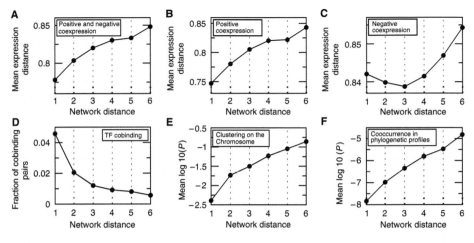

Figure 24.4 Coexpression and functional associations on the scale of the whole metabolic network. Mean expression distance is plotted as a function of the metabolic network distance separating the metabolic gene pairs for (A) all metabolic gene pairs; (B) positively coexpressed pairs; and (C) negatively coexpressed pairs. (D) Fraction of metabolic gene pairs that share at least one transcription factor binding site in their promoter region as a function of metabolic network distance. (E) Mean chromosome clustering gene pair association score dependency. (F) Mean phylogenetic profile cooccurrence association score dependency. From [202].

case (Figure 24.4A–C). Further analysis shows that network distance is also predictive of transcription factor usage, closeness of location in the genome, and phylogenetic profiles. The pristine nature of a stoichiometrically structured network underlies the success of such predictions.

24.3 Omics Data as Constraints

Condition-specific omics data sets can be used as constraints (see Chapter 17). A pioneering paper [8] described mapping omics data as constraints as a proxy for the transcriptional regulatory network to improve the prediction of growth rates of yeast. Since then, a number of methods have been developed to integrate systematically genome-scale omics data as constraints into GEMs.

The use of these algorithms falls into two main categories. The first is to use omics data as constraints on quantitative models that predict phenotypic functions. This use has been evaluated critically and these methods are still in need of development [253].

The second use is a semi-quantitative approach to customize global reconstructions for multicellular organisms to particular cell or tissue types based on expression profiling data from the target cells. This use has led to a number of successful studies. Some have already been described in Chapter 6.

Discovering novel drug targets for cancer chemotherapy [133] In Chapter 23 we learned about synthetic lethal pairs of genes, where the removal of one gene is not lethal, but the removal of both is. The observation that primary cancers have metabolic lesions has led to the search for a synthetically lethal paired gene. Thus, if this paired

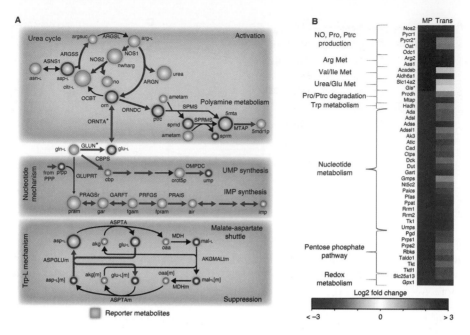

Figure 24.5   High-throughput data support model predictions in macrophage activation. (A) Reporter metabolites provide a global analysis of the expression data. Major changes pertained to predicted pathways of activation and suppression. Green nodes are scaled by degree of enrichment. Circled metabolites in red and blue represent significantly changed metabolites detected. (B) Directionality of computational predictions was in high accordance with the transcriptional and proteomic response of lipopolysaccharide (LPS) stimulated cells. Pycr2, Oat, and Gls expression contradicted model predictions, but the proteomics data confirmed the predictions. Expression profiling (TRANS) data after 24 h of stimulation and Model Prediction (MP) are shown. From [50].

gene is the target of drug intervention then only cancer cells are affected, since normal cells do not express the metabolic lesion.

Candidate synthetic lethalities for cancer were first determined on a large scale [124]. This led to the identification of synthetically lethal pairs to fumarate dehydrate deficiency in the biosynthetic pathway for heme [133]. Experimentation found such a synthetic pair in a hereditary leiomyomatosis and renal cell cancer. Reactions in the heme biosynthetic pathway were thus discovered as targets for the development of cancer chemotherapy. These highly unexpected targets were discovered by using genome-scale models as a context for mapping multiple different data types.

Identifying the role of isoleucine and a transcription factor in virulence [239]   Intracellular pathogens adapt metabolism to their host environment during pathogenesis. One study [239] generated transcriptional profiling data of pathogenic intracellular growth to investigate the relationship between metabolism and pathogenesis of *Listeria monocytogenes*. The data were analyzed both through traditional pathway enrichment analysis and through integration with a genome-scale model of *L. monocytogenes* metabolism.

An algorithm was used that computes a flux distribution which best uses reactions that are associated with up-regulated genes and that avoids using reactions that are associated with down-regulated genes (see Figure 24.2). This leads to a prediction of differential reaction use between conditions. This analysis led to a focused experimentation on highly active pathways. The activity of these pathways was confirmed experimentally by generating conditional knock-out strains. Prospective experiments that were based on the identified pathways showed that limiting concentrations of branched-chain amino acids induced virulence activator genes and elucidated the role of amino acid metabolism in pathogenesis. Isoleucine and a transcription factor (codY) were in this way implicated in virulence and pathogenesis.

24.4 Omics Data and Validation of GEM Predictions

GEMs can compute overall phenotypes. They can predict the detailed expression and activity state of a network by forcing the synchronization of all cellular parts needed to generate the overall phenotypic state. The biological predictions that such computations make can subsequently be validated using omics data.

The most parsimonious flux map [234]   After hundreds of generations of adaptive evolution at exponential growth, *E. coli* grows as predicted using genome-scale metabolic models (Chapter 26). Given the number of alternative solutions (Chapter 20), is the predicted pathway usage in the solutions consistent with gene and protein expression in the wild-type and evolved strains? Analyzing the optimal solutions for wild-type and evolved strains, it was found that:

- more than 98% of active reactions from the optimal growth solutions are supported by transcriptomic and proteomic data;
- the evolved strains up-regulate genes within the optimal growth predictions, and down-regulate genes outside of the optimal growth solutions;
- bottlenecks from dosage limitations of computationally predicted essential genes are overcome in the evolved strains; and
- regulatory processes were identified that may contribute to the development of the optimal growth phenotype in the evolved strains, such as the down-regulation of known regulons and stringent response suppression.

Thus, differential gene and protein expression from wild-type and adaptively evolved strains supported the observed growth phenotype changes, and was consistent with GEM-computed optimal growth states. Adaptive laboratory evolution is discussed in more detail in Chapter 26.

Metabolic junctions in macrophage activation [50]   Macrophages are central players in immune response, manifesting divergent phenotypes to control inflammation and innate immunity through the release of cytokines and other signaling factors. Metabolism plays a critical role in macrophage activation. Genome-scale modeling and polyomic data (transcriptomics, proteomics, and metabolomics) analysis can be used to assess metabolic features that are critical for macrophage activation. The murine RAW 264.7 cell line was used experimentally to determine metabolic modulators of activation. Metabolites well-known to be associated with immunoactivation (glucose and arginine) and immunosuppression (tryptophan and vitamin D3) were among the most critical effectors found. Intracellular metabolic mechanisms were assessed, identifying a suppressive role for *de novo* nucleotide synthesis. The underlying metabolic mechanisms of macrophage activation were identified by analyzing multi-omic data obtained from LPS-stimulated RAW cells in the context of our flux-based predictions. The glutamine junction was found to be a balancing point between activated and suppressed state.

Promiscuous and specific enzymes [277] Enzymes are thought to have evolved highly specific catalytic activities from promiscuous ancestral proteins. By analyzing the *E. coli* iAF1260 genome-scale metabolic model, it was found that 37% of its enzymes act on a variety of substrates and catalyze 65% of the known metabolic reactions. However, it is not apparent why these generalist enzymes remain after an extended period of evolution. An analysis of multiple disparate data types using the genome-scale model as a context found that there are marked differences between generalist enzymes and specialist enzymes, known to catalyze a single chemical reaction on one particular substrate *in vivo*. Specialist enzymes: (i) are frequently essential, (ii) maintain higher metabolic flux, and (iii) require more regulation of enzyme activity to control metabolic flux in dynamic environments than do generalist enzymes. Thus, by using the metabolic network as a context, one can show how environmental conditions influence enzyme evolution toward high specificity, and how some enzymes are not under such selection pressure.

24.5 Summary

- A variety of omic data types are now available that allow for genome-wide measurements of biological components.
- These high-dimensional data sets can be mapped onto network reconstructions.
- Network reconstructions allow the mapping of multiple omics data types simultaneously, thus enabling their mechanistic integration and fundamental interpretation.
- A number of COBRA methods have been developed for omic and polyomic data analysis, and a growing number of basic and applied studies show the utility of this approach.
- Reconstructions thus increase the 'resolution' of understanding of omic data as they can relate all the measurements against a mechanistic context rather than treating the measurements as independent.

25 Model-Driven Discovery

If observed facts of undoubted accuracy will not fit any of the alternatives it leaves open, the system itself is in need of reconstruction – Talcott Parsons

A number of genome-scale networks have been reconstructed based on available data for the target organism. At present, however, there is no organism for which a complete data set exists. GEM-derived predictions based on incomplete reconstructions will fail when the expressed phenotype relies on a missing component or interaction. The mismatch between prediction and observation can be used to build hypotheses about the missing components. Given the high-dimensional models and data sets in genome-scale science, computer algorithms facilitate hypothesis generation greatly [297,478].

25.1 Models Can Drive Discovery

Missing information As detailed in Chapter 3, metabolic models are constructed from a biochemically, genetically, and genomically structured knowledge base. This knowledge base couples the components of a metabolic reaction with its catalytic enzyme and the gene(s) that encodes it. The information used to create the knowledge base is incomplete, and Figure 25.1 summarizes how we might be missing information. A reconstruction can have missing reactions, protein, or genes. This chapter discusses how genome-scale models can be used to discover systematically some of the missing information using a combination of computation and experimentation.

The iterative nature of high-dimensional model-building Network reconstructions represent a knowledge base about the target organism that leads to the construction of predictive models of organism function. Because, at any given point in time, the information available for the target organism is incomplete, the genome-scale model cannot compute all organism functions correctly. Some of the model-derived predictions will be inconsistent with the data obtained and some will be consistent (see step two in Figure 25.1G). The consistent computational outcomes can be considered validations of the GEM (Chapter 4), while the inconsistent ones represent failure of prediction. The analysis of these failures leads to model updates and a re-versioning of the model. The model-building process is therefore iterative and constantly incorporates and

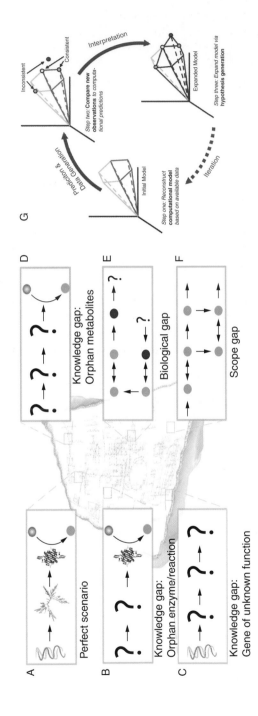

Figure 25.1 Overview of missing information in metabolic reconstructions. The case of complete information (Panel A). Knowledge of metabolism is incomplete and contains gaps in the form of known unknowns that arise due to orphan enzyme activities (Panel B), genes with unknown functions (Panel C), and orphan metabolites (Panel D). Network gaps can be identified in metabolic reconstructions as blocked reactions and dead end metabolites (Panel E) and missing pathways or scope gaps (Panel F). Prepared by Ottar Rolfsson. Reconstructions and model-building drive biological discovery in an iterative manner (Panel G). Prepared by Jeff Orth, based on [83].

reconciles more information. The additional information incorporated may result from bibliomic data or consultations with experts on the target organism, or from new experiments. How, then, are new experiments designed to reconcile inconsistencies? First, one needs to classify the types of prediction failures that occur, analyze the types of gaps in a network reconstruction, and devise a way to fill those gaps.

Failure modes When building GEMs, organism functions are predicted and then measured and the outcomes are compared. Some of the most common predictions are those associated with the ability of an organism to grow. Thus, growth is predicted in a given environment and then measured. This prediction is qualitative – either growth or no growth. The outcome of a growth/no growth comparison between prediction and experiments can be organized into a table (see Table 23.2 and Figure 23.5).

False positives occur when the model predicts growth incorrectly, and therefore has unrealistic capabilities or missing down-regulation of key gene products that enable growth. False negatives occur when the model fails to predict growth, indicating that the model is missing content. False-positive and false-negative predictions open the door to gap-filling and iterative model-driven discovery opportunities.

The commonly found types of missing information in metabolic network reconstructions are (Figure 25.2): (1) *gaps in the network* where a reaction is missing, or (2) *orphan reactions* that are known, required reactions in the network with no gene association.

Three types of network gaps The first category of missing information is represented by gaps (sometimes called holes) in the network where a reaction that should occur in the organism is absent. These gaps are manifested in network reconstructions

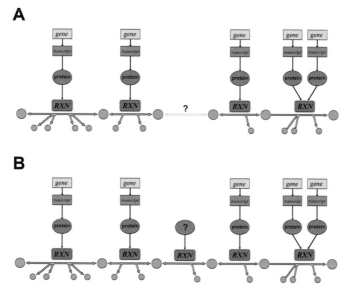

Figure 25.2 Two overarching gap categories: gap reactions (A) and orphan reactions (B). From [297].

as *dead-end metabolites*, which have a producing reaction but no consuming reaction (Figure 25.2A). It is also possible for a metabolite to be consumed in a reaction but not produced by any reactions. These two types of dead ends have been termed *root no-consumption metabolites* and *root no-production metabolites* [370]. When simulating steady-state flux maps in these networks, the reactions producing or consuming dead-end metabolites can never carry flux under any conditions. These reactions are said to be *blocked*. Further, any reactions upstream or downstream from them will also be unable to carry flux under steady-state conditions.

There are several reasons why there may be gaps in a network. First, it is possible that the actual biochemical network is missing an enzyme that is part of a completed pathway in related organisms. An example of this situation is in the O-antigen synthesis pathways of many *E. coli* K-12 strains, in which an IS5 insertion element has disrupted a rhamnosyltransferase gene, inactivating it [238, 367]. These strains are therefore unable to produce a functioning rhamnosyltransferase enzyme, blocking the entire downstream pathway that produces O-antigen. In such cases, there really is a gap in the biochemical network. The *E. coli* K-12 MG1655 reconstruction *i*AF1260 [115] contains a biologically realistic gap that is unable to carry flux. This gap is a *true biological gap*, as the blocked metabolites and reactions in the model are blocked in the actual organism as well; *E. coli* strains with this mutation are unable to produce O-antigen.

A second reason for gaps in network models relates to the scope of the models themselves. To date, most genome-scale network reconstructions do not include other systems, such as signaling or transcription and translation. Metabolites that are produced in metabolism but then enter these other systems may be left as gaps in models, even though their biological functions are known. These are known as *scope gaps*. An example of scope gaps are the tRNAs in *i*AF1260. This reconstruction contains tRNA charging reactions, but there are no consuming reactions for these tRNAs, even though it is well-known how charged tRNAs are used in the process of translation.

Finally, there may be a gap because it is not known what biochemical reaction produces or consumes a certain metabolite. In this case, the gap is not biologically realistic as it is the result of limited knowledge. These are called *knowledge gaps*, and to fill them, new biological discoveries must be made.

Orphan reactions The second category of missing information in a metabolic network is orphan reactions. These are biochemical reactions that are known to occur but are catalyzed by an unknown gene product (Figure 25.2B). Such reactions can be identified without genes using several different types of evidence, including biochemical assays of crude cell extracts, a known phenotype such as uptake or secretion of a particular substrate, or the implied presence of a reaction due to the presence of other genes in a conserved pathway.

Even the most well-studied organisms have many genes with unknown functions, and many of these genes may code for orphan reactions. *E. coli* K-12 MG1655, for example, still has 981 partially or fully uncharacterized genes out of a total of 4495 genes, according to version 13.6 of EcoCyc [200]. However, it is also possible that genes with currently known functions may also catalyze orphan reactions, so these genes must not be ignored when attempting to identify the gene or genes associated with an orphan.

Orphan reactions can be *local*, where the gene is unknown in one organism but known in at least one other. They can also be *global* orphan reactions, in which there are no known genes in any organism that code for a catalyzing enzyme. It has been determined that 30–40% of all known enzymatic activities are global orphans [197, 231]. Some apparent orphans may be due to inadequate or incorrect information in databases, but most are truly unknown [338]. Global orphans are evenly distributed across the different types of known biochemical reactions, and most have been characterized experimentally in more than one organism [72]. A database called ORENZA (http://www.orenza.u-psud.fr/) lists currently known global orphan reactions [230].

Directing discovery To address the challenge of missing information, algorithms have been developed to determine the probable gene candidates that fill knowledge gaps in network reconstructions (Figure 25.3). These algorithms utilize global network topology and genomic correlations, such as genome context and protein fusion events [71], as well as local network topology and/or phylogenetic profiles [71, 201]. Similar tools have been developed that utilize mRNA coexpression [203] and can evaluate more general metabolic pathway databases [150].

In addition to these network topology-based methods, an optimization-based procedure has also been developed to fill network gaps and evaluate reaction reversibility along with adding additional transport and intracellular reactions from databases of known metabolic reactions [370]. These studies produce specific targets for drill-down experiments needed for confirmation of these computationally generated hypotheses. We will now describe the algorithms and give examples of drill-down experiments that have resulted in the discovery of new biological information. Table 25.1 summarizes some of the published experimental discoveries resulting from drill-down studies.

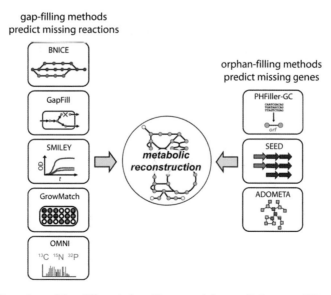

Figure 25.3 Overview of the different algorithms used for predicting gap-filling reactions and orphan-filling genes. From [297].

25.2   Predicting Gap-filling Reactions

SMILEY   False-negative predictions lead to gap-filling. An early algorithm called SMILEY was created to predict which reactions are likely missing from a network when the model predicts no growth but growth is indicated experimentally [357] (see Figure 25.4). A matrix **U**, containing an extensive list of known metabolic reactions along with a matrix **X**, containing reactions for exchanging intracellular metabolites, was constructed. SMILEY uses mixed-integer linear programming (MILP) to attempt to identify a flux distribution that leads to growth on the substrate of interest while minimizing the total number of reactions added from **U** and **X** to the stoichiometric matrix for the target organism. It thus predicts which reactions should be added to the model in the most parsimonious fashion (i.e., adding the smallest number of reactions necessary) to reconcile *in silico* and *in vivo* growth predictions.

A drill-down example   Analyzing high-throughput growth screens for *E. coli*, 54 conditions were identified in which growth was observed but the *i*JR904 GEM predicted no growth [357]. These 54 cases were analyzed leading to drill-down experiments. Below is one such example.

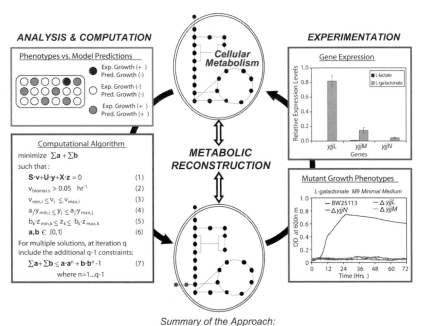

Summary of the Approach:
1. Use the model to find positive growth environments not explained by the model.
2. Use the model to hypothesize what transport and/or enzymatic reactions are missing.
3. Use bioinformatics tools and experimentation to find genes responsible for missing activities.

Figure 25.4   The SMILEY algorithm for reconciling computation and observation. There are basically four steps: (1) determining discrepancies, (2) finding candidate missing reactions using a MILP algorithm, (3) knocking-out the candidate genes to see if discrepancy disappears, and (4) if it does, then clone and express the gene to confirm the predicted biochemical function. From [357].

Table 25.1 Summary of applications of gap-filling and orphan-filling methods and new discoveries made. From [297].

Discovery/application	Gap-filling method used	Validation methods used	Refs.
putP as *E. coli* propionate transporter	SMILEY	Gene knock-out phenotype, up-regulation of gene detected (RT-PCR)	[357]
idnT as *E. coli* 5-keto-d-gluconate transporter	SMILEY	Gene knock-out phenotype, up-regulation of gene detected (RT-PCR)	[357]
iolC as a possible *E. coli* thymine transporter (not verified)	SMILEY	Measured thymine excretion (HPLC), gene knock-out phenotype	[357]
dctA, yeaU, yeaT as D-malate uptake genes	SMILEY	Gene knock-out phenotypes, up-regulation of genes detected (RT-PCR), assay of purified YeaU enzyme, ChIP analysis of YeaT	[357]
Pyrimidine catabolism pathway in *Y. enterocolitica, Acinetobacter* sp. AP1, *P. syringae, C. crescentus,* and *A. tumefaciens*	SEED	Gene knock-out phenotypes in *E. coli,* detection of pathway products in *E. coli* (GC/MS)	[241, 302]
dipA and *dipB* in DIP synthesis pathway of *T. maritima*	SEED	Assay of purified DipA enzyme, assay of DipB and entire pathway in cell extracts	[365]
ynel as *E. coli* succinate semialdehyde dehydrogenase	ADOMETA	Gene knock-out phenotypes, assay of purified YneI enzyme	[135]
86 new reactions for *i*PS189 *M. genitalium* reconstruction	GapFill and GrowMatch	Improved accuracy of model gene essentiality predictions	[409]
Refinement of *i*MA945 *Salmonella* reconstruction	GapFill and GrowMatch	Improved accuracy of model growth phenotype predictions	[5]
32 new reactions for *i*Bsu1103 *B. subtilis* reconstruction	Modified GrowMatch	Improved accuracy of model gene essentiality and growth phenotype predictions	[162]

413

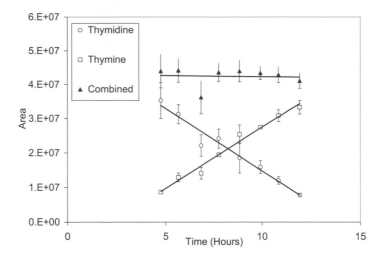

Figure 25.5   SMILEY algorithm predicts adding thymidine transport reaction to the *E. coli* model will enable growth and secretion of thymine. HPLC analysis shows that thymine concentration in the media increases at the same rate as thymidine concentration decreases. From [357].

E. coli can grow on thymidine, but the GEM computed no growth. SMILEY predicted that *E. coli* would be able to grow on thymidine if a transport reaction for thymine is added to the network reconstruction. The GEM then predicts that thymidine is consumed, split into deoxyribose-1-phosphate and thymine by the thymidine phosphorylase reaction, and thymine is then excreted while deoxyribose-1-phosphate is used as a carbon source.

HPLC analysis of *E. coli* growing on thymidine confirmed this mechanism by showing that thymine accumulation in the media increased as the concentration of thymidine decreased in the media (Figure 25.5). Thus, the biochemical prediction was verified.

Over two dozen gene knock-out strains were screened for potential thymine transport activity, and only the ΔtolC (b3035) strain showed a reduced growth rate on thymidine, indicating that this gene may encode a part of a thymine transporter. Transporters are generally composed of many subunits, and other gene encoding subunits of the transporter would have to be identified by constructing double, and even triple, knock-out strains. Multiple KO strains were not constructed and the full identity of the transporter was not established, thereby leaving the possibility for further biological discovery open to future research.

BNICE   When larger gaps exist in a biochemical network between structurally unrelated chemical species, there are potentially many possible sets of reactions that might fill the gap and restore connectivity. Several *pathway predictor* algorithms have appeared that compute a series of reactions connecting two compounds.

An early algorithm in this category called BNICE was developed to identify all biochemical reactions that could link two metabolites realistically [158]. In BNICE, molecules are represented as bond–electron matrices and reactions are represented by matrix addition. BNICE then searches iteratively for all possible sequences of chemical

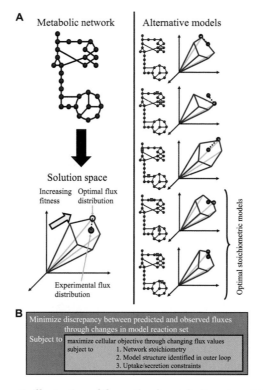

Figure 25.6 (A) Schematic illustration of the optimal metabolic network identification approach. Changes in the model reaction set lead to changes in the FBA predicted optimal flux distribution (yellow) that can be compared to the experimental fluxes (red). (B) Bi-level optimization scheme for optimal metabolic network identification. From [166].

transformations from one metabolite to another, offering a pathway prediction as a gap-filling hypothesis between chemically different molecules. Although gap-filling predictions generated by BNICE have not yet been verified experimentally and published, pathway-finding algorithms are clearly a valuable tool.

OMNI False-positive predictions can also be dealt with algorithmically. OMNI is a MILP-based algorithm that compares *in silico* predictions to experimental measurements in an attempt to improve a constraint-based model (Figure 25.6 [166]). OMNI uses measured metabolic flux data, which is obtained through ^{13}C labeling experiments. OMNI compares fluxes measured experimentally to those that were predicted, and then seeks to minimize the total difference between measured and predicted fluxes by adding or removing reactions while maintaining a predicted growth rate above a defined minimum. It uses a matrix **F** that contains fixed reactions which cannot be deleted and a matrix **D** that contains reactions that may be deleted. To improve a model by predicting missing reactions, a library of candidate reactions such as the **U** matrix used by SMILEY can be provided as **D** to the algorithm, which will then add reactions as needed to achieve a more realistic flux distribution.

The OMNI algorithm has many potential uses other than gap-filling. Because it can remove reactions as well as add them, it can be used to remove reactions corresponding to poorly annotated genes if they do not match the experimental data. OMNI can also be used to identify alternative reaction mechanisms or bottleneck reactions in evolved strains or metabolically engineered strains. To date, no gap-filling predictions made by OMNI have been verified experimentally, but the algorithm has been used successfully to identify the reasons why some evolved strains did not reach their computationally predicted optimal states [166].

GapFind and GapFill   While SMILEY and OMNI both compare computational predictions to experimental data to add missing reactions to the model, they do not specifically target and attempt to eliminate gaps in networks. GapFind and GapFill are algorithms that do not require any experimental data, and simply attempt to minimize the total number of gaps in a model [370]. GapFind is a MILP algorithm that can identify every gap in a network by identifying blocked metabolites, i.e., those that cannot be produced or consumed at steady-state under any conditions.

GapFill is another MILP algorithm, and its objective is to minimize the total number of gaps by reversing the directionality of existing reactions, adding new reactions (as in SMILEY and OMNI), adding transport reactions for blocked metabolites, or adding intracellular transport reactions between compartments for multi-compartment models. GapFill attempts to reduce the number of gaps with the smallest number of model modifications possible.

GrowMatch   GrowMatch uses experimentally determined gene essentiality data to identify incorrect model predictions [217]. The GrowMatch algorithm begins with the simulation of the effects of a gene knock-out in the model. Two types of inconsistencies are possible: false positive (FP) and false negative (FN) (see Table 23.2). GrowMatch attempts to correct these two types of disagreements differently. In the case of a FP mutant, the model has some extra capabilities that are not realistic and should be removed or constrained. FN mutants are missing capabilities. Corrections to FP or FN mutants made by GrowMatch can be either global or conditional. Global corrections resolve at least one inconsistency while creating no new inconsistencies, and conditional corrections resolve one inconsistency while creating inconsistencies in other mutant strains.

25.3   Predicting Metabolic Gene Functions

PathoLogic pathway hole filler   PathoLogic is a program for constructing metabolic networks from annotated genomes automatically. It uses EC numbers, Gene Ontology terms, or annotated gene names to map reactions to genes. It then assembles the reactions into pathways by comparing the reactions to the reference database MetaCyc and adding any missing reactions [196, 308]. After performing these steps, however, many of the reactions in the new pathways may be orphans.

PathoLogic includes a *hole-filler* program that attempts to identify the genes associated with these reactions [150]. The first step is to identify genes in other organisms that code for enzymes that catalyze an orphan reaction from the Swiss-Prot and PIR [459] databases. The sequences of these genes are then compared to the genome of the organism of interest using BLAST [16], and genes with similar sequences are identified. After identifying candidate genes, the hole-filler program then uses a simple Bayesian network to calculate the probability that each gene actually encodes an enzyme that catalyzes the orphan activity. By testing the hole-filler on known reactions in the Pathway/Genome Databases for *Caulobacter cresentus*, *Mycobacterium tuberculosis*, and *Vibrio cholerae*, a precision of 71% was achieved with a *p*-value cutoff of 0.9.

A subsystems approach   Given the increasing number of fully sequenced microbial genomes, the Fellowship for Interpretation of Genomes (FIG) launched the Project to Annotate 1000 Genomes. This large-scale project uses a subsystems approach to annotate these genomes [305]. Briefly, a *subsystem* is defined as a set of *functional roles* that act together to carry out a specific biological process or form a complex. Traditional biochemical pathways are examples of subsystems.

One subsystem is annotated at a time across many different organisms [304], in contrast to the traditional approach of annotating one full genome at a time. The organization of genes in many different genomes into common subsystems can be beneficial in attempts to characterize genes and fill gaps. First, subsystems can help indicate the presence of gaps even in organisms for which formal metabolic network reconstructions have not yet been built. When one organism is missing a particular subsystem component (a reaction) that most other organisms with the same subsystem have, it is an indication of missing content.

Bioinformatics methods can then be used to analyze the genes comprising the subsystems in these other organisms to identify which genes in the target organism likely encode the missing reactions [303]. Analysis of chromosomal clustering has proven to be the most useful analysis method [302]. Genes that encode enzymes in the same pathways are often clustered close together on the genome, so genes and their orthologs that are found near each other in multiple genomes are likely to have related functions. This approach followed by experimental verification has already led to several new discoveries.

A drill-down example   Comparative genomics methods in the context of subsystems were used to fill a gap in the metabolism of *Thermotoga maritima* [365]. This organism, like other hyperthermophilic bacteria and archaea, produces di-*myo*-inositol 1,1'-phosphate (DIP) as an osmoprotecting metabolite. The four reaction pathway that produces DIP in many of these organisms has been characterized partially, and the genes encoding the first two steps in *T. maritima*, *ips* (TM1419) and *imp-* (TM1415), were already known. However, the genes encoding the enzymes for the final two steps had not been discovered in any organism.

Chromosomal clustering and phylogenetic profiles were used in the SEED platform to predict which genes may be responsible for the two reactions. It was found that the genes *dipA* (TM1418a) and *dipB* (TM1418b) tend to cluster with *ips* and *imp* in several types of thermophilic bacteria and archaea (Figure 25.7). These genes also had similar phylogenetic profiles and were even combined into one gene (*dipAB*) in several organisms. *DipA* was expressed in *E. coli* and purified, while *dipB* was cloned into *E. coli* but could not be purified. Purified *dipA* was then confirmed to carry out its predicted CTP:inositol monophosphate cytidylyltransferase reaction *in vitro*, and

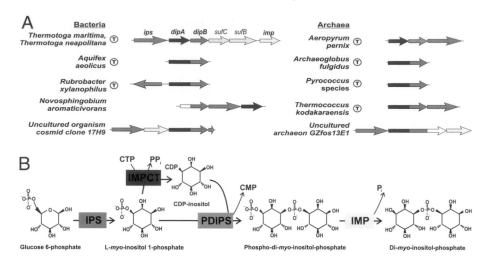

Figure 25.7 The SEED algorithm allows for cross-species genome-structure analysis for the identification of missing reactions in *T. maritima* (A). Chromosomal clustering and phylogenetic profiling predict which genes are necessary to fill the last two steps in the reaction mechanism (B). Copyright 2005, National Academy of Sciences, U.S.A. From [365].

crude extracts of *E. coli* with *dipB* expressed were assayed for DIP synthase activity. This reaction was found to produce DIP along with a P-DIP intermediate. Finally, the entire pathway was cloned from *T. maritima* into *E. coli* and was confirmed to function as expected. Among the other organisms that were predicted to contain this pathway is *Aeropyrum pernix*, which was shown by ^{31}P NMR to produce DIP.

ADOMETA A bioinformatics framework called ADOMETA has been developed that analyzes gene coexpression, phylogenetic profiles, chromosomal clustering, and protein fusions to find association between similar reconstruction components from many species [71, 201, 203]. ADOMETA uses the local network structure around orphan reactions in a metabolic network reconstruction to generate hypotheses for which genes are associated with these orphans. The first level of the local network consists of the genes associated with the reactions adjacent to an orphan in the model, i.e., reactions that share a common metabolite. The second level of the network consists of the genes associated with reactions adjacent to the first level, and so on. The genes in the organism with unknown functions (and genes with known functions as well) can then be compared to the genes in the local network surrounding an orphan reaction using different types of functional association evidence. Genes that are close to each other in a metabolic network are more likely to have similar coexpression profiles, phylogenetic profiles, and so forth.

Analysis of gene coexpression gives functional association evidence. Gene expression data are available for many different organisms under many conditions. When two genes have similar expression patterns, they are likely to have related functions because they may be used in the same functional pathway. The expression patterns of different genes in an organism can be correlated easily, and it is also possible to compare expression of orthologous genes in different organisms, which improves the effectiveness of this method.

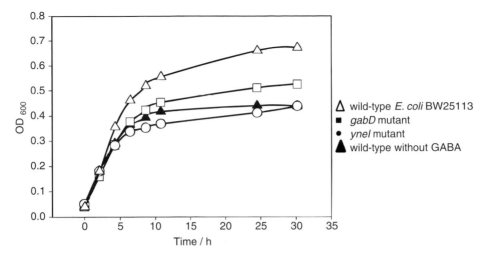

Figure 25.8   ADOMETA algorithm tests all *E. coli* genes with unknown function for SSADH activity and identifies two promising candidates. Mutant strains of these genes show inhibited growth, possibly due to succinate semialdehyde toxicity. From [135].

Phylogenetic information can also be used for inferring associations. A phylogenetic profile consists of a vector of ones and zeros, with a one for every organism that contains an ortholog of a certain gene and a zero for every organism that does not. Genes with related functions evolve together, and thus tend to be found in the same organisms. Genes with the most similar phylogenetic profiles are therefore more likely to exist in the same pathways.

A drill-down example   For over 25 years, it has been known that *E. coli* is able to carry out the NAD^+-dependent succinate semialdehyde dehydrogenase (SSADH) reaction, but it was not known which gene encoded the enzyme [98]. This reaction converts the toxic intermediate succinate semialdehyde to succinate. Using the *i*JR904 GEM, ADOMETA was used to test all *E. coli* genes with unknown function for potential SSADH activity. The genes *ynel* (b1525) and *ydcW* (b1444) were identified as the top two candidates [135]. As *ydcW* had been identified recently as a γ-aminobutyraldehyde dehydrogenase, it was not investigated.

A Δ*ynel* strain along with a Δ*gabD* (b2661) strain (missing the known gene encoding $NADP^+$-dependent SSADH) were grown in M9 glycerol media with GABA as the only nitrogen source. Both strains grew to lower concentrations than wild-type, presumably due to succinate semialdehyde toxicity (Figure 25.8). These strains were also grown in media containing succinate semialdehyde, and again did not grow as well as wild-type. NAD^+- and $NADP^+$-dependent SSADH activities were assayed in crude cell extracts, and this activity was not detected in the Δ*ynel* strain but was detected in Δ*gabD* and wild-type. Finally, the YneI and GabD proteins were expressed and purified, and YneI was found to accept both NAD^+ and $NADP^+$ as co-factors. All of this evidence confirmed the ADOMETA prediction that *ynel* encodes the NAD^+-dependent SSADH of *E. coli*.

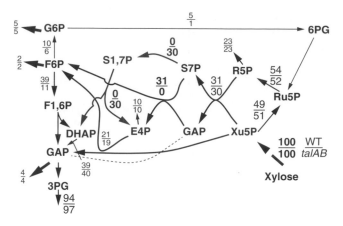

Figure 25.9 Discovery of new reactions catalyzed by two glycolytic enzymes, phosphofructokinase and aldolase. Flux distribution of the wild-type and *talAB* knock-out mutant growing on xylose. The novel reactions are shown in red and the transaldolase reaction that is removed from the KO strain is shown in blue. The relative flux through each reaction (with consumption of xylose set to 100) is shown (upper number, wild-type; lower number, *talAB* KO strain). From [275].

Discovering new reactions carried out by characterized enzymes GEM-guided comparison of growth and metabolism between knock-out and wild-type strains can lead to the discovery of new functions. The pathways of central carbon metabolism are considered to be well known. However, a detailed study of multiple KO strains of *E. coli* genes involved in central carbon catabolism grown under 12 different nutrient conditions revealed new functions of familiar glycolytic enzymes [275].

Differences between model predictions and experimental data indicated that unreported reactions existed within this extensively characterized metabolic network. Novel reactions involved in the breakdown of sedoheptulose-7-phosphate (a pentose pathway intermediate) to erythrose-4-phosphate and dihydroxyacetone phosphate were detected in transaldolase KO stain mutants, without any noticeable changes in gene expression. Two reactions, triggered by the accumulation of sedoheptulose-7-phosphate, were catalyzed by two universally conserved glycolytic enzymes, phosphofructokinase and aldolase (Figure 25.9). Thus, new GPRs result in the *E. coli* GEM. This example underscores the challenge of finding all connections between the proteome and the reactome.

25.4   Summary

- Discrepancies between experimental outcomes and prediction from genome-scale models can lead to the discovery of missing parts of the underlying network reconstruction.
- The two principal types of missing information are (1) reactions that create gaps in the network, and (2) a missing genetic basis for a known reaction; so-called orphan reactions.

- There are a number of algorithms now available for identifying candidate gap-filling reactions and genes to associate with orphan reactions. These different methods rely on a number of different types of data as input.
- A growing number of drill-down studies show the utility of gap-filling algorithms.
- As new reconstructions are built, one can expect the use of these gap-filling methods to become a standard part of the reconstruction process. They are already a part of some automated reconstruction protocols.

26 Adaptive Laboratory Evolution

Darwin would be amazed to see where his ideas have led – Richard Lenski

Evolution is fundamental to biology. We now have the potential to observe it in the laboratory, to define its dynamics, and to determine the genetic bases that enable new phenotypes. What happens in the laboratory may not be directly applicable to what happens in natural habitats, however, unless that habitat can be reproduced accurately in the laboratory setting. Nevertheless, adaptive laboratory evolution (ALE) is opening up new possibilities for the fundamental research of biology. The term *distal causation* is used to describe changes in biological properties over many generations. We can now not only control short-term evolutionary processes in the laboratory, but through inexpensive whole genome re-sequencing, also determine the genetic basis for distal causation.

26.1 A New Line of Biological Inquiry

ALE can be used to study the genotype–phenotype relationship   ALE has been used extensively to study the genetic and biochemical basis for bacterial adaptation. Using whole-genome re-sequencing, the mutations that are selected during ALE can be identified readily. The introduction of these mutations into the starting strain allows for the determination of causality of the identified mutations. This determination is performed by first introducing one mutation at a time, then the pairwise combinations, then triple combinations, and so forth until the full complement of the mutations found has been introduced. Newer genome editing methods allow the introduction of multiple genetic changes simultaneously.

Experience to date has shown that if the dominant mutations are relatively few, then the determination of the genetic basis for adaptation is possible. For a more complex genetic basis, new methods like MAGE [450] may enable one to delineate the effect of many mutations, where each one contributes in a small way to the overall phenotypic change. However, finding the biochemical functional changes in the gene products and how they affect the phenotype has proven to be a challenging task. When successful, the elucidation of the underlying molecular mechanisms leads to the discovery of new cellular processes and a deeper understanding of the functions of the mutated gene products.

ALE is thus a process that can be used to study the genetic basis for the generation of new phenotypes under a chosen selection pressure in a controlled laboratory setting. Furthermore, with recent advances in experimental and genome-scale modeling methods, this approach is likely to become increasingly used for a variety of basic and applied research purposes.

The role of GEMs ALE experimentation can be guided using genome-scale models. GEMs are based on the genetic composition of the target organism and the predicted phenotypic states are based on an optimization criterion. If the optimization criterion used is consistent with the selection pressure applied, then predictions of outcomes of ALE experiments can be made. In other words, one can compute the outcome of an ALE experiment before performing it.

GEMs can be used to compute a variety of phenotypic states. Early on, GEMs were used to predict optimal growth rates of laboratory strains under defined growth conditions. Such predictions worked well for some substrates, but not all. The failure to predict optimal growth rates under certain substrate conditions led to the hypothesis that the organism had the latent ability to grow at the predicted rates but was not adapted to do so. This hypothesis has been addressed using ALE [177]. The procedure of GEM formulation (see Part I), failure of predicting growth optimal states in a production strain through the use of phenotypic phase planes (see Chapter 22), and performing ALE to an optimal state is illustrated in Figure 26.1.

Performing ALE ALE experiments are carried out in a variety of ways. Figure 26.2 shows two commonly used approaches to ALE experiments: (a) they can be done through serial passing, with or without entering stationary phase; or (b) cells can be grown in a chemostat for extended periods of time. Other culture methods can also be used. For example, cell cultures can be continued based on the properties of individual colonies observed on agar plates, and only the cells from a particular colony are streaked onto the next plate. Population-based selection of the fastest growth rate phenotype is perhaps best done through serial passage in mid-exponential phase. This procedure puts the cell population into extended exponential growth in which a mutant with a slight growth advantage will increase exponentially in abundance in the population. It has proved to be very efficient to evolve fast-growing strains.

The trait being selected for improves over time. If cells are stored during an ALE experiment, then one can create a fossil record for the experiment. Frozen cells can then be revived after the experiment to obtain a time profile of how the genotype and the phenotype change over the course of the ALE. The time coordinate in these experiments is normally scaled in terms of generations, although the total number of cell divisions during the experiment might be a better ordinate as the generation of mutations presumably happens during cell division due to sequence errors associated with the DNA polymerase [221]. Significant adaptation has been shown to occur during the first 500 to 2000 generations. However, bacteria can continue to improve their growth rate over tens of thousands of generations. The definition of an 'endpoint' thus refers to the state of the population at the time the experiment was terminated.

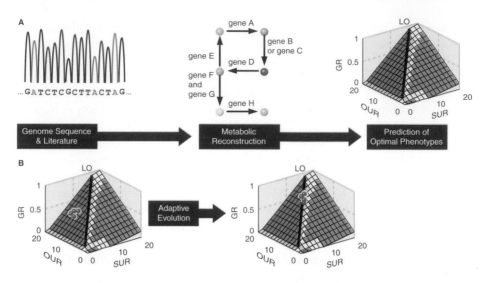

Figure 26.1   GEMs can be used to predict the outcome of ALE experiments. (A) Metabolic reconstructions allow for the assessment of phenotypic potential, as, for example, described by PhPPs (see Chapter 22) (SUR, substrate uptake rate; OUR, oxygen uptake rate; GR, growth rate). On the 3D graph, the optimal phenotype occurs along the line of optimality (LO). (B) In cases where microorganisms have been found to grow suboptimally initially (such as on an unusual substrate, or after gene knock-outs which disrupt the wild-type metabolic network) the microorganisms acquire mutations during ALE to optimal phenotypes. From [316].

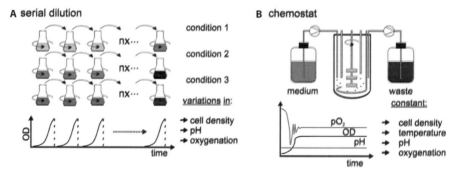

Figure 26.2   Illustration of the two most common ALE methods: (A) serial dilution and (B) chemostat. From [99].

26.2   Determining the Genetic Basis

Complexity of the genetic basis   After a new phenotype has been developed through ALE, a clone from the endpoint population can be isolated and its DNA sequenced. The full genome sequence of the starting and the endpoint strains are obtained, and the sequence differences can be verified by small-scale sequencing of the region where a sequence difference is found. Sometimes, the number of mutations found in an endpoint strain of ALE is relatively low – on the order of 2 to 6 – and the same genes are mutated in replicate ALE experiments. Conversely, there are sometimes

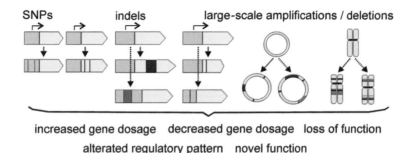

Figure 26.3   Various types of sequence changes occur during an ALE experiment. From [99].

many mutations found and different gene products are mutated in replicate ALE experiments. In some cases, a mutator strain has appeared and a large number of mutations can be found in the endpoint [400].

Types of genetic changes found   A variety of different mutations and genomic changes are found in the endpoints (Figure 26.3). Typically, most of the sequence changes found are point mutations. In many cases, however, up to a third of the differences are found in the form of small *indels* (sequence insertions or deletions) between 1 and 80 bp in length. Occasionally, large-scale duplications or rearrangements are found. Thus, a variety of sequence changes are found in the endpoints. The number, type, and extent of genetic changes will vary from one selection pressure and duration of selection to another.

Determining causality   For strains where genetic manipulation tools are available, the mutations found at the end of the ALE can be introduced into the starting strain. In *E. coli*, for which a good genetic manipulation system is available, an allelic replacement takes up to three weeks, and serial introductions of up to six mutations have been carried out. Thus, the endpoint strains can be 'reconstructed' by introducing all the mutations found in the endpoint of ALE. Clearly, with an increasing number of mutations, this task becomes more and more difficult and can take more than half a year to do all strain reconstructions of interest. Fortunately, methods have been developed to introduce multiple mutations into cells simultaneously [450], which will make this task much easier in the future and make the reconstruction of complex genetic bases feasible for new phenotypes.

The strains in which one or more mutations have been introduced can be phenotyped along with the starting and the endpoint strains. Such comparison allows for the identification of the causal mutations and any epistatic interactions between them (Figure 26.4). In several cases, a full reconstruction of the phenotype is achieved, allowing for the determination of the genetic basis for the phenotypic change.

Dynamics of adaptation   With the causal mutations found in the endpoint strains and the stored strains from the intermediate points of the ALE, one can monitor the dynamics of fixation of the mutations (Figure 26.4). Interestingly, the endpoints in ALE

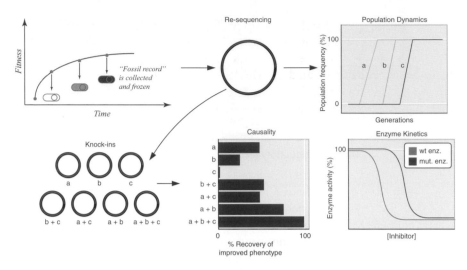

Figure 26.4   Tools for understanding the genetic basis for the phenotypes developed during ALE. Sequencing technologies are now available to determine readily and inexpensively mutations acquired during ALE. Known mutations can be screened for in frozen stocks of evolutionary time-points to determine the dynamics of mutation fixation on the population level. Mutations can be reintroduced singly and in combination into the ancestor strain by allelic replacement, allowing the mutation contribution to phenotype, genetic interactions, and molecular mechanisms to be determined. Modified from [273,316].

sometimes have a 100% allelic frequency of the causal mutations found in the end-point population.

At a population level, one can trace the abundance of a mutation over the time course of the ALE experiment. The time of appearance of the mutation can be determined, as well as the rate at which it fixes and how completely it 'sweeps' the culture (Figure 26.4). Mutations begin to appear fairly early in a typical ALE experiment where there is strong selection pressure. In general, the most causal mutations tend to fix first, followed by those that are less influential, somewhat in order of their contribution to the phenotype. Thus, one sees a gradual decrease in phenotypic change with time of ALE. This sequence of appearance of mutations is expected if they are relatively independent, with the most causal fixing first. If positive interactions between mutations occur, known as epistatic effects, the progress towards the new phenotype could accelerate as the new mutation appears against the mutational background that has formed up to that point during the ALE experiment.

At the individual cell level, it may now become possible to sequence a large number of clones obtained from an intermediate time point. In this case, one can monitor how mutations pair up and generate variants that are selected for and what other variants they win over. Once we are able to perform such experiments in a cost-effective manner, we should be able to get a fine-grained view of how competition amongst sequence variants and and epistatic interactions takes place.

Ranking the fitness of ALE endpoints   ALE experiments can be repeated over and over again, and the endpoints do not always contain an identical set of mutations.

In some cases, the same genes show up mutated, but with non-identical mutations in replicate endpoints. This result suggests that there are many similar mutations possible in the gene products that generate strong effects on the phenotype. In some cases, the phenotypes generated are similar in replicated endpoints, but the sets of mutations that occur are in different sets of genes. This difference may be because there are many approximately equivalent ways to generate the phenotype being selected for.

The reconstructed strains can be competed against one another directly, providing a fine-resolution view of what mutations and mutation combinations are superior. These experiments have to be sufficiently short-term so that new mutations do not appear and change the fitness of the variants in competition. Outcomes of competition experiments can be compared to results of *in vitro* assays of the mutated gene products. For instance, the relative competitive fitness of the mutants may correlate positively with an increase in the binding affinity of the mutated enzyme for substrate, providing a clue that increased fitness in the mutants may arise from the increased binding affinity.

26.3   Interpretation of Outcomes

The interpretation of the outcome of an ALE experiment is a hierarchical process. There are four principal levels of interpretation (Table 26.1): (1) changes in DNA sequence, (2) changes in gene product biochemical properties, (3) changes in the functional state of a network, and (4) changes in phenotypic function. Interpretation of genetic changes is done by re-sequencing as described above and interpretation of phenotypes is done based on the selection pressure used. For growth rate selection, characterization is performed by the usual assessment of growth and metabolic properties. Thus, the first and fourth level in this hierarchy are easy to study.

Interpretation through molecular biology   *In vitro* activity assessment can then be performed for the mutated gene products and compared to the wild-type to try to understand how the function of a gene product changed. It is difficult to extrapolate the understanding of altered functions determined for an isolated protein to how such changes in a component in a network alter its functional state.

Table 26.1   Multi-level interpretation of outcome of ALE experiments.

Level	Difficulty	Approach
Genetic	Easy	Whole-genome re-sequencing
Molecular biology	Very difficult	Cloning, expression, gene product characterization
Systems biology	Moderate	Omics data analysis
Phenotype	Easy	Growth, metabolism, and other phenotypic measures

The interpretation of the mechanistic effects of individual mutations is difficult. Such difficulty arises from the complex intracellular environment and the difficulty in tracing the effects of the mutation through multiple different processes that can be altered inside the cell as a result of the mutation. Further, the gene product can have other functions than the primary functional assignment given to it, some of which can be unknown. This interpretation may thus be more of a challenge in systems biology than molecular biology.

Interpretation using omics data sets   Starting and endpoint strains can be compared on a global level by generating (poly)omic data sets. Comparison of hundreds, let alone thousands, of data points is difficult without having some systematic context for the interpretation. GEMs provide such a context. Genome-scale reconstruction not only enables the computation of the optimal phenotype selected for, but also gives information about the relative activity state of all the processes underlying the phenotype. Initial results with such genome-scale measurements and their interpretation using GEMs suggest that this is a productive path forward and perhaps not entirely surprising, as such analysis is close to phenotypic function, and not focused on individual molecular mechanisms [234].

ALE as a 'hypothesis generator'   Biological experiments are always focused on comparisons to a set of well-defined conditions – the 'controls.' Well-defined perturbations from these reference states are implemented and the results are interpreted. Perturbations are typically single variables, such as a change in an environmental condition (i.e., a nutrient or quorum sensing molecule of interest) or a genetic parameter (i.e., a knock-out of a gene of interest). The causal mutations identified by re-sequencing endpoints of ALE may be viewed as *system perturbation* variables. We know that they have a major effect on the phenotype and can be interpreted in terms of the effects on the cell (or system) as a whole. A notable difference is that in the case of the single environmental or genetic perturbation variable, the variables are chosen by the investigator, whereas the causal mutations from an ALE experiment are not known ahead of time. However, once known, a causal mutation represents an important perturbation from the starting sequence. ALE can thus be viewed as a *hypothesis generator,* that represents a path to biological discoveries that may not be attainable in other ways.

Selection pressures   The future prospects of ALE seem significant. We can now perform the ALE experiments, do the re-sequencing, establish causality, and interpret the results with genome-scale models. The bottleneck may turn out to be the conceptualization and implementation of the selection pressure, i.e., specify the objective function. Initial experiments have focused on the obvious and easily implemented growth rate selection pressure. Metabolic engineers are beginning to use ALE to determine the genetic basis for tolerance to the fermentation product, i.e., ethanol, butanol, etc., that they wish to make. Similarly, resistance to antibiotics can be studied this way, and its basis elucidated. The more elusive, complex, and newly discovered electrogenic phenotype has been selected for using varying voltage on electrodes where the cells grow and deliver their electrons [144, 245].

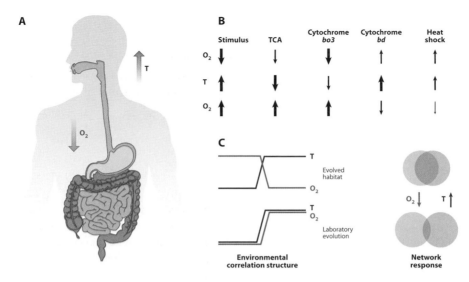

Figure 26.5 Predictive behavior during entry into the mammalian gut. (A) Diagram of the environmental changes encountered by *E. coli* upon ingestion by a mammal. Upon ingestion, bacteria encounter increased temperature (T) in the mouth; subsequently, the bacteria are swallowed and reach a low-oxygen (O_2) environment in the stomach. (B) Changes in expression of specific genes or groups of genes during temperature or oxygen perturbations. The cytochrome *bo3* oxidase complex is utilized in high-oxygen environments, whereas cytochrome *bd* oxidase is utilized in low-oxygen environments. As expected based on the predictive–dynamic interpretation, responses to oxygen downshift and temperature upshift are correlated strongly. TCA, tricarboxylic acid cycle genes. (C) Evolved decoupling of temperature and oxygen responses. The left side of the panel shows the correlation structure for temperature and oxygen in the ecologically relevant transition and in the artificial decoupling experiment; the right side shows the degree of overlap in the responses of wild-type or laboratory-evolved cells to a drop in oxygen (blue) or rise in temperature (red). From [132].

Complex adaptations that can be mimicked in the laboratory ALE experiments can be performed using a constant environment to determine optimal growth properties. Most often, life cycles of bacteria are more complicated. For instance, the life cycle of commensal *E. coli* strains revolves around four principal shifts: first a heat shock to 37°C, then a pH shock to gastric pH levels, then an anaerobic and community environment shock, and finally a cold shock back to ambient temperatures.

Such simple shifts can be mimicked in the laboratory [132]. *E. coli* normally experiences anaerobic shock after the heat shock associated with entering a mammal's mouth (although there is a pH shock in between). This sequence of two shocks is illustrated in Figure 26.5A. The associated gene expression changes are shown in Figure 26.5B. An evolution to a sequence of a temperature shock followed by plenty of oxygen, Figure 26.5C, leads to a systematic change in the expression profile. Such temporal environmental changes can be implemented in ALE experiments and the reprogramming of the regulatory system can be studied.

Challenges to mimicking adaptations in the laboratory Organisms can have complex life cycles that are a challenge to mimic in the laboratory. Pathogens such as

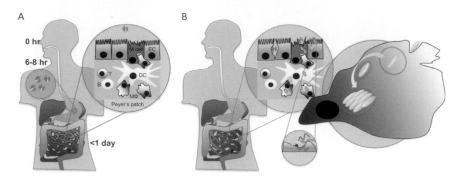

Figure 26.6   (A) Following ingestion, salmonellae transit through the stomach and reach the intestine within one day. (B) Once the intestinal lumen is reached, serovars of *Salmonella enterica* subsp. *enterica* capable of mediating a systemic infection use a number of strategies to invade additional host tissues. A key tactic employed by *Salmonella* is the manipulation of its environmental niche by establishing a *Salmonella*-containing vacuole in host macrophages. Figure courtesy of Brian Schmidt.

Salmonella have a complicated path to traverse. Bacteria of the genus *Salmonella* include pathogenic variants that cause diseases such as typhoid fever and foodborne illnesses. Similar to *E. coli*, ingested salmonellae must survive multiple environmental niches if they are to colonize a human host successfully.

Salmonella colonization is a complex process with several well-characterized steps (Figure 26.6). Ingested salmonellae must survive the transit through the stomach, and surviving bacteria reach the small intestine within a day. The most pathogenic *Salmonella* strains are capable of escaping the gastrointestinal lumen and invading additional tissues. They establish an intracellular infection in immune cells, including macrophages and dendritic cells, and use them to spread quickly to additional tissues [186, 451]. *Salmonella* reach the mesenteric lymph nodes within days, and spread to additional sites such as the liver, gall bladder, and spleen within about a week.

Genomic pathogenicity islands encoding virulence factors such as adhesins, invasins, secretion systems, and toxins are critical to successful *Salmonella* colonization [294]. For example, *Salmonella* Pathogenicity Island (SPI) 1, 4, and 5 facilitate epithelial invasion. SPI-4 encodes 7 genes, including *siiE*, an adhesin used to adhere to epithelial cells [186]. SPI-1 encodes a secretion system that injects intestinal cells with factors such as *sipA* and *sopE* that can cause cytoskeletal actin to polymerize, resulting in bacterial engulfment and uptake [260]. Survival in macrophages is facilitated by SPI-2, 3, and 5. For example, SPI-2 encodes 45 genes, including a secretion system and effectors such as *spiC*, which protects the *Salmonella*-containing vacuole by inhibiting fusion with host lysosomes [260].

There is evidence from both experimental and GEM studies to support the hypothesis that *Salmonella* adapts to the host tissue environments through a diverse metabolism that can be supported by alternative substrates taken from the host, and that *Salmonella* modulates cellular metabolism under conditions that induce virulence genes [210, 404]. To fully analyze the colonization process one would have to build a multi-step objective function that would mimic the environmental queues that *Salmonella* responds to at the various stages. Mimicking them may prove difficult in the lab if one wishes to study the adaptation process and find mutations with gain or loss of function.

Thus, time-dependent objective functions are important in the survival of organisms and patterns of changing environmental conditions, only some of which have been implemented in ALE experiments. One can look ahead and imagine that the social properties of bacteria, such as synergistic, parasitic, and competitive lifestyles, can also be studied with ALE.

Synthetic biology The recent emphasis on the design of new biological and phenotypic functions has generated much attention and is broadly referred to as 'synthetic biology.' ALE plays a role in this field. It turns out that the predictive power of GEMs can be used to compute *a priori* new metabolic phenotypes of interest. There is an entire field of metabolic engineering that designs strains for bioproduction purposes. As will be discussed in Chapter 27, various computational methods have been developed to couple metabolic functions of interest to a selection pressure [328].

26.4 A Specific Example of Nutrient Adaptation

ALE as a solution to failure of prediction GEMs have been used successfully to predict optimal growth rates of *E. coli* K-12 MG1655 on a variety of commonly used substrates, such as glucose, malate, succinate, and acetate [104], as shown in Figure 26.1. Interestingly, while enterobacteria can grow on glycerol, the GEMs failed to predict optimal growth rates of *E. coli* K-12 MG1655 on glycerol. This observation led to the hypothesis that this *E. coli* strain was not adapted to grow optimally on glycerol and that ALE should produce fast-growing strains on glycerol. This proved to be the case [177], and by 2006, full re-sequencing became possible and the causal mutations were identified.

Analysis of mutations found Two commonly mutated genes in replicate glycerol ALE experiments were *glpK* (encoding glycerol kinase) and *rpoC* (encoding the subunit of RNA polymerase) [167] (see Figure 26.7). Subsequently, the mutations in *glpK* were studied through *in vitro* assays and alterations in V_{max} and allosteric binding constants were determined [20]. The growth advantage of single-mutation knock-in (KI) strains did not correlate with changes in either kinetic parameter of the enzyme. Growth of these single-mutation KI strains on other substrates, such as succinate, was altered, even if glycerol kinase is not used as a metabolic enzyme under such conditions. Further investigation showed that this mutation in *glpK* altered cAMP levels, through a yet-to-be-determined mechanism. This experience demonstrates the difficulty in determining the underlying mechanism through which the mutations act, as there can be secondary functions of proteins that are unknown.

The mutations in *rpoC* lead to well-defined kinetic changes in the RNA polymerase as determined by *in vitro* assays [78]; most notably that the pause time on the promoter was reduced 15-fold. When the expression states of the wild-type and the single-mutation KI strain were compared, it was discovered that this mutation led to a genome-wide and systematic change in the expression state of *E. coli*. Whole classes of genes associated with well-defined functions (such as flagella, TCA cycle, the electron transport system, etc.) were affected. It is as if the RNA polymerase itself acts as a global transcription factor.

Systems biology of mutations   The analysis of omics data (proteomics and transcriptomics) from the wild-type and the ALE strains using GEMs proved to be productive. The genome-wide changes in the expression were found to be consistent with the optimal growth solutions computed from the GEM [234]. Often, computation with GEMs does not yield a single optimal growth solution and the omics data were found to be consistent with the most parsimonious set of optimal solutions, i.e., those that achieved the optimal growth rate with the minimal use of gene products and the minimization of the total flux load throughout the metabolic network. This satisfying result brings together evolutionary biology and mathematics through the common denominator of optimization.

The causal mutations found are systemic perturbation variables. Thus, ALE is a way to discover mechanisms and functions that would be hard to identify through other means.

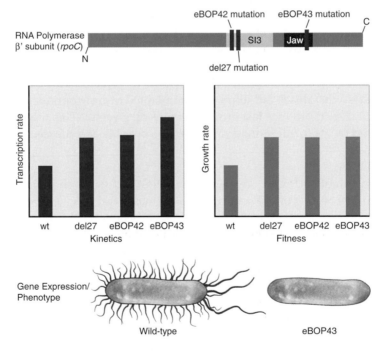

Figure 26.7   Mutations to the *rpoB* and *rpoC* genes encoding the RNA polymerase were repeatedly found when *E. coli* was adapted to growth in glycerol minimal medium. *In vitro* transcription assays showed that the mutants increase the elongation rate and other kinetic parameters. Knock-ins of the mutations into the wild-type strain showed drastically improved fitness; some are attributed to down-regulation of fimbriae (pili) and flagella genes as found by gene expression and motility studies. From [316].

Table 26.2   Collection of ALE experiments. Adapted from [99].

Organism	Type of stress	Detailed conditions
E. coli	Nutrient	Glucose, glycerol, lactate, lactose, 1,2 PDO, phosphate, cellobiose, xylose
	Environmental	High temperature, UV light, freeze–thaw, osmotic stress, toxicity (ethanol, isobutanol, *n*-butanol, H_2O_2), low pH
Saccharomyces	Nutrient	Glucose, xylose, maltose, arabinose, sulfate, galactose, lactate
	Environmental	Multiple abiotic stress, NaCl, $CuSO_4$, NaCl, ethanol

26.5   General Uses of ALE

Over the past couple of decades, ALE has been used for a growing number of applications. Reviews of ALE studies [326] show that GEMs have been used successfully for several broad categories of applications: distinguishing between essential and non-essential genes across environmental conditions [401]; identifying epistatic interactions [157]; predicting growth properties [104]; guiding metabolic engineering [290]; and charting the functional dependence (coupling) between genes [56]. Adaptation to nutrients and environmental stress has received much attention. *E. coli* and *Saccharomyces* have been evolved to grow under a variety of nutritional conditions (see Table 26.2). Tolerization to environmental stress has also received a fair amount of attention.

Detailed mutational analysis has not been performed in many cases yet and much work lies ahead to get a broad view of where mutations take place in a genome and what mechanisms get altered. A summary of many of the early *E. coli* ALE studies is found in Figure 26.8, showing that the mutations occur throughout the genome. This figure is in effect an early version of an ALE-based annotation of a genome.

In other words, perturbations in the survival conditions of the organism followed by adaptation will lead to the identification of *systemic perturbations*. The genetic basis for them is likely to become an important part of how we annotate genomes, and help us understand how they evolve through multi-dimensional spaces defined by the physical, chemical, and biological challenges that an organism needs to meet as is illustrated schematically in Figure 26.9.

26.6   Complex Examples of Adaptive Evolution

As interesting as ALE is, it may not have direct relevance to evolution in a natural environment, except, of course, for those environments that can be reproduced accurately in a laboratory environment. ALE experiments have now been performed in

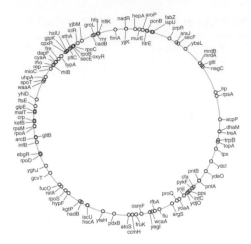

Figure 26.8 Intragenic mutations identified in a series *E. coli* ALE studies prior to 2011. Single-nucleotide substitutions, insertions, and deletions found within the open reading frames by whole-genome sequencing in multiple *E. coli* ALE studies are shown on a circular representation of the *E. coli* chromosome. From [80].

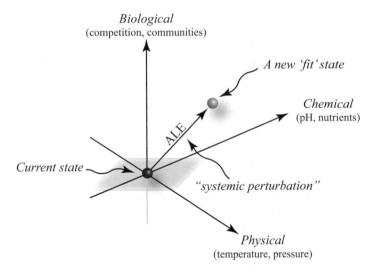

Figure 26.9 A schematic illustration showing that an organism adapted to one condition needs to adjust to new conditions and can do so through ALE.

live animals, with complex organisms, and longitudinal monitoring of pathogenesis can be analyzed in a similar fashion.

Evolution during colonization   Germ-free animals can be inoculated with a defined bacterial strain and colonization observed. Studies with the colonization of sterile mice with *E. coli* have been reported [94, 142]. Rapid genetic diversification was observed upon colonization, characterized by the systematic selection of mutations in three

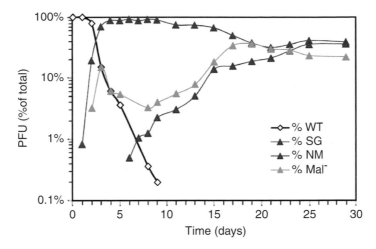

Figure 26.10 Rapid diversification occurs during gut colonization of *E. coli* into a sterile mouse. Evolution of colonies collected from the feces of one representative mouse (out of 8 studied). Phenotypes of bacteria observed in motility agar: small granulous (SG), totally non motile (NM) and colonies unable to use maltose (Mal⁻). From [94].

different pathways: in the global regulator EnvZ/OmpR controlling outer membrane permeability, in the flagellar operon, and in the maltose regulon. The three strains then coexist in the animal (see Figure 26.10).

The studies concluded that the cohabitation of *ompB* and *flhDC* mutants is due to a tradeoff between bile salt resistance and nutritional competence in the very crowded gut environment. Diversification of *ompB* and *flhDC* mutants could be reproduced in a chemostat that mimicked ecological parameters in the mouse distal gut. The results indicate that a tradeoff between stress resistance and nutritional competence is sufficient to mediate the diversification of bacteria. These results illustrate how experimental evolution in natural environments allows for the identification of the selective pressures that organisms face in their natural environment, as well as the diversification mechanisms.

Adaptation of pathogens Samples of pathogens can be collected over the time course of infection in patients. Strain collections of *Pseudomonas aeruginosa* are available from cystic fibrosis patients that have been collected over a long time and from multiple patients. The approaches outlined above for ALE can be applied to such series of examples from a 'patient adaptive evolution' (see Figure 26.11).

Analysis of the samples show that there is a linear accumulation of mutations with colonization and adaptation time, but a non-linear improvement in measured phenotypic changes [125]. These data are an example of what was mentioned above; namely, the most causal mutations tend to fix first followed by those that cause less-pronounced phenotypic changes. Note the time scale and the length of time it takes for the microbe to adapt to the host and eventually become pathogenic.

With the ability to re-sequence strains cheaply, such carefully documented longitudinal collections of strains from patients are great resources to detail the colonization

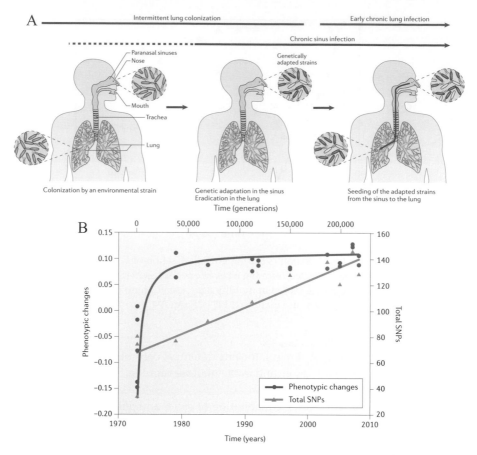

Figure 26.11   (A) A model for the typical course of *Pseudomonas aeruginosa* colonization and infection in the cystic fibrosis (CF) airway. (B) Rates of mutation accumulation and phenotypic changes in CF-associated isolates of *P. aeruginosa*. From [125].

and infection process. Having many outcomes of this adaptation from many patients should lead to comparative analysis of the genomes of the pathogens and how they change over time. Such data should not only help the development of the understanding of disease progression, but also help with devising interventions and therapies.

Fruit flies adapt to low-oxygen concentrations   ALE experiments have been performed with multicellular organisms. Long-term laboratory selection for over 200 generations has been performed with the fruit fly, *Drosophila melanogaster*, to tolerate severe, normally lethal, levels of hypoxia [474]. Deep re-sequencing of the adapted strains and comparison to the wild-type flies led to the identification of a number of DNA regions under selection. Several of the hypoxia-selected regions contained genes encoding or regulating the Notch signaling pathway.

ALE experiments are likely to progress from microorganisms to more complicated organisms. With DNA sequencing becoming cheaper and cheaper, the focus will be on the data analysis. Such analysis will not only rely on increasingly sophisticated bioinformatics tools, but also on network reconstructions and systems biology models that can describe and analyze the phenotypic changes that are observed.

Recap The convergence of several experimental and computational technologies has now allowed us to design and implement evolutionary experiments in the laboratory. This possibility opens the prospect of obtaining a fine-grained genome-scale understanding of the nature of genotype–phenotype relationships, and how they are formed and selected. Clearly, this allows us to address the very core principles of biology. One would expect that, for bacteria, this field would quickly develop and allow us to formulate the fundamentals of prokaryotic biology at a new and deep fundamental level. Several detailed reviews on ALE have appeared [80, 99, 108, 125, 168, 326].

26.7   Summary

- The objective function used to compute phenotypic states from GEMs can be used to predict the outcomes of ALE experiments.
- Evolution experiments in the laboratory under controlled conditions and whole-genome re-sequencing allows us to determine the genetic basis for adaptation. With the rapid adaptation of bacteria, the time scale of such experiments can be just a few weeks.
- Introduction of the mutations found into the starting strain allows for the determination of causality of the mutations.
- Samples can be saved at various time points during the adaptive evolution, and with the known causal mutations we can study the dynamics of their selection over time.
- Mechanistic analysis of how the altered biochemical properties of the mutated products lead to phenotypic change is challenging, but the use of omics data in the context of genome-scale models leads to insights into how successful phenotypes are produced.
- Adaptive evolutions are now being performed in creative ways and a broad set of questions are being addressed.

27 Model-driven Design

DNA reveals the past, and predicts the future. Get used to it, learn how to read it, and use it practically – Henrik Wegener

Organisms used for bioprocessing are engineered to achieve the production of the desired chemical compound. Such compounds can be natural or non-natural. In the former case, the existing gene portfolio of the organism is altered in its function to change the production phenotype of the cell. In the latter case, a series of heterologous genes are introduced to confer the capability on the host to carry out a new biochemical function. Once such a function is established, the host functions are modified to achieve an optimal function of the pathway incorporated. Metabolic engineering and microbial factory design represent the most ambitious efforts at directed and systematic construction of a new phenotypic function. Over the past decade, GEMs have played an increasing role in this process [207, 223, 224, 328].

27.1 Historical Background

Many compounds can be made biologically The chemical and pharmaceutical industries produce products with sales in excess of \$3.5T per year, representing one of the world's largest industries. A broad variety of chemical compounds are manufactured in large quantities. Most of these products are derived currently from petroleum as a feedstock. With technological advances in the life sciences, it has become clear in recent years that a fair fraction of these compounds can be made through biological means (Figure 27.1). Given the significant economic impetus for producing chemicals from sustainable feedstocks, major effort is going into designing cells to produce industrially, commercially, and pharmaceutically valuable compounds.

Cell factories The notion of a cell factory is illustrated at the center of Figure 27.1. Substates are fed to an engineered organism that can produce and secrete chemical compounds that it does not produce naturally. Substrates are typically sugars (both five- and six-carbon sugars) derived from renewable biomass, although there is an increasing interest in the use of one-carbon substrates (methane, carbon dioxide, and synthetic gas). The most commonly used production hosts are *E. coli* and yeast as they

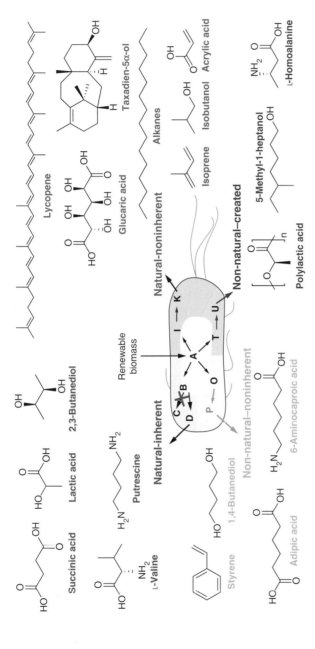

Figure 27.1 Classes of chemical compounds produced by microbial cell factories. Natural-inherent compounds are made by cells naturally and their production can be enhanced through metabolic engineering. In contrast, the production of natural-noninherent, non-natural-noninherent, and non-natural-created chemicals that are not synthesized via native pathways begins with the design of appropriate metabolic pathways, constructed via heterologous and/or combinatorial expression of known genes or the creation of new genes. Inherent metabolites and pathways are indicated in black, and noninherent metabolites and pathways are represented by different colors according to the category: natural-inherent, green; natural-noninherent, blue; non-natural-noninherent, orange; non-natural-created, red. A, carbon source; B, I, O, and T, metabolic intermediates; C, by-product; D, K, P, and U, target products. From [223].

439

are easy to manipulate genetically, although there is a range of other production hosts in use.

There are three significant overall parameters used to describe a bioprocess.

- The titer of the final product is a key performance metric. For fuels or bulk chemicals, titers in excess of 100 g/liter are desired typically. For specialty chemicals, 1 g/liter may be sufficient.
- The conversion yield of the substrate to a product is an important economic variable. It is normally described on a mass product produced per mass substrate consumed basis. For fuels and bulk chemicals the yield is a key parameter for economic success, and typically needs to exceed 30–40%.
- The productivity of a process is measured in g/liter/h, and for a high efficiency process the productivity should be in the range of 3 to 5 g/liter/h, or higher.

There will clearly be great economic opportunities arising from bioprocessing in the future. Success is based on a combination of computational and experimental methods. GEMs can play an important role in cell factory design [207,223,224,328,460].

The three-phase history of metabolic engineering   Microbial cell factories have been used for centuries to produce fermentation products for human consumption. When antibiotics were discovered, microbial cell factories took on a role to produce medicinal compounds. The performance of production strains was improved through a mutagenesis process followed by selection. This tedious process, over time, led to the identification of strains with better and better performance characteristics. This effort may be described as phase one in the history of microbial factory design (see Figure 27.2).

When genetic engineering was developed in the 1970s and 1980s, the genetic makeup of a production host could be altered directly, in contrast to random changes induced by mutagenesis. Genetic engineering opened the door for *rational* engineering of the host. However, it was quickly learned that it was hard to predict the consequences of direct gene manipulation on host production performance, and such targeted changes only came with 'local' prediction. Thus, only limited predictability was possible and the field continued to use trial and error in the progression of the design of the host strain.

When whole-genome sequencing developed in the mid to late 1990s, gene knock-out procedures become easier to perform, and when GEMs became available in 2000, multiple genome-scale technological drivers were in place to enable metabolic engineering procedures at a systems or 'global' level. These procedures gave rise to an iterative workflow that should accelerate the design procedure (Figure 27.2C). Throughout the mid to late 2000s, a plethora of additional computational and molecular tools were developed, broadening the ability to practice multi-scale pathway design, metabolic engineering, and cell factories design.

The plethora of methods for pathway design   A suite of computational methods are now available for pathway design and incorporation into a production host (Figure 27.3). Using methods from physical chemistry, one can now compute all the chemically possible pathways from one compound to another. This allows

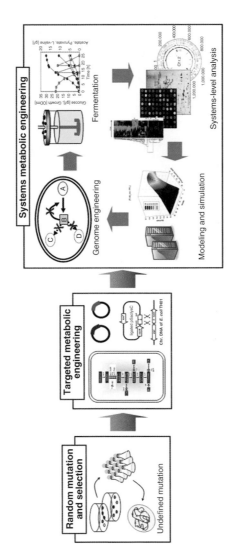

Figure 27.2 An illustration of the three-phase history of metabolic engineering. From [327].

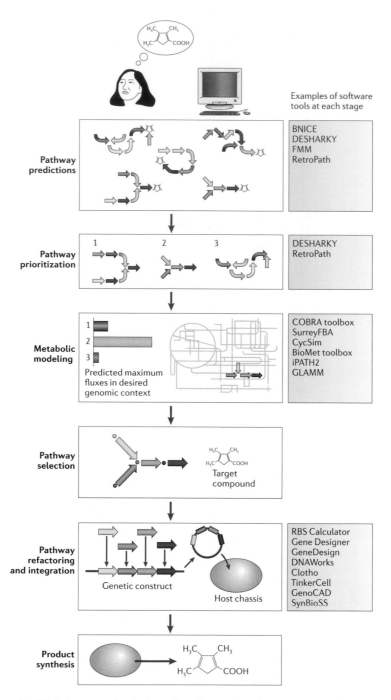

Figure 27.3 Multiple issues in the design of biochemical pathways can now be addressed. From [264].

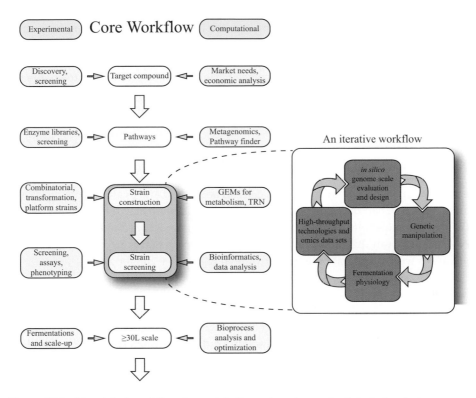

Figure 27.4   The global workflow for metabolic engineering and cell factories design.

for the computational generation of a large number of possible pathways. Such a collection of pathways can then be filtered based on thermodynamic criteria, co-factor coupling requirements, EC numbers for the chemical transformations, the length and complexity of the pathways design, and other criteria.

Candidate pathways can then be placed into a GEM of the production host's metabolic network. Comparative *in silico* evaluation is then performed to predict theoretical production capacities of the pathway. Similarly, a comparative evaluation of possible hosts can be performed as each host organism has slightly different metabolic capabilities and characteristics. The implementation of the pathway can then be evaluated from a synthetic biology standpoint based on which synthetic expression constructs can be designed for the host and the required enzymes. This comes down the refactoring process, described in Chapter 8.

Workflows for cell factory design   These computational tools for pathway design are a part of an even larger workflow (Figure 27.4). This workflow goes from the selection of the target compound, to the pathway design, to the initial construction of a production strain, to the screening of production strain properties. This workflow is not linear, and the construction and screening steps in particular are a part of an iterative workflow shown in the figure. For each one of these steps, there are required computational and experimental tools. A larger integrated workflow is needed to coordinate all the steps in the process through the implementation of laboratory information management systems (LIMS) and metadata generation and analysis.

Currently it takes > 5 years and well over \$50M to get a high-performance production strain. Clearly, this performance needs significant improvement to enable the broad deployment of cell factories. The use of systems biology and GEMs within the overall workflow and process modeling should lead to an improvement in the field. Some examples are found below.

27.2   GEMs and Design Algorithms

The integration of a pathway into a host cell can be studied and evaluated using GEMs. As described in Chapter 19, one can state a single-valued objective for the production of the desired metabolite and then compute its optimal yield from a given substrate. The ME models can then compute the optimal proteome that is needed to enable the pathway, as well as estimate the transcriptional regulatory challenges.

These computations can then be put into the context of other required physiological functions, such as growth and maintenance. The latter can be computed by elevating the ATP load, but assessing the interaction between growth and product production is more complex. If the objective of a cell is to grow as fast as possible, it goes counter to the use of a large fraction of the substrate consumed to produce a secreted product. Ingenious and intricate design algorithms have been devised to reconcile these conflicting objectives.

Bi-level optimization procedures   Bi-level optimization involves the nesting of objectives. In this way, one can synchronize the growth objectives of a cell and the metabolic engineering objectives of product production [58, 334, 335]. An outer optimization problem is put over the *E. coli* biomass optimization problem discussed in Chapter 21. This outer problem is used to find the minimum number of gene knock-outs or additions that maximizes the secretion of a by-product of interest while the inner problem maximizes the biomass formation. In other words, the outer problem manipulates the gene portfolio of the host such that the optimal biomass formation problem comes with the secretion of the desired compound. Growth and production become coupled.

Growth-coupled designs   The natural tradeoff between a forced product secretion and cellular growth is shown in Figure 27.5. The faster a cell grows, the less product it makes, and vice versa. If growth is maximized, no product is produced as all the substrate is used for cellular growth.

This tradeoff can be altered through a bi-level design algorithm. The first such algorithm, called OptKnock [58], modified the reactome in a GEM to achieve growth coupling and the best product production characteristics. The algorithm 'shapes' the solution space so that the optimal growth rate comes with a finite secretion rate of the desired product (Figure 27.5). A pathway design can be built into the production host, then the genes removed that were specified algorithmically by OptKnock, and, if the secretion is truly growth-coupled, then adaptive laboratory evolution would generate the desired design automatically.

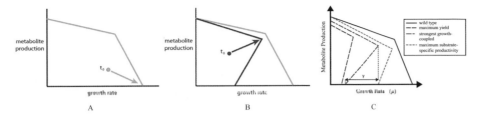

Figure 27.5 The concept of a growth-coupled design. (A) A wild-type strain will evolve from t_o to suppress all metabolite production and devote all its resources to growth. (B). A growth-coupled strain will evolve to a maximum growth rate that comes with metabolite production (red curve). (C) Different growth-coupled design criteria. Three criteria can be used to select a 'best' OptKnock design. From [116].

The notion of growth coupling is intuitive. It has been used in practice where strain designers have enough insight into the metabolism of the host to find growth-coupled design through reasoning. This approach has been demonstrated for natural-inherent metabolites [21, 62, 63, 152, 185, 443, 472, 475]. For more complex designs and intricate co-factor coupling scenarios, genome-scale algorithms can find all possible growth-coupled designs. Some simple growth-coupled designs have been implemented using OptKnock [126]. Many more exist in the patent literature and in proprietary files of bioprocessing companies.

Growth coupling at the genome-scale The bi-level optimization procedure has been used to compute comprehensively optimal growth-coupled designs based on the *E. coli* model GEM [116]. Three substrates were considered (glucose, xylose, and glycerol) and the production of 10 products was considered (ethanol, lactate, glycerol, alanine, pyruvate, fumarate, malate, succinate, 2-oxoglutarate, and glutamate). Optimal host strain designs were reported for designs which possess maximum yield, substrate-specific productivity, and strength of growth coupling. These criteria are illustrated in Figure 27.5C. Because the OptKnock algorithm can run for an undetermined number of gene alterations, the complexity of the genetic design was limited to 10 gene knock-outs. In total, growth-coupled designs could be identified for 36 of the total of 54 conditions considered. There were 17 different substrate/target pairs for which over 80% of the theoretical maximum potential yield could be achieved. Thus, growth coupling cannot be found for all desired products.

Design calculations can be redone allowing for the addition of new reactions and not restricting the computations to gene knock-outs. The bi-level optimal solutions will be sensitive to the reactome of the host, and thus updated and expanded network reconstructions would change the predicted designs. No comprehensive experimental evaluation of growth-coupled designs exist. It should be noted that thousands of growth-coupled designs exist in the patent literature and in unpublished company studies.

Other growth-coupling strain design algorithms COBRA methods are thus a useful tool for predicting the target genes for metabolic engineering. This design use of GEMs is based on their capability to analyze and predict the re-distributed metabolic fluxes following genetic modifications (see Chapter 23). Non-growth-coupled production strains typically exhibit a decrease in product yield over time, whereas growth-coupled strains can enhance product yield (Figure 27.6A). The conception of OptKnock and its productive uses has spurred the development of a series of algorithms for metabolic engineering analysis and design. The characteristics of these algorithms are comparatively summarized in Table 27.1.

Most COBRA methods for strain design systematically identify reactions that, when perturbed, lead to growth-coupling. As demonstrated above, the capabilities of a production strain can be shown with a production envelope (Figure 27.6B). Selection of wild-type cells for fast growth will minimize product secretion (Figure 27.6B, white circle), while OptKnock-designed cells will evolve to a maximum growth rate where the production rate is finite (Figure 27.6B, blue circle). However, OptKnock strain designs occasionally have alternative optima with other products being secreted, potentially leading to a low product secretion rate (Figure 27.6B, green circle). To eliminate alternative optima, the product can be added to the biomass function (called objective tilting [116]) or MILP can be applied (using RobustKnock [416]) to find designs that provide the maximum lower bound on product yield while maximizing growth, leading to the red circle in Figure 27.6B.

Those strain design methods perturb the reactome. However, genetic modifications change the gene portfolio. The GPRs relate the two. It turns out that strain designs that are based on reactome modification can require additional gene deletions to remove all relevant isozymes. Experience also shows that the predictions made are occasionally not feasible when they require the removal of one reaction that is catalyzed by a multi-specific enzyme (Figure 27.6C). To avoid such predictions, heuristic approaches such as OptGene [330] and genetic design through local search (GDLS) [252] identify growth-coupled production strains. Such strain designs may be more realistic and easier to test *in vivo*.

Strain design algorithms are not limited to manipulations of the gene portfolio of the host. Various methods have been developed to add gene functions to the host cell and to use other design considerations. These are detailed in [478] and summarized in Table 27.1. Thus, COBRA approaches allow the coupling of non-native product synthesis to a cellular objective (see the three categories of products in Figure 27.1). However, there is a limited number of cases published in the peer-reviewed literature [126, 467].

27.3   GEMs and Cell Factory Design

The number of case studies showing the use of GEMs for pathway design and selection, for metabolic engineering, and for host cell design is growing. There are different uses of GEMs (Figure 27.7). These studies include the identification of knock-out and/or amplification target genes, as discussed above, and polyomic data analyses, as discussed in Chapter 24. A few select cases are shown below.

Tuning expression   Engineering strategies found using model-driven analysis can often be nonintuitive and highlight some of the most interesting recent findings using GEMs. For instance, researchers used the GEM to not only determine a gene

Table 27.1 Optimization algorithms to redesign metabolic networks. From [478].

Name	Type of optimization problem	Type of intervention	Accessibility
OptKnock	Bi-level, MILP	Knock-outs	GAMS
RobustKnock	Multi-level, MILP	Knock-outs	MATLAB
OptGene	Evolutionary	Knock-outs	Online (as part of OptFlux)
Objective tilting	Bi-level, MILP	Knock-outs	MATLAB (via COBRA toolbox)
OptStrain	Bi-level, MILP	Addition of non-native reactions/pathways	N/A
SimOptStrain	Bi-level, MILP	Knock-outs and addition of non-native reactions/pathways	N/A
BiMOMA	Bi-level, MINLP	Knock-outs	N/A
OptReg	Bi-level, MILP	Knock-outs, up-regulations and down-regulations	N/A
GDLS	Heuristic	Knock-outs, up-regulations and down-regulations	N/A
FSEOF	LP	Up-regulations and down-regulations	N/A
OptORF	Bi-level, MILP	Knock-outs and over-expressions (of both metabolic and regulatory genes)	N/A
OptForce	Bi-level, MILP	Knock-outs, up-regulations and down-regulations	GAMS
EMILiO	Bi-level, MILP	Knock-outs, up-regulations and down-regulations	N/A

447

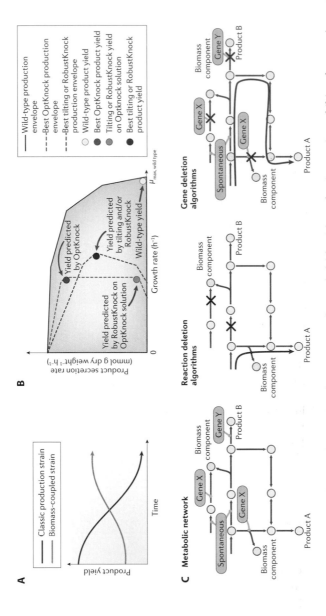

Figure 27.6 Principles of GEM-guided strain design. Panel A: non-growth-coupled (classic) production strains exhibit a decrease in product yield over time, whereas growth-coupled (biomass-coupled) strains can enhance product yield. Panel B: designs for growth-coupled strains are predicted to force product secretion while allowing optimal growth of the organism. Several methods have been developed to predict strains that undergo growth-coupled production, and these methods model reaction deletion, gene deletion, or reaction addition. Different reaction deletion algorithms, such as OptKnock [58], objective tilting [116], and RobustKnock [416], can provide different optimal growth-coupled strain designs owing to algorithmic differences. Panel C: many algorithms predict the set of reactions that must be blocked (or deleted) to obtain a desired product. However, methods such as OptGene [330] and genetic design through local search (GDLS) [252] provide a more realistic view by modeling genetic modifications, as some genes catalyze multiple reactions and some reactions are spontaneous. μ_{max}, predicted maximum growth rate. From [236].

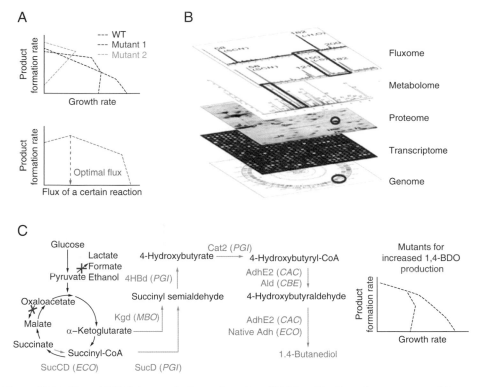

Figure 27.7   Use of GEMs in metabolic engineering. (A) Identification of gene targets. (B) Polyomic data analysis. Marked in red are target genes that show significant changes under different conditions. WT, wild-type. (C) A representative example of the development of an *E. coli* strain for the production of 1,4-butanediol by systems metabolic engineering. Orange arrows indicate increasing fluxes by the over-expression of corresponding genes. Red crosses indicate the knocking-out of the corresponding reactions identified by OptKnock simulation. CAC, *C. acetobutylicum*; CBE, *Clostridium beijerinckii*; ECO, *E. coli*; MBO, *Mycobacterium bovis*; PGI, *Porphyromonas gingivalis*. Adh, alcohol dehydrogenase; AdhE2, bifunctional alcohol-aldehyde dehydrogenase; Ald, aldehyde dehydrogenase; Cat2, 4-hydroxybutyryl-CoA transferase; 4HBd, 4-hydroxybutyrate dehydrogenase; Kgd, a-ketoglutarate decarboxylase; SucCD, succinyl-CoA synthetase; SucD, CoA-dependent succinate semialdehyde dehydrogenase. From [223].

that needed to be up-regulated, but were able to tune the expression level of this gene after subsequent GEM analysis of a deleterious over-expression event (recall Figure 22.4C,D).

Co-factor coupling   The highest flavanone production was predicted and determined experimentally by knocking-out genes strategically to not only increase the production of the redox carrier (NADPH) to drive the heterologous flavanone catalyst, but also to maintain the optimal redox potential of the cell (i.e., the ratio of NADPH to $NADP^+$) [70, 131].

Production of a non-native metabolite   Aided by a GEM, a design team improved the production potential of the non-native metabolite 1,4-butanediol in *E. coli* by over three orders of magnitude [467]. The team was able to rewire the host cell to produce the compound via native and non-native pathways by ensuring that the production of 1,4-butanediol was the only means by which the host cell could maintain a redox balance and grow anaerobically.

27.4   Summary

- Metabolic engineering and cell factories design represent the most ambitious designs of phenotypes undertaken.
- Dozens of genetic manipulations are required, which have to function harmoniously and lead to an optimal functional state of the cell factory and thus optimal bioprocessing outcome.
- Historically, such designs have proven to be laborious, lengthy, and very costly.
- COBRA methods coupled with a plethora of pathway design and experimental methods form the basis for an integrated and iterative workflow to produce cell factory designs.
- This field is young, but if developed successfully will have significant positive worldwide economic consequences.

PART V
Conceptual Foundations

Never trust an experiment that is not supported by a good theory – Jacques Monod

After reading the four main parts of the book, one naturally begins to think about what the broader implications of COBRA are and what lies ahead for this field. This fifth part of the text begins to address these issues. First we talk about the core values that the COBRA approach is based on. We delineate the educational values associated with the material and try to outline the background that is needed to practice in this field. We finish the book with a brainstorming chapter that discusses the broader implications of genome-scale models and how they are likely to affect the life sciences going forward.

28 Teaching Systems Biology

We are drowning in a sea of data and thirsting for knowledge. Most biology today is low input, high-throughput, and no output biology
– Sydney Brenner

Driven by high-throughput technologies, systems biology has emerged and grown as a new discipline over the past decade and a half. With vast quantities of data being generated, the need to analyze them to generate biological knowledge has emerged. There is an overarching need to integrate various omics data types into a coherent whole and view cellular functions as systems. The fundamental paradigm of a bottom-up mechanistic approach to systems biology that has evolved to meet this challenge is based on the following core sequence of events (Figure 28.1):

$$\text{components} \xrightarrow{reconstruction} \text{networks} \xrightarrow{math\ representation} in\ silico \text{ models} \xrightarrow{COBRA\ tools} \text{physiology}$$

Each one of these steps relies on different disciplines. One needs to have command of the underlying concepts and how they are applied in the practice of systems biology. This chapter summarizes the intellectual and disciplinary bases for the core paradigm of the bottom-up approach to systems biology, one that overcomes the challenge that Sydney Brenner stated so succinctly.

28.1 The Core Paradigm

1. Measuring biological components Cells are made up of thousands of different molecules in many different classes. Over the last 10–20 years, technologies have developed to enumerate these molecular components. The generation of omics data sets is based on high-throughput (HT) technologies that, in turn, are based on *miniaturization, automation, and multiplexing*. These measurements are the enablers for systems biology, as the principal use for these measurements is for network reconstruction. Systems biology is thus enabled by technological developments.

2. Reconstruction of networks Omics data sets, along with a detailed and comprehensive review of the literature (i.e., bibliomic data) lead to the reconstruction of a network that underlies a biological process of interest. During this reconstruction process, there are defined procedures for determining which data are included and which

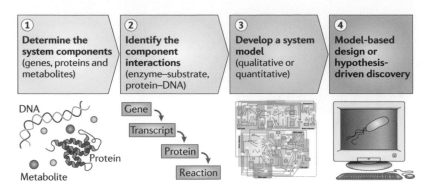

Figure 28.1   The systems biology paradigm. From [254].

are not, depending on the level of validation of the components and their interactions in the target cell. Network reconstruction can take place in two different ways.

Bottom up: this is a mechanistic way of network reconstruction; reaction by reaction, gene by gene, molecule by molecule, protein by protein. The result of this process is a curated *knowledge* base. This information results in a biochemically, genetically, and genomically (BiGG) structured knowledge base. A network reconstruction of a particular genome or target organism basically represents everything that is known about that organism or genome.

Top down: this approach involves going through the omics data sets to make inferences of interactions between biological components by using statistical methods.

Both bottom-up and top-down approaches are practiced and are complementary. The former leads to the formulation of mechanistic models and is thus emphasized in this text.

3. Conversion to a mathematical model enabling *in silico* analysis   A structured organism-specific BiGG knowledge base is then converted into a mathematical format and thus becomes an *in silico* model of the knowledge that it represents. Once it has been represented mathematically, the capabilities of the network can be interrogated. This activity is foundational to systems biology, because it enables the computation of 'functional states' of a network based on its components. It is thus possible to determine the properties of the BiGG knowledge base and what physiological states are consistent, or inconsistent, with the data from which it is comprised. Detailed curation and validation of a genome-scale model against available data is performed.

4. Prospective uses: discovery, understanding, and design   Once an organism-specific knowledge base has been formed, it can be used in a prospective fashion. A few key uses and applications have emerged to date. The first set of uses is related to omics data-mapping and analysis. The second is for systematic discovery that is focused on any missing pieces in the knowledge base. Third, an *in silico* model can be used to develop an integrated understanding of a complex biological process, such as adaptation or disease progression. Fourth, a predictive *in silico* model can be used for design and to enable synthetic biology at the network scale. In essence, the systems biology paradigm runs the gamut from analytical chemistry all the way to design.

28.2   High-throughput Technologies

Significant creativity and innovation has gone into the development of technologies that allow for the comprehensive measurement of the chemical composition of a cell. All of these technologies contain the ability to *multiplex, automate, and miniaturize*. These technologies generate so-called omics data types. They enable measurements to be carried out on a genome, cell, tissue, or organismic scale. What is being measured is presence/absence of a molecule, or its abundance, requiring one to understand analytical chemistry as well as precision vs. accuracy.

Sometimes millions of measurements are made in one experiment. This data-generation rate immediately raises the issues of databasing, bioinformatics, statistical assessment, and so forth. It is becoming clear that there are more data being generated than exists the capability to analyze it. Therefore, the analysis problem grows much faster than the data-generation problem [319]. It is for this reason that large quantities of data exist that have yet to be fully analyzed.

Conceptual foundations   Although the HT technologies are varied and are being developed continually, they still have a set of underlying conceptual foundations.

- Advancement in omics technologies come down to *automation, miniaturization, and multiplexing*. The resulting measurements are foundational to systems biology.
- As HT technologies are based on analytical chemistry, those core values need to be understood. Such values include reproducibility, precision, accuracy, and detectability.
- HT technologies are basically instruments that perform chemical assays. The basics of instrumentation need to be understood, such as detection technology, molecular probes, fluorescence, and microscopy.
- Enormous amounts of data are generated by HT instruments. The field of bioinformatics grew out of the need to analyze the large amounts of data being produced. Thus, there are conceptual values associated with software development and use, such as LIMS, statistics, data storage, data retrieval, visualization, and unique specifiers.
- Traditionally, analytical chemistry results in just a few measurements, but the HT technologies can generate hundreds to thousands, or even more. Omics data provide a chemical component *parts list* of the cell under investigation.

28.3   Network Reconstruction

Mathematically speaking, a parts list is a one-dimensional (1D) object. It is simply a string of items. For instance, an annotation of a genome sequence is a list of all the genes in the genome and their corresponding functions. A network reconstruction includes not only these components, but all their interactions as well, leading to a two-dimensional (2D) object, a table or a matrix (recall Figure 1.8).

Approach   As mentioned earlier, network reconstruction can be approached in either a top-down or bottom-up fashion. This text will focus on the latter. In addition, there are two ways to think about the networks being reconstructed. One way is to have a disciplinary bias, i.e., to focus *a priori* on metabolism or transcriptional regulation, or signaling; the alternative way is to be blind and just look at the results from molecular interaction screens (Figure 28.2). We will focus on the former.

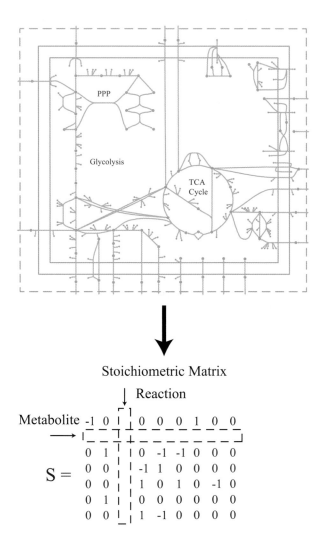

Figure 28.2   A map representing a set of curated chemical reactions can be converted into a mathematical format.

Determining links The information about links in a network can be at different levels of detail and specificity (see Figure 16.2). The information can be abstract, as in the form of a statistical correlation with no mechanistic information, all the way to a specific mechanistic description. To build computational models that can compute functional states of networks, the information needs to be at a chemical mechanistic level, or at least at the level of having causal relationships.

In order to reconstruct networks in a bottom-up fashion, all of the links, which will be chemical interactions, must first be identified. To establish just one link in one of these networks requires extensive work and extensive knowledge [427]. In a step-by-step fashion, a list of reactions that compose a network is built. The table shown in Figure 1.8 contains stoichiometric coefficients that are found in the corresponding list of chemical equations. This table also represents an example of quality controls ensuring that the chemical reactions presented are accurate. If this step is ignored and computation is based on an unchecked list of reactions, they may violate elemental balancing, for instance, and then the computations of network properties could lead to nonsensical results. Therefore, quality control becomes a very important issue.

Four steps to the reconstruction process There is a basic workflow that is followed when creating a network reconstruction that is detailed in Chapter 3.

1 **Draft reconstruction** This step involves the automatic retrieval of information from the genome annotation of the target organisms and other databases. Thus, an understanding of bioinformatics is needed, i.e., what the confidence scores mean, what gene location on the genome means, etc.

2 **Manual curation** This step is the most laborious. All the links and nodes need to be examined manually, their existence verified, and any independent literature data to support them uncovered.

3 **Genome-scale model** Once a reconstruction has been curated, it can be turned into a genome-scale model. This step requires the definition of system boundaries (typically the periphery of a cell), inputs, and outputs.

4 **Evaluation and validations** The functions of a system can then be looked at as a whole. These functions include growth rate, knock-out sensitivity, etc. Retrospective comparison with existing data is performed.

Once a reconstruction is completed, a working *in silico* genome-scale model is in place that can be used prospectively as a platform for design and discovery. Ultimately, a reconstruction is a knowledge base. It is a formal representation of all the knowledge that exists about the target organism.

Conceptual foundations The conceptual foundations associated with bottom-up network reconstructions can now be identified.

- To reconstruct networks, an intermediate level of knowledge of the basic life sciences is needed. This includes biochemistry, genetics, genomics, cell biology, and physiological functions.
- Reconstructions are organism-specific and they represent a structured knowledge base about the target organism. They *are* knowledge bases.

- To build a computational model (not just generate a network map), the information that goes into a reconstruction has to undergo a careful QA/QC procedure. An understanding of quality control engineering thinking is useful.
- An understanding of fundamental chemical and physical principles is important as this enables the conversion of data into a structured mathematical format.
- A computational model based on a BiGG knowledge base for the target organism can be used to relate genotype to phenotype, and can thus be used to reconcile physiological functions with all the underlying components.
- It is essential to understand the difference between a network and its reconciliation against an actual physiological function or its functional state. These are two very different issues. If something is missing from a network you may not be able to compute a functional state.

28.4   Computing Functional States of Networks

The computational interrogation of a reconstructed network constitutes a core activity in systems biology. It involves converting a reconstruction into a mathematical format and characterizing it. Characterization is focused on the topological properties of the network and the computation of allowable functional states.

28.4.1   Conversion to a computational model

A BiGG knowledge base can be converted into a mathematical format (Figure 28.2). Once in this format, the data can be used to compute network, or physiological states. This mathematical format is nothing more than a restatement of the data – a mathematical representation of what is known. The steps involved in the conversion to a functional model were detailed in Chapter 18.

Conceptual foundations   The conversion and ultimate representation of all the reactions, metabolites, and their interactions into a mathematical format requires a specific set of conceptual foundations.

- Representing basic data mathematically requires knowledge of logistical format, basic chemistry, and the physics of chemical transformations.
- The formation of a computational model converts a network into a system. This requires an understanding of the basic concepts of systems science, such as the definition of a systems boundary, what is inside and what is outside, the definition of inputs and outputs, and the statement of physico-chemical constraints, such as mass and energy balances.
- Because computational models are specific to the target organism, specific parameters for that organism have to be measured. This requirement calls for an understanding of the measurement of uptake and secretion rates as well as phenomenological parameters, such as maintenance energy and membrane permeability.

28.4.2   Topological properties

Once a network has been represented mathematically, its topological properties are described with a matrix (Figure 28.2). While a network map may be drawn in many different ways, often representing the bias of its creator, the stoichiometric matrix is a unique representation. This matrix conveys network properties such as the connectivity of a compound. As an example we will briefly discuss equivalence here, but it is detailed in Chapter 20.

Equivalence   For a given system with the same input and output, there are many different internal states that can produce that input/output relationship and that phenotypic state. Biologists call these *silent phenotypes* – they cannot be seen, but may be known to exist. However, now that computational models of cells exist, these silent phenotypes can be computed. Equivalence really boils down to topological robustness; something previously thought to be important can be clipped out of the cell and the cell finds an alternative way of functioning in an equivalent manner. Ultimately, there may be several different ways of achieving the same physiological state. This property was detailed in Chapter 20.

Conceptual foundations   In order to understand network topology, another set of conceptual foundations is required.

- Because a pathway map is a mathematical representation, the key to this representation is the stoichiometric matrix, which is also a connectivity or an incidence matrix. The mathematics are unique, while the map is not.
- Linear algebra is needed to analyze topological and steady-state properties of reconstructed networks. Therefore, a basic knowledge of matrix properties is needed, such as knowing the fundamental subspaces associated with a matrix, finding good basis sets for such spaces, knowing singular value decomposition, and other fundamental concepts of linear algebra.
- While linear algebra is the mathematics used to analyze topological and steady-state properties, it is important to realize that it deals with spaces that are infinite in all directions. Of course, no physical process has variables that are infinite in all directions, so convex analysis is needed to limit the spaces so that they become finite. Convex analysis deals with edges of spaces and vertices.

28.4.3   Determining the capabilities of networks

Once a network has been represented mathematically and converted into a computational model, functional states of a network under governing constraints can be computed. One fundamental concept to keep in mind is that one network can have many different functional states.

Functional states   Networks have different functions even if their topology is the same. For example, core metabolic reactions in the human liver form a network.

However, the *functional state* of the network in which these reactions participate is very different depending on whether someone is fasting or eating. And it actually makes a difference what one is fed with; whether it's a high protein diet, a high fat diet, or a high carbohydrate diet – different functional states will occur for each of these scenarios. The network, however, remains the same. Network topology versus its functional state is illustrated in Figure 28.3.

To relate a network representation to a physiological function, one must be able to compute the functional state. Such computation will illuminate how material is flowing through a given network, i.e., glucose consumption, rate of ATP production,

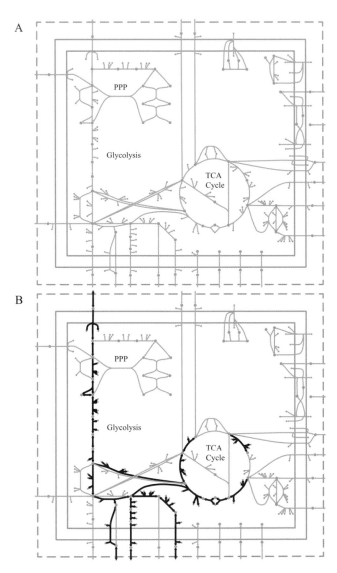

Figure 28.3   Topological representation of a metabolic network (Panel A) and the display of its functional state in the form of a flux map (Panel B).

and so forth. There is currently a lot of interest in computing functional states for different purposes, such as changing the metabolic fluxes in *E. coli* so that it produces more ethanol.

Biological causation   In biology, there exist *proximal* and *distal* causation. Therefore, models built in biology have to account for both. They need to recognize explicitly chemical causation and biological causation. Anything that is alive has to obey physico-chemical laws. These laws can take two different forms: they are either *balances*, i.e., osmotic balance, mass balance, energy balance, etc., or *bounds*, i.e., all concentration variables have to be positive, all fluxes inside a cell tend to be bounded by a v_{max}.

In the process of forming a model based on the reconstruction, these physical laws need to be applied and not violated by the network. However, applying all of the physico-chemical laws is not enough to compute a unique state. Biological systems can 'choose' their own functional states and can evolve under the governing constraints. Thus, living systems obey a genetic program that evolves and changes over multiple generations. Biological causation is thought of in terms of willful generations of alternatives on replication or reproduction to create new variants that then get selected over time.

Building models   Chemical and biological causation must be respected in order to develop a computational model that relates gene products to phenotypic functions (see Figure 1.5). In other words, if a mechanistic genotype–phenotype relationship is to be developed, both forms of causation must be accounted for. Systems biology aims to bridge the two.

The functional states that a model computes correspond to an observable physiological state. To go from all the components and all the physical and chemical laws that they obey and try to calculate biological outcomes is a challenging endeavor. Such states can be computed by using a constraint-based approach, which is conceptually and philosophically very different from a theory-based approach. This approach is detailed in Part III of this text.

In biology, a particular unique solution cannot be calculated. Non-uniqueness represents a departure from traditional educational values in engineering and physics where the focus is on finding *the* right solution. However, the range of allowable, or feasible, solutions can be bracketed. It is possible to tell what is infeasible based on a network reconstruction and the constraints under which it must operate. Thus, a *solution space* can be formed, in which every point is a candidate functional state of a network. Therefore, a solution space becomes mathematically, what is known as the *reaction norm* in evolutionary biology – all of the different phenotypes that can be achieved from a given genotype.

Conceptual foundations   The conceptual foundations associated with the determination of functional states are as follows.

- It is very important to understand the difference between functional states and topological properties. A network with a fixed topological structure can take on many different functional states.

- A modeling framework consistent with dual causation in biology can be developed using constraint-based approaches. Living systems function and evolve under governing constraints. Governing constraints do not specify the state of a living system, they bracket it.
- There is a significant difference between theory and constraint-based thinking. One example is equivalent states and the lack of uniqueness in biology. The constraint-based approach allows for biological uncertainties in experimentation and observation, but contains mathematical rigor.
- Constraint-based optimization methods need to be used.
- Hierarchical thinking needs to be deployed in systems biology.

28.4.4   Dynamic states

Networks can have dynamic states during which they either respond to overcome external disturbances or move from one homeostatic state to another. In the language of control theory, these are called the *disturbance rejection* and *servo* problems. These states can be described by continuum models on which classical kinetic modeling approaches are based, or a discrete formalism where individual molecules are considered leading to sophisticated stochastic simulations. The properties of these trajectories can be analyzed by methods of linear or non-linear analysis. Some of these topics are treated in a companion book [317].

Conceptual foundations   Detailed models require an increasing amount of information and mathematical sophistication.

- For the analysis of dynamic states, a large amount of kinetic information is needed to formulate the dynamic mass balance equations.
- Simulations of dynamic states require knowledge of ordinary differential equations.
- The properties of sets of ordinary differential equations can be linear (where the spectrum of time scales and modularization in time is analyzed) and non-linear (where the stability of the steady state and the types of dynamic states that exist are analyzed; such as multiple steady states, sustained oscillations, and chaos).
- For the analysis of stochastic effect, complicated algorithms are needed for simulation. Such issues are necessarily the subject of advanced course work.

28.5   Prospective Experimentation

Uses of network models   The use of computational models for several classes of basic and applied problems is growing.

- First, models are very useful for the analysis of multiple omic data sets.
- Second, models can systematize the discovery of missing pieces in a network reconstruction. No network reconstruction to date is complete or without knowledge gaps.

- Third, the use of models can help one to understand complicated physiological processes. For example, the process of adaptation of bacteria, in particular, has been studied through the use of genome-scale models. Now that genomes can be fully re-sequenced following adaptation, all the mutations that underlie the adaptation process can be found. Thus, a model, along with those identified mutations, help to understand how and why a bacteria took a certain developmental path.
- Fourth, predictable models can be used for design and to implement synthetic biology at a network scale. Metabolic engineering applications of genome-scale metabolic models are numerous.

These applications were discussed in Part IV of this text.

Perturbation experiments in systems biology   The experiments that go into validating network models are *perturbation experiments*, where certain parameters are perturbed and the system response with and without the perturbation is examined. Both states are also computed. Perturbation experiments come in many forms:

- Perturbations can be performed by changing individual environmental or genetic parameters.
- Dual perturbation experiments can be performed where both an environmental and genetic parameter is changed, leading to a 2×2 experimental design.
- Another type of perturbation is a systemic change resulting from an adaptation of an organism. The adapted and starting organism can then be compared.
- One can use a viral infection as a perturbation parameter. Crossing viral infections with gene knock-outs can reveal the host factor that a virus relies on, as the host will become refractory to infection without the gene present.
- Biochemical and network perturbations can be made. For instance, the co-factor coupling of an enzyme can be changed and the network-wide consequences of changing the coupling can be assessed.

Conceptual foundations   The conceptual foundations associated with these uses are classical.

- The discovery process comes down to classical hypothesis-testing and evaluation with the exception that the hypothesis is computer-generated (see Chapter 25). Associated with the experimental process can be HT screens of phenotypic states calling for an understanding of cell and tissue physiology. Hypothesis testing may also rely on classical molecular biology through the cloning of a gene product under investigation and the testing of its biochemical functions.
- The development of an understanding of a biological process based on a simulation of a network model falls under classical systems science. Here, one should be familiar with the notion of input/output relationships and how they are tested with external disturbances.

- Network design comes down to the implementation of synthetic thinking. This often requires *in silico* computational procedures, prototyping, and testing. The thought process that goes into engineering design is similar.

28.6   Building a Curriculum

A curriculum and educational materials need to be built around the systems biology paradigm, generally viewed as a four-step process, each having different disciplinary underpinnings that need to be mastered as a part of understanding the overall systems biology process (Figure 28.1).

The challenge   This book is based on a series of core educational values and material with measurable learning tools that have evolved over the past decade. What has become clear is that the background material for systems biology is quite diverse and its core educational values differ from existing disciplines. To help orient the reader, the conceptual foundations inherent in the core systems biology paradigm are summarized in this chapter. These values are likely to evolve and expand over the coming years, and gain granularity and detail as the research in the field is distilled into a standard set of educational values that is commonly accepted.

To begin with, *systems biology* is a slightly misleading name. A more accurate term would be *molecular* systems biology. While this discipline grew out of the relatively recent developments of genome sequencing and expression profiling, it is incorrect to think that systems analysis is new to biology. Systems analysis has been practiced in biology for a long time by disciplines such as ecology, immunology, and physiology. What is new is the existence of genome sequences with which one can identify all the genomes and molecules and how they interact and create cellular function. This is what systems biology is about. In a way, it helps realize the promise of molecular biology as stated in Chapter 1.

Educational values   Every curriculum is built around core educational values. For example, in chemical engineering, the educational values revolve around the processing of chemicals, thus one must learn thermodynamics, mass transfer, process control, etc. There are core values associated with each of these segments, but they all share a common focus that leads back towards chemical processing. Other disciplines like analytical chemistry, electrical engineering, and physics are all well-defined in terms of their core educational values, and courses have been structured around those values. Of course, these disciplines have been around a very long time, and these values and courses have been developed over their long history.

Systems biology, on the other hand, has only been around for about a decade as a discipline. It has become clear that in order to practice systems biology, you need to learn a broad spectrum of basic disciplines. These disciplines must then be put together in a logical fashion to develop a core sequence of educational values along with the material to support them to eventually form the educational blueprint for systems biology. One way to do so is illustrated in Figure 28.4.

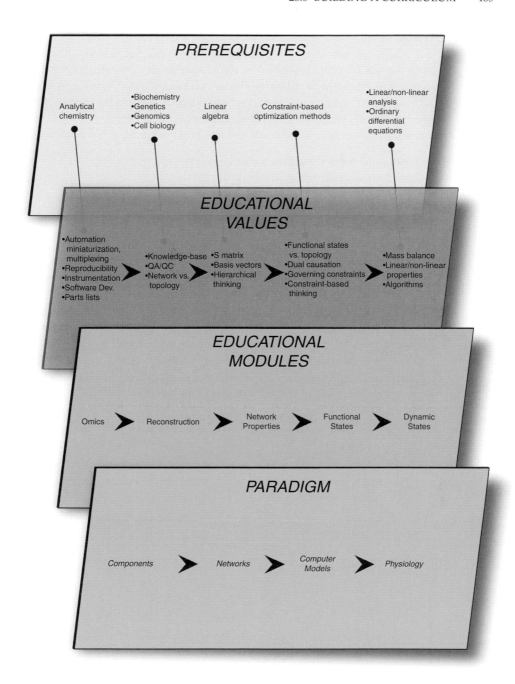

Figure 28.4 An overview of the educational issues that support the core systems biology paradigm.

28.7 Summary

- A four-step paradigm for systems biology has emerged. The disciplinary material underlying each of the four steps can be defined. The educational values associated with this material is being delineated and the outline of a core curriculum with core educational values is emerging.
- Educational values associated with systems biology can be summarized as:

    1 **component determination:** technology development, analytical chemistry, informatics, statistics, and parts lists;

    2 **network reconstruction:** knowing the basic life sciences, understanding QC/QA procedures, building knowledge bases, and the mathematical representation of data;

    3 *in silico* **modeling:** conversion of a network reconstruction into a computational model, network topology and connectivity matrices, and the basics of linear algebra. Functional states of networks can be computed using constraint-based optimization, and dynamic states of networks require complex analysis and simulation procedures; and

    4 **prospective experimentation:** discovery through hypothesis testing, developing an integrated understanding of a process through computation and comparison of component properties to integrated processes, and fundamentals of the design procedure and integrated thinking.

29 Epilogue

How can the events in space and time which take place within the spatial boundary of a living organism be accounted for by physics and chemistry? The preliminary answer ... can be summarized as follows: The obvious inability of present-day physics and chemistry to account for such events is no reason at all for doubting that they can be accounted for by those sciences – Erwin Schrödinger

It is becoming increasingly clear that making non-trivial predictions of phenotypic functions from genomic information requires a thorough understanding of the molecular processes involved in the expression of genes into functional proteins, the organization of these proteins with lipids and nucleic acids in spatio-temporal structures, and the assessment of the role of such structures in carrying out metabolic processes, creating ion currents involved in membrane excitability, maintaining electro-neutrality and osmotic homeostasis, and performing other cellular functions. This statement represents a grand challenge for biological and medical research for the twenty-first century. The COBRA methods covered in this text can address some of these issues. They represent a first step in addressing this grand challenge.

29.1 The Brief History of COBRA

Constraint-based reconstruction and analysis has been applied to biochemical reaction networks for over two decades. Research articles that utilize COBRA methods for interpreting and predicting biological phenotypes appearing from 1986 to mid-2013 have been collected and the cumulative rate of progress in the field can be graphed (Figure 29.1). The analysis of this published literature shows that the history of deployment of COBRA methods can be divided into four phases [52].

Initial studies (1986–1998) Early interest in constraint-based models was directed towards the determination of theoretical pathway yields and metabolite overflows [121, 258]. Then, experimental metabolic fluxes and growth rates were found to be consistent with computation based on optimization of physiologically meaningful objective functions, including minimal production of reactive oxygen species for hybridoma cells [373] and maximal growth rate for laboratory strains of *E. coli* [438]. The quantitative match between model predictions and measured cellular behavior opened up the possibility of predicting cellular phenotypes from a biochemically reconstructed network.

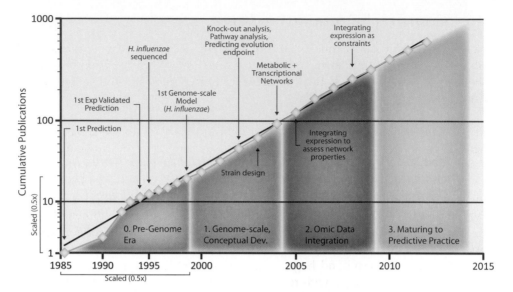

Figure 29.1 The use of COBRA methods to analyze or predict biological performance, 1986–2013. The history of the field can be divided into 4 phases as described in the text. Notice that the doubling time in the number of publication is 3 years (knowledge generation), twice as long as Moore's Law (speed of processing information). The detailed list of publications is hosted on http://sbrg.ucsd.edu/cobra-predictions. Prepared by Aarash Bordbar.

Formulating genome-scale models (1999–2004) The generation of whole-genome sequences and their annotation in the mid 1990s enabled the formulation of genome-scale models (GEMs). GEMs allowed for the representation of the complete metabolic gene portfolio in a target organism in the assessment of its phenotypic functions. Importantly, metabolic reactions in a GEM were now linked directly to genes and their genomic location, enabling the prediction of the consequences of gene knock-outs (Chapter 23). GEMs facilitated the study of the global organization of cellular behavior, such as pathway structure, adaptive evolution endpoints, metabolic fluxes, and bacterial evolution.

Integrating omics data (2005–2009) As the generation of 'omics' data became cheaper and large data sets appeared, researchers began to analyze these data sets using GEMs (Chapter 24). Initially, the metabolic network was used as a scaffold to interpret transcriptional changes, similar to pathway analysis. Subsequently, omics data were used directly by constraining individual metabolic reactions to increase the context specificity of GEMs.

Maturing to predictive practice (>2010) These efforts resulted in highly curated and validated GEMs that are now enabling the research community to obtain meaningful predictions of biological functions. This capability has brought on a degree of predictability of phenotypic functions for a target organism.

29.2 Common Misunderstandings

The COBRA approach is thus relatively new and its fundamental premise and features are not widely known. Although several review articles have been published on COBRA methods, their uncommon conceptual basis has led to a few misunderstandings.

Use of the steady-state assumption The ability to estimate and analyze possible phenotypic states generated by a reconstructed network has proven to be quite useful. In their simplest form, such computations can answer questions about network capabilities. They can answer questions like 'is a steady state (i.e. a physiological state) achievable in a given environment?' Such computations form the basis for the prediction of the consequences of gene knock-outs, auxotrophies, synthetic lethals, drug interventions, comparative strain capabilities, consequences of horizontal gene transfer, and so forth. The computation of the answers to such questions represents one of the most powerful and large-scale sets of biological predictions made from computer models in biology, with remarkably good outcomes. It is not so much the assumption of a steady state that these predictions rely on, but the characteristics of the reconstructed network to function at all in a given environment or without a particular gene product.

Dynamic versus steady states Biology is inherently dynamic, thus much discussion about GEMs has focused on how steady-state analysis can be so informative. All complex systems come with a spectrum of time scales and thus many characteristic response times. In engineering sciences the use of temporal composition readily leads to the identification of fast subsystems that are put into a steady state relative to the slower processes taking place. This approach has been present from the beginning of COBRA applications. The first paper using the pre-genome *E. coli* metabolic model described the analysis of fast cellular metabolism relative to a slowly changing nutritional environment [138]. The steady-state assumption in this situation simply states that cellular metabolism adjusts rapidly to a changing environment. This COBRA method is now called dynamic FBA, or dFBA.

   Temporal decomposition is a standard procedure in many areas of physics and engineering but seems conceptually new to many in the life sciences. Hierarchies exist in biology, in time, in space, in complexity, and in DNA sequence. The judicious use of the steady-state assumptions is one way to move up and down the time-scale hierarchy.

The objective function Much has been said about the objective function, and I suspect that the discussion will continue for some time to come. Perhaps the simplest description is that the objective function makes up for our ignorance, as we do not know enough about the target cell to compute full solutions to the equations of a genome-scale model. However, a careful consideration reveals that the objective function can represent distal causation; a fundamental feature of biology (see Chapter 21). Therefore, it is likely that the study of and understanding of the objective function is of fundamental importance to address key biological issues and questions. In a way, biologists can claim it as 'their function' that describes biological phenomena instead of physical or chemical phenomena. In fact, it can now be examined experimentally through controlled evolution in the laboratory or other settings (see Chapter 26).

The scope of applications Flux balance analysis is often thought to be the maximization of microbial growth rate based on metabolic inputs and the optimal use of its reconstructed network. Perhaps this is understandable, as this well-posed maximization problem was one of the first successes of COBRA. It is formulated mathematically based on data and a clear hypothesis. However, many more applications of COBRA have emerged since, e.g., biomarker discovery, inborn errors in metabolism, gap-filling, etc., as described throughout this book. In particular, one should note that there are unbiased COBRA methods that do not require the use of an objective function to pursue the analysis and answers sought. The majority of key papers from phase 3 in Figure 29.1 did not need to use an objective function.

Multiple possible solutions A core conceptual value in the physico-chemical and engineering sciences is to find 'the solution' to a problem. COBRA generates a range solutions that are not unique, so what's the use? Well, in biology, multiple solutions represent reality. Recall that in a bio-population there are many alternative solutions that represent the population and its properties. And we now know through ALE studies that distal causation produces multiple different solutions to the evolutionary problem presented.

Few parameters are needed The first GEMs had a surprising ability to provide understanding, reconcile data, and to generate predictions. This ability seemed confusing at first, as the expectation was that many more parameters were needed to describe genome-scale behavior. Acceptance that biological predictions could indeed be made with essentially 'parameter-free models' came slowly [25]. A reason for needing few key parameter values lies in the fact that there are few governing constraints in a given situation and those that are redundant are not critical.

29.3   Questions in Biology and in Systems Biology

General types of questions asked in biology There are fundamentally three types of questions that are asked in biology: 'what,' 'how,' and 'why.'

What is there?   We have made substantial strides in answering this type of question in cell and molecular biology. We can sequence entire genomes and use bioinformatic analyses to determine what is in a genome. We can expression-profile an organism under various conditions. We thus now have extensive information about genomes, cells, and organisms, and are in a position to continue to generate much more data. It is indeed this impressive availability of data that has made biology 'data-rich' and has been the driving force for the emergence of systems biology.

How does it work?   Science seeks to generate mechanisms and formulate theories to explain the world around us. Functional genomics tries to assign function to individual gene products and segments of a genome. The large number of interactions that need to be taken into account to explain cellular functions has grown substantially with our growing knowledge of cellular components. The drive to reconstruct genome-scale networks and to assess their functional states is a response to this question.

Why does it work the way it does?   The answers to such questions are generally very difficult to obtain. In biology, they are often based on the understanding of evolution and making teleological arguments.

Some questions in systems biology   There are several types of questions that can now be addressed with the methods of systems biology, the experimental tools, and the algorithms that we now have at our disposal. A few such questions are as follows.

What are the limits of reconstruction?   Genome-scale reconstruction started with metabolism, primarily due to the extensive knowledge available about the properties of the metabolic gene products. Every time the scope of the *E. coli* reconstruction was expanded (Chapter 4), more biological functions could be addressed, studied, and explained. The initial efforts to move past metabolism to include protein synthesis and structures produced surprisingly promising results. It therefore seems that the boundaries of reconstruction will grow steadily until we have put as many cellular processes as possible into a chemically consistent format. The increasing scope of reconstructions puts the function of more and more cellular components into context and provides the mutual constraints that cellular components put on one another; they must function in harmony to get cellular functions balanced.

How constrained are biological systems?   This question is quite interesting and hard to answer. On one hand, looking at the images of cells and their content, it is surprising at first blush that they can function at all, given all the constraints by which the cell has to abide. On the other hand, evolution clearly shows that a beautiful diversity of functional states can be achieved with these myriad constraints.

Is there a set of 'practical constraints' that can be defined?   In the early days of COBRA, this was a pressing question. Could any meaningful solutions be obtained from GEMs? The notion of governing constraints in a given condition proved valid as meaningful solutions could be obtained with almost 'parameter-free' models. With time, we have steadily grown in our ability to define an increasingly practical and useful set of constraints.

How malleable are the constraints?   As discussed in Chapter 17, there are hard and adjustable constraints. The latter can be adjusted using ALE under laboratory conditions. Some ALE experiments show a relatively narrow set of outcomes while others show a surprising diversity in the genetic changes in the evolved strains. So on one hand, there may be many 'equivalent' solutions to an evolutionary challenge, while for others there may be relatively few outcomes.

Will we ever know the objectives of a cell?   A key question. As stated above, the use of an objective function to study mechanisms is unique to biology as it describes distal causation. But how can we know what the objectives are that drive the evolution of better phenotypes? Fundamentally, objective functions are in some sense analogous to diffusion that macroscopically seems purposeful, but is a result of an underlying stochastic process of randomly moving non-identical molecules.

Can we develop a multi-scale theoretical framework?   Biological systems are organized hierarchically in time, space, complexity, and other characteristics. Clearly, to be able to move up and down these hierarchies is important to get a global understanding of biological phenomena. Methods to decompose systems in time and space have been developed in the physical sciences, and as discussed above, the steady-state assumption for a fast subsystem is one such approach. The challenge is to do the same for biological functions and properties.

What is 'self?'   A key biological property is that which distinguishes 'self' from 'non-self'. This distinction spans the biological spectrum from molecules to human societies. A heterologous gene expressed in a bacteria does not fit in, and the host knows it. The new protein does not belong; it is 'non-self.' Similarly, when individuals move from one culture to the next, even if the communities are indistinguishable ethnically, the behavior of the newcomer is recognized by the native population as foreign. The question of self and non-self is a fundamental scale-independent property in biology. It is exciting to find out if GEMs can address this question. Presumably, a part of the answer lies in the development of biological information theory.

Are functional states a reflection of awareness?   Biological systems are aware and awareness is a property found at all scales in biology. One can argue that the simplest organism, such as a bacterium with its two-component sensing systems that measure all the important variables in the environment, is thus 'aware' of its environment. In response to sensing ('measuring') these variables it responds; a response that includes adjustment of gene expression. Such adjustment leads to a particular expressed functional network state. This functional state can thus be considered a readout of the awareness of the cell.

29.4   Why Build Mathematical Models?

Mathematical modeling has been practiced in various branches of science and engineering and is key to answering the questions raised in systems biology. The generic purpose and utility of model-building has been succinctly summarized and discussed [24]; (1) to organize disparate information into a coherent whole; (2) to think (and calculate) logically about what components and interactions are important in a complex system; (3) to discover new strategies; (4) to make important corrections to the conventional wisdom; and (5) to understand the essential qualitative features. All of these issues are addressed directly or indirectly with GEMs.

Ultimately, GEMs are mechanistic genotype–phenotype relationships. They are in some sense analogous to the physical laws (Figure 29.2). They describe a functional

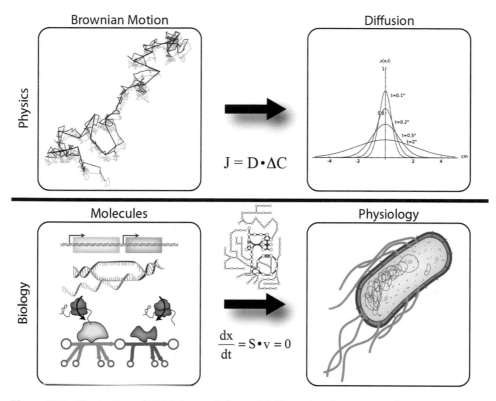

Figure 29.2 Illustration of Fick's law and the multi-dimensional genotype–phenotype relationship.

relationship between the microscopic (different genes and molecules versus a homogeneous population of molecules) and a microscopic behavior (phenotype versus molecular diffusion). The physical laws are low-dimensional, well-structured, and require a modest amount of data to apply. They contain fundamental parameters (such as the speed of light or the unit charge on an electron). Genotype–phenotype relationships need to be genome-scale and thus are high-dimensional. The equations are general and multiple large-scale data sets may be required to work them out. They involve many parameters, which are heavily constrained in their numerical values, but are not fundamental. Such is the fundamental complexity of the living process.

29.5   What Lies Ahead?

There are many practical consequences of the complexity of genome-scale science. Its practice requires a combination of numerous experimental and computational methods, all of which are high-dimensional and multi-variate. Otherwise, some of the key questions of systems biology cannot be addressed and answered. The use of myriad experimental and computational methods leads to the formulation of *workflows* (sometimes called *pipelines* if only data analysis is needed) that describe the order in which such methods are deployed. Workflows can be linear, have parallel paths, or be iterative.

Iterative workflows   The genotype–phenotype relationship is complex. It requires the analysis and modeling of high-dimensional data sets. To improve the rate of progress of genome-scale science, iterative workflows are beginning to appear. Such iterative workflows, as illustrated in Figure 29.3A, comprise reconstruction, modeling, prediction, and experiment. Each cycle of the workflow should lead to a biologically meaningful result. Several studies that have successfully used the *E. coli* GEM for generating iteratively new knowledge include: assignment of ORF function, discovery of antibiotics, metabolic engineering, and discovering new activities of genes that already have functional assignment (see Figure 29.3).

Strengths and limitations of current GEMs   Although GEMs have been able to deal successfully with several basic and applied questions in biology, they have their limitations. The analysis of a large number of applications of GEMs has led to the summary of their strengths and weaknesses (Table 29.1). The weaknesses mostly come down to the missing kinetics and regulation in the current GEMs. Clearly, these attributes need to be included in future models. The third generation of GEMs should arise from time-dependent polyomic data sets where such parameters are estimable [312].

When kinetic parameters become available in GEMs, we can begin to characterize them with additional state variables, over and above the fluxes that are currently used. We will be able to find aggregate variables, or 'pools,' through a hierarchical analysis [317]. The state of such pools will become important, such as what fraction of a regulatory enzyme is in an active state and what fraction is rendered inactive by an inhibitor. Such analyses will expand to focus on variables such as the capacity to carry a particular property (such as the adenylates to carry high-energy bonds) and how fully occupied such capacities are [361]. Key dynamic characteristics of transition states can be analyzed, and the drivers for change in the row space of the stoichiometric matrix can be identified, just as the pathway vectors are used to characterize its orthogonal null space (Chapter 12).

Expanding the scope of GEMs   The scope of GEMs will continue to expand, as more cellular processes are detailed mechanistically and added to the formal structure that GEMs provide. The ME models show that incorporating and integrating protein synthesis with metabolism mechanistically opens up new vistas in systems biology. Detailing operon structures on a genome-scale using metastructure data should follow. The incorporation of DNA structure and transcription binding as a ready-to-compute biochemical network in a mathematical format would overcome the limitations presented by the current Boolean formulation of the TRN (Transcriptional Regulatory Network), and allow for complex regulatory interactions to be modeled and predicted mechanistically. It is likely that DNA synthesis, post-translational modifications, and other cellular processes that involve biochemical interactions that can be described by a biochemical interaction network can also be incorporated into GEMs. In brief, what lies ahead for GEMs is the iterative expansion to include other cellular processes well beyond metabolism.

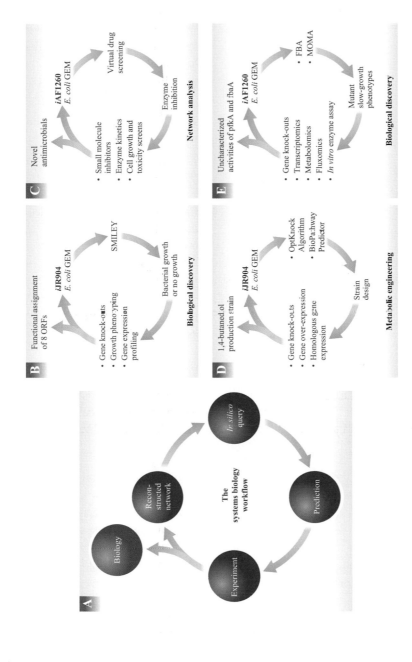

Figure 29.3   Illustration of iterative workflows. From [263]. References: Panel B [357], Panel C [393], Panel D [467], and Panel E [275].

Table 29.1   Strengths and limitations of the metabolic GEM applications. From [263].

Application	What the model can do (Strengths of the *E. coli* GEM)	What the model cannot do (Areas for future progress)
Metabolic engineering	Gene deletion (combinatorial)	Limited coverage of molecular biology
	Gene addition	Predicting the effects of perturbations to regulatory elements
	Gene over- and under-expression	Predicting allosteric inhibition
	Rapidly test the systemic effects of heterologous pathway additions	There is no explicit representation of metabolite concentrations
	Design biomarkers/biosensors for characteristic function	Account for enzyme kinetics
	Determine media supplementation strategies. Map high-throughput data to identify bottlenecks	Cannot predict the performance of nonnative genes/proteins in *E. coli* accurately
	Design strains through evolution	
Biological discovery	Predict growth on different carbon sources/media conditions	Predict the regulation of isozymes/parallel pathways
	Guide the functional assignment of network gaps	Predict enzyme promiscuity
	Guide the discovery of previously uncharacterized gene product functions (graph theory analysis)	Predictive power is inherently limited, because the model is not complete in scope
	Guide the reannotations of incorrectly annotated genes	Predict the expression of genes
	Connect orphan metabolites to known reactions	Predict the functional state of proteins (e.g., posttranslational modification)
Phenotypic behavior	Predict optimal cellular behavior	Differentiate between computed alternate optimal flux distributions of the cell *a priori*
	Understand energetics and occurrence of suboptimal behavior	Explain the reasons for suboptimal performance *a priori*
	Infer impact of regulation. Provide a context for which experimental data can be interpreted	Provide a framework for incorporating additional regulatory interactions that are currently under development

Table 29.1 *(cont.)*

Application	What the model can do (Strengths of the *E. coli* GEM)	What the model cannot do (Areas for future progress)
	Predict and understand absolute and conditional gene essentiality	
	Predict and understand shifts in growth conditions	
Network analysis	Evaluate metabolic networks from a systems view through node and link dependencies, essentialities, overall network robustness	Does not always include the biological mechanisms behind the network connections. Few predictions can be validated experimentally
	Describe the complex interactions of the components of the metabolic network	
	Evaluate modularity of function	
	Evaluate regulation based on network structure	
Bacterial evolution	Predict essential genes	Account for changes in regulatory elements
	Predict the endpoint of evolution	Predict the time-course of evolution
	Understand the basis for epistatic interactions and mutational effects	Predict location of mutations in the genome. Predict the effects of mutations in the genome
	Provide insights into evolutionary trajectories	Account for strain-specific genomic differences
Interspecies interaction	Model the exchange of metabolites	Model interactions that affect metabolic regulation
	Analyze high-throughput data from different strains	Inability to measure flux exchange in multi cell-type systems
	Determine the cost/benefit ratio for different types of commensalism	There are still too many unknowns to accurately build an interactions network
		Limited ability to define individual genetic content in large communities
		Limited spatial knowledge in large communities

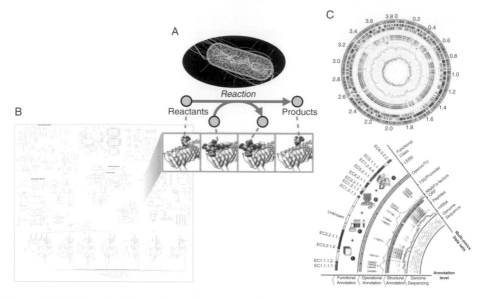

Figure 29.4 Three dominant views of genome-scale science. (A) The biophysical view, (B) the network view, and (C) the genome, or metastructure view (image from [148] used with kind permission from Springer Science+Business Media B.V.), top image from [269], bottom image from [349].

The payoff for this increased complexity will be more accurate predictions of the molecular phenotype, and new hierarchical analysis methods should enable a better understanding of the cellular phenotypes that result. Detailed reconstruction of a strain will be followed by multi-strain reconstruction, and through comparative analysis will help to define and understand a species. Reconstruction of microbial communities is likely to follow. Comparative analysis of human cell and tissue types, and multi-organ models are beginning to appear, and perhaps some of the differences between animals can be analyzed at this level, such as the difference between carnivores and herbivores. How high COBRA methods can reach in the biological hierarchy is uncertain at present, but the possibilities seem numerous.

Merging the three dominant views in cell and molecular biology We think of cell and molecular biology in terms of three paradigms that come with some disciplinary biases (Figure 29.4). The biophysical view of molecules, protein, DNA, and membrane is familiar. Images of this representation at the genome scale appeared in the early 1990s and provided an early view of genome-scale science. In the mid 1990s, whole-genome sequences appeared and images of whole genomes became commonplace. We now understand that genomes are much more complicated than their base pair sequence and 3-dimensional arrangement, and the metastructure point of view emerged in the late 2000s. Genome-scale views of networks appeared in the late 1990s as the first genome-scale reconstructions began to appear.

These three dominant viewpoints are familiar. For systems biology to advance, we have to learn how to reconcile them, think about them as an integrated entity, and be able to formulate computational descriptions of their simultaneous functions. This will be challenging to achieve, and especially to visualize the results simultaneously.

Towards hierarchical theory and modularization Systems and network analysis in biology is at an early stage of development. It seems clear that one of the challenges that the field is faced with is the development of a multi-scale analysis framework. Conceptual foundations and some issues can be identified.

Dealing with physical vs. biological causation: as illustrated in Figure 15.3, these two issues are at opposite ends of a hierarchy of events that we have to deal with. It is clear that this relationship is hierarchical, and calls for multi-scale analysis.

Consequently, one challenge is to determine the information content in the various omics data types as they address different layers in the hierarchy. Constraints at the lowest level must hold at all higher levels. However, there will be additional constraints and considerations that arise as we move up the hierarchy. Thus, there may be measurable changes at a lower level that are inconsequential at a higher level. The existence of hitchhiker mutations is one example, and we can expect to find similar examples with other omics data types. *In silico* analysis methods that deal with these issues explicitly are needed.

In the intracellular environment there is much *thermal noise*, as biological components are constantly bouncing into one another. The question of 'who talks to whom?' has received much attention, particularly in the study of signaling networks, and should perhaps more clearly be asked as 'who listens to whom?' as biological components can be bouncing into one another without any resulting chemical reaction or any influence on the network as a whole. Chemical interactions that lead to no further consequences would be 'dead ends' in a network. Dead ends can turn into 'live contacts' if new links are established. Robust and properly functioning networks within the constraints of thermal noise is clearly important in maintaining basic cellular functions.

A molecule within the complex intracellular environment will, by virtue of physico-chemical and topological constraints, only be able to interact with a finite number of potential partners. Thus, every molecule and every component seems to function in what might be called a *small world*. What is happening in one small world or in one locality inside a cell may not be aware of what is happening a few locales over, at least not at every point in time. Therefore, what is happening at one location inside a cell may be only loosely connected to what is occurring elsewhere. In terms of a computer programming language, this may be thought of as parallel processing, which every so often needs to be synchronized by some higher-level organization. The process of spontaneous self-assembly will play a role.

This necessarily leads to the consideration of the 3D arrangements of cells and the morphogenic properties of groups, or modules, executing integrated functions in molecular biology. Therefore, the 2D representations of biochemical reaction networks that are discussed in Part I will eventually have to become three-dimensional. The consideration of the architecture of cellular processes at the

~100 nm length scale are likely to lead to an exciting new dimension in molecular systems biology, both in terms of *in silico* analysis as well as new generation of measurement tools.

Hierarchical analysis will come down to aggregating or combining the elementary variables (e.g., concentrations and fluxes) into new quantities that will take us from chemistry to biology systematically. Mathematical definitions of aggregate variables in terms of pools and pathways have appeared [361]. The systematic decomposition of pools appear through a combination of network topology and kinetic values through the use of temporal decomposition and modal analysis [198, 310].

As we become better at biologically driven hierarchical analysis of large networks, we will begin to be able to formulate mathematical definitions of key properties like redundancy, robustness, causality, and so on. Potentially, such analyses may graduate to definitions of network properties that may relate to fundamental biological properties such as 'what is self?'

Closing   Hopefully, this text will be useful to those that are interested in network reconstruction, the biochemically and genomically accurate representation of such reconstructions, and methods to interrogate the functional states of networks. After 15 years of development, it seems like the COBRA field is just in its infancy and that it has so many exciting developments ahead of it.

References

[1] http://ecocyc.org/.

[2] http://systemsbiology.ucsd.edu.

[3] http://www.wadsworth.org/databank/ecoli.htm.

[4] http://bd2k.nih.gov.

[5] M. AbuOun *et al*. Genome scale reconstruction of a *Salmonella* metabolic model: Comparison of similarity and differences with a commensal *Escherichia coli* strain. *Journal of Biological Chemistry*, 284:29480–29488, 2009.

[6] R. Agren *et al*. Reconstruction of genome-scale active metabolic networks for 69 human cell types and 16 cancer types using INIT. *PLoS Computational Biology*, 8:e1002518, 2012.

[7] R. Agren *et al*. The RAVEN toolbox and its use for generating a genome-scale metabolic model for *Penicillium chrysogenum*. *PLoS Computational Biology*, 9:e1002980, 2013.

[8] M. Akesson, J. Forster, and J. Nielsen. Integration of gene expression data into genome-scale metabolic models. *Metabolic Engineering*, 6:285–293, 2004.

[9] R. A. Alberty. *Thermodynamics of Biochemical Reactions*. Wiley-Interscience, Hoboken, NJ, 2003.

[10] W. J. Albery and J. R. Knowles. Evolution of enzyme function and development of catalytic efficiency. *Biochemistry*, 15:5631–5640, 1976.

[11] W. J. Albery and J. R. Knowles. Efficiency and evolution of enzyme catalysis. *Angewandte Chemie International Edition in English*, 16:285–293, 1977.

[12] T. E. Allen and B. O. Palsson. Sequenced-based analysis of metabolic demands for protein synthesis in prokaryotes. *Journal of Theoretical Biology*, 220(1):1–18, 2003.

[13] E. Almaas, B. Kovacs, T. Vicsek, Z. N. Oltvai, and A. L Barabasi. Global organization of metabolic fluxes in the bacterium *Escherichia coli*. *Nature*, 427:839–843, 2004.

[14] E. Almaas, B. Kovacs, T. Vicsek, Z. N. Oltvai, and A. L. Barabasi. Network biology: Understanding the cell's functional organization. *Nature Reviews Genetics*, 5:101–113, 2004.

[15] O. Alter, P. O. Brown, and D. Botstein. Singular value decomposition for genome-wide expression data processing and modeling. *Proceedings of the National Academy of Sciences of the United States of America*, 97:10101–10106, 2000.

[16] S. F. Altschul *et al*. Basic local alignment search tool. *Journal of Molecular Biology*, 215:403–410, 1990.

[17] J. Amberger *et al*. Mckusick's online Mendelian inheritance in man (OMIM). *Nucleic Acids Research*, 37:D793–D796, 2009.

[18]   R. T. Anderson *et al.* Stimulating the *in situ* activity of *Geobacter* species to remove uranium from the groundwater of a uranium-contaminated aquifer. *Applied Environmental Microbiology*, 69:5884–5891, 2003.

[19]   M. R. Antoniewicz, J. K. Kelleher, and G. Stephanopoulos. Elementary metabolite units (EMU): A novel framework for modeling isotopic distributions. *Metabolic Engineering*, 9:68–86, 2007.

[20]   M. K. Applebee *et al.* Functional and metabolic effects of adaptive glycerol kinase (GLPK) mutants in *Escherichia coli. Journal of Biological Chemistry*, 286:23150–23159, 2011.

[21]   S. Atsumi, T. Hanai, and J. C. Liao. Non-fermentative pathways for synthesis of branched-chain higher alcohols as biofuels. *Nature*, 451:86–89, 2008.

[22]   G.J. Baart *et al.* Modeling *Neisseria meningitidis* metabolism: From genome to metabolic fluxes. *Genome Biology*, 8:R136, 2007.

[23]   T. Baba *et al.* Construction of *Escherichia coli* K-12 in-frame, single-gene knockout mutants: The Keio collection. *Molecular Systems Biology*, 2:2006.0008, 2006.

[24]   J. E. Bailey. Mathematical modeling and analysis in biochemical engineering: Past accomplishments and future opportunities. *Biotechnology Progress*, 14:8–20, 1998.

[25]   J. E. Bailey. Complex biology with no parameters. *Nature Biotechnology*, 19:503–504, 2001.

[26]   A. Bar-Even *et al.* The moderately efficient enzyme: Evolutionary and physicochemical trends shaping enzyme parameters. *Biochemistry*, 50:4402–4410, 2011.

[27]   A. L. Barabasi. *Linked: the New Science of Networks*. Perseus, Cambridge, 2002.

[28]   C. L. Barrett, M. J. Herrgard, and B. O. Palsson. Decomposing complex reaction networks using random sampling, principal component analysis, and basis rotation. *BMC Systems Biology*, 3:30, 2009.

[29]   B. L. Bassler and R. Losick. Bacterially speaking. *Cell*, 125:237–246, 2006.

[30]   D. Baycin-Haydal *et al.* Proteomic analysis of Chinese hamster ovary (CHO) cells. *Journal of Proteome Research*, 11:5265–5276, 2012.

[31]   M. S. Bazaraa, H. D. Sherali, and C. M. Shetty. *Nonlinear Programming: Theory and Algorithms*, 2nd Edition. John Wiley and Sons, New Jersey, 1993.

[32]   D. A. Beard and H. Qian. Thermodynamic-based computational profiling of cellular regulatory control in hepatocyte metabolism. *American Journal of Physiology*, 288:E633–E644, 2005.

[33]   S. A. Becker, N. D. Price, and B. O. Palsson. Metabolite coupling in genome-scale metabolic networks. *BMC Bioinformatics*, 7(111), 2006.

[34]   S. A. Becker *et al.* Quantitative prediction of cellular metabolism with constraint-based models: The COBRA toolbox. *Nature Protocols*, 2:727–738, 2007.

[35]   S. A. Becker and B. O. Palsson. Genome-scale reconstruction of the metabolic network in *Staphylococcus aureus* N315: An initial draft to the two-dimensional annotation. *BMC Microbiology*, 5:8–19, 2005.

[36]   S. A. Becker and B. O. Palsson. Context-specific metabolic networks are consistent with experiments. *PLoS Computational Biology*, 4:e1000082, 2008.

[37]   S. A. Becker, N. D. Price, and B. O. Palsson. Metabolite coupling in genome-scale metabolic networks. *BMC Bioinformatics*, 7:111, 2006.

[38]   S. L. Bell and B. O. Palsson. ExPa: A program for calculating extreme pathways in biochemical reaction networks. *Bioinformatics*, 21:1739–1740, 2005.

[39]   D. J. Beste *et al*. GSMN-TB: A web-based genome-scale network model of *Mycobacterium tuberculosis* metabolism. *Genome Biology*, 8:R89, 2007.

[40]   F. R. Blattner *et al*. The complete genome sequence of *Escherichia coli* K-12. *Science*, 277(5331):1453–74, 1997.

[41]   A. S. Blazier and J. A. Papin. Integration of expression data in genome-scale metabolic network reconstructions. *Frontiers in Physiology*, 3:doi10.3389/fphys.2012.00299, 2012.

[42]   A. Boiteux, B. Hess, and E. E. Sel'kov. Creative functions of instability and oscillations in metabolic systems. *Current Topics in Cell Regulation*, 17:171–203, 1980.

[43]   H. P. J. Bonarius, G. Schmid, and J. Tramper. Flux analysis of underdetermined metabolic networks: The quest for the missing constraints. *Trends in Biotechnology*, 15:308–314, 1997.

[44]   B. K. Bonde *et al*. Differential producibility analysis (DPA) of transcriptomic data with metabolic networks: Deconstructing the metabolic response of *M. tuberculosis*. *PLoS Computational Biology*, 7:e1002060, 2011.

[45]   R. Bonneau *et al*. A predictive model for transcriptional control of physiology in a free living cell. *Cell*, 131:1354–1365, 2007.

[46]   F. C. Boogerd *et al*. ATP mutants of *Escherichia coli* fail to grow on succinate due to a transport deficiency. *Journal of Bacteriology*, 180:5855–5859, 1998.

[47]   A. Bordbar. *Utilizing genome-scale models to enhance high-throughput data analysis: from pathways to dynamics*. PhD thesis, University of California, San Diego, 2014.

[48]   A. Bordbar *et al*. Insight into human alveolar macrophage and *M. tuberculosis* interactions via metabolic reconstructions. *Molecular Systems Biology*, 6:422, 2010.

[49]   A. Bordbar *et al*. A multi-tissue type genome-scale metabolic network for analysis of whole-body systems physiology. *BMC Systems Biology*, 5:180, 2011.

[50]   A. Bordbar *et al*. Model-driven multi-omic data analysis elucidates metabolic immunomodulators of macrophage activation. *Molecular Systems Biology*, 8:558, 2012.

[51]   A. Bordbar, N. Jamshidi, and B. O. Palsson. IAB-RBC-283: A proteomically derived knowledge-base of erythrocyte metabolism that can be used to

simulate its physiological and patho-physiological states. *BMC Systems Biology*, 5:110, 2011.

[52] A. Bordbar, J. M. Monk, Z. A. King, and B. O. Palsson. Constraint-based models predict metabolic and associated cellular functions. *Nature Reviews Genetics*, 15:107–120, 2014.

[53] A. Bordbar and B. O. Palsson. Using the reconstructed genome-scale human metabolic network to study physiology and pathology. *Journal of Internal Medicine*, 271:131–134, 2012.

[54] S. Bordel, R. Agren, and J. Nielsen. Sampling the solution space in genome-scale metabolic networks reveals transcriptional regulation in key enzymes. *PLoS Computational Biology*, 6:e1000859, 2010.

[55] A. P. Burgard and C. D. Maranas. Optimization-based framework for inferring and testing hypothesized metabolic objective functions. *Biotechnology and Bioengineering*, 82:670–677, 2003.

[56] A. P. Burgard, E. V. Nikolaev, C. H. Schilling, and C. D. Maranas. Flux coupling analysis of genome-scale metabolic network reconstructions. *Genome Research*, 14:301–312, 2004.

[57] A. P. Burgard and C. D. Maranas. Optimization-based framework for inferring and testing hypothesized metabolic objective functions. *Biotechnology and Bioengineering*, 82:670–677, 2003.

[58] A. P. Burgard, P. Pharkya, and C. D. Maranas. OptKnock: A bilevel programming framework for identifying gene knockout strategies for microbial strain optimization. *Biotechnology and Bioengineering*, 84:647–657, 2003.

[59] V. Busskamp *et al*. Optogenetic therapy for retinitis pigmentosa. *Gene Therapy*, 19:169–175, 2012.

[60] M. W. Calhoun *et al*. Energetic efficiency of *Escherichia coli*: Effects of mutations in components of the aerobic respiratory chain. *Journal of Bacteriology*, 175:3020–3025, 1993.

[61] R. Caspi *et al*. The MetaCyc database of metabolic pathways and enzymes and the BioCyc collection of pathway/genome databases. *Nucleic Acids Research*, 36:D623–D631, 2008.

[62] T. B. Causey, K. T. Shanmugam, L. P. Yomano, and L. O. Ingram. Engineering *Escherichia coli* for efficient conversion of glucose to pyruvate. *Proceedings of the National Academy of Sciences of the United States of America*, 101:2235–2240, 2004.

[63] T. B. Causey, S. Zhou, K. T. Shanmugam, and L. O. Ingram. Engineering the metabolism of *Escherichia coli* W3110 for the conversion of sugar to redox-neutral and oxidized products: Homoacetate production. *Proceedings of the National Academy of Sciences of the United States of America*, 100:825–832, 2003.

[64] R. L. Chang *et al*. Drug off-target effects predicted using structural analysis in the context of a metabolic network model. *PLoS Computational Biology*, 6:e1000938, 2010.

[65] R. L. Chang *et al.* Metabolic network reconstruction of *Chlamydomonas* offers insight into light-driven algal metabolism. *Molecular Systems Biology*, 7:518, 2011.

[66] R. L. Chang *et al.* Structural systems biology evaluation of metabolic thermotolerance in *E. coli*. *Science*, 340:1220–1223, 2013.

[67] P. Charusanti *et al.* Genetic basis of growth adaptation of *Escherichia coli* after deletion of *pgi*, a major metabolic gene. *PLoS Genetics*, 6:e1001186, 2010.

[68] A. K. Chavali *et al.* Systems analysis of metabolism in the pathogenic trypanosomatid *Leishmania major*. *Molecular Systems Biology*, 4:177, 2008.

[69] A. K. Chavali *et al.* A metabolic network approach for the identification and prioritization of antimicrobial drug targets. *Trends in Microbiology*, 20:113–123, 2012.

[70] J. A. Chemler *et al.* Improving NADPH availability for natural product biosynthesis in *Escherichia coli* by metabolic engineering. *Metabolic Microbiology*, 12:96–104, 2010.

[71] L. Chen and D. Vitkup. Predicting genes for orphan metabolic activities using phylogenetic profiles. *Genome Biology*, 7(2):R17, 2006.

[72] L. Chen and D. Vitkup. Distribution of orphan metabolic activities. *Trends in Biotechnology*, 25:343–348, 2007.

[73] B. K. Cho *et al.* Genome-scale reconstruction of the Lrp regulatory network in *Escherichia coli*. *Proceedings of the National Academy of Sciences, USA*, 105:19461–19466, 2008.

[74] B. K. Cho *et al.* The transcription unit architecture of the *Escherichia coli* genome. *Nature Biotechnology*, 27:1043–1049, 2009.

[75] B. K. Cho *et al.* Deciphering the transcriptional regulatory logic of amino acid metabolism. *Nature Chemical Biology*, 8:65–71, 2012.

[76] V. Chvatal. *Linear Programming*. W.H. Freeman, New York, 1983.

[77] W. W. Cleland. What limits the rate of an enzyme catalyzed reaction? *Accounts of Chemical Research*, 8:145–151, 1981.

[78] T. M. Conrad *et al.* RNA polymerase mutants found through adaptive evolution reprogram *Escherichia coli* for optimal growth in minimal media. *Proceedings of the National Academy of Sciences, USA*, 107:20500–20505, 2010.

[79] T. M. Conrad *et al.* Whole-genome resequencing of *Escherichia coli* K-12 MG1655 undergoing short-term laboratory evolution in lactate minimal media reveals flexible selection of adaptive mutations. *Genome Biology*, 10:R118, 2009.

[80] T. M. Conrad, N. E. Lewis, and B. O. Palsson. Microbial laboratory evolution in the era of genome-scale science. *Molecular Systems Biology*, 7(509):509, 2011.

[81] International Human Genome Sequencing Consortium. Finishing the euchromatic sequence of the human genome. *Nature*, 431:931–945, 2004.

[82] J. B. Courtright and U. Henning. Malate dehydrogenase mutants in *Escherichia coli* K-12. *Journal of Bacteriology*, 102:722–728, 1970.

[83]   M. Covert, E. M. Knight, M. J. Herrgard, J. L. Reed, and B. O. Palsson. Integrating high-throughput and computational data elucidates bacterial networks. *Nature*, 429:92–96, 2004.

[84]   M. W. Covert, I. Famili, and B. O. Palsson. Identifying constraints that govern cell behavior: A key to converting conceptual to computational models in biology? *Biotechnology and Bioengineering*, 84:763–772, 2003.

[85]   M. W. Covert and B. O. Palsson. Transcriptional regulation in constraints-based metabolic models of *Escherichia coli*. *Journal of Biological Chemistry*, 277:28058–28064, 2002.

[86]   M. W. Covert and B. O. Palsson. Constraints-based models: Regulation of gene expression reduces the steady-state solution space. *Journal of Theoretical Biology*, 221:309–325, 2003.

[87]   M. W. Covert *et al.* Metabolic modeling of microbial stains *in silico*. *Trends in Biochemical Sciences*, 26:179–186, 2001.

[88]   M. W. Covert, C. H. Schilling, and B. O. Palsson. Regulation of gene expression in flux balance models of metabolism. *Journal of Theoretical Biology*, 213:73–88, 2001.

[89]   I. T. Creaghan and J. R. Guest. Succinate dehydrogenase-dependent nutritional requirement for succinate in mutants of *Escherichia coli* K12. *Journal of General Microbiology*, 107:1–13, 1978.

[90]   F. Crick. Project k: The complete solution of *E. coli*. *Perspectives in Biology and Medicine*, 17:67–70, 1973.

[91]   F. Crick. *What Mad Pursuit: A Personal View of Scientific Discovery*. Basic Books, New York, 1988.

[92]   M. A. Croxen and B. B. Finlay. Molecular mechanisms of *Escherichia coli* pathogenicity. *Nature Reviews Microbiology*, 8:26–38, 2010.

[93]   A. Danchin. *The Delphic Boat: What Genomes Tell Us*. Harvard University Press, New Jersey, 2003.

[94]   M. De Paepe *et al.* Trade-off between bile resistance and nutritional competence drives *Escherichia coli* diversification in the mouse gut. *PLoS Genetics*, 7:e1002107, 2011.

[95]   R. De Smet and K. Marchal. Advantages and limitations of current network inference methods. *Nature Reviews Microbiology*, 8:717–729, 2010.

[96]   K. A. Dill, K. Ghosh, and J. D. Schmit. Physical limits of cells and proteomes. *Proceedings of the National Academy of Sciences, USA*, 108:17876–17882, 2011.

[97]   P. D. Dobson *et al.* Further developments towards a genome-scale metabolic model of yeast. *BMC Systems Biology*, 4:145, 2010.

[98]   M. I. Donnelly and R. A. Cooper. Two succinic semialdehyde dehydrogenases are induced when *Escherichia coli* K-12 is grown on gamma-aminobutyrate. *Journal of Bacteriology*, 145:1425–1427, 1981.

[99]   M. Dragosits and D. Mattanovich. Adaptive laboratory evolution – principles and applications for biotechnology. *Microbial Cell Factories*, 12:64, 2013.

[100] N. C. Duarte *et al.* Global reconstruction of the human metabolic network based on genomic and bibliomic data. *Proceedings of the National Academy of Sciences, USA*, 104:1777–1782, 2007.

[101] N. C. Duarte, B. O. Palsson, and P. Fu. Integrated analysis of metabolic phenotypes in *Saccharomyces cerevisiae*. *BMC Genomics*, 5:63, 2004.

[102] I. Dubchak and K. Frazer. Multi-species sequence comparison: The next frontier in genome annotation. *Genome Biology*, 4:122, 2003.

[103] J. S. Edwards, R. Ramakrishna, and B. O. Palsson. Characterizing the metabolic phenotype: A phenotype phase plane analysis. *Biotechnology and Bioengineering*, 77:27–36, 2002.

[104] J. S. Edwards, R. U. Ibarra, and B. O. Palsson. *In silico* predictions of *Escherichia coli* metabolic capabilities are consistent with experimental data. *Nature Biotechnology*, 19:125–130, 2001.

[105] J. S. Edwards and B. O. Palsson. Systems properties of the *Haemophilus influenzae* Rd metabolic genotype. *Journal of Biological Chemistry*, 274:17410–17416, 1999.

[106] J. S. Edwards and B. O. Palsson. The *Escherichia coli* MG1655 *in silico* metabolic genotype; Its definition, characteristics, and capabilities. *Proceedings of the National Academy of Sciences, USA*, 97:5528–5533, 2000.

[107] J. S. Edwards and B. O. Palsson. Robustness analysis of the *Escherichia coli* metabolic network. *Biotechnology Progress*, 16:927–939, 2000.

[108] S. F. Elena and R. E. Lenski. Microbial genetics: Evolution experiments with microorganisms: The dynamics and genetic bases of adaptation. *Nature Reviews Genetics*, 4:457–469, 2003.

[109] I. Famili, J. Forster, J. Nielsen, and B. O. Palsson. *Saccharomyces cerevisiae* phenotypes can be predicted using constraint-based analysis of a genome-scale reconstructed metabolic network. *Proceedings of the National Academy of Sciences, USA*, 100:13134–13139, 2003.

[110] I. Famili, R. Mahadevan, and B. O. Palsson. K-cone analysis: Determining all candidate values for kinetic parameters on a network-scale. *Biophysical Journal*, 88:1616–1625, 2005.

[111] I. Famili and B. O. Palsson. Systemic metabolic reactions are obtained by singular value decomposition of genome-scale stoichiometric matrices. *Journal of Theoretical Biology*, 224:87–96, 2003.

[112] K. Fang *et al.* Exploring the metabolic network of the epidemic pathogen *Burkholderia cenocepacia*. *BMC Systems Biology*, 5:83, 2011.

[113] X. Fang *et al.* Development and analysis of an *in vivo*-compatible metabolic network of *Mycobacterium tuberculosis*. *BMC Systems Biology*, 4:160, 2010.

[114] R. P. Faynman. *The Character of Physical Law*. MIT Press, Cambridge, Mass, 1965.

[115] A. M. Feist *et al.* A genome-scale metabolic reconstruction for *Escherichia coli* K-12 MG1655 that accounts for 1260 ORFs and thermodynamic information. *Molecular Systems Biology*, 3:121, 2007.

[116] A. M. Feist *et al.* Model-driven evaluation of the production potential for growth-coupled products of *Escherichia coli*. *Metabolic Engineering*, 12:173–186, 2010.

[117] A. M. Feist, M. J. Herrgard, I. Thiele, J. L. Reed, and B. O. Palsson. Reconstruction of biochemical networks in microorganisms. *Nature Reviews Microbiology*, 7(2):129–143, 2009.

[118] A. M. Feist and B. O. Palsson. The growing scope of applications of genome-scale metabolic reconstructions using *Escherichia coli*. *Nature Biotechnology*, 26:659–667, 2008.

[119] A. M. Feist and B. O. Palsson. The biomass objective function. *Current Opinion in Microbiology*, 13:344–349, 2010.

[120] A. Feizi *et al.* Genome-scale modeling of the protein secretory machinery in yeast. *PLoS One*, 8:e63284, 2013.

[121] D. A. Fell and J. R. Small. Fat synthesis in adipose tissue. an examination of stoichiometric constraints. *Biochemical Journal*, 238:721–786, 1986.

[122] E. Fischer and U. Sauer. A novel metabolic cycle catalyzes glucose oxidation and anaplerosis in hungry *Escherichia coli*. *Journal of Biological Chemistry*, 278:46446–46451, 2003.

[123] R. D. Fleischmann *et al.* Whole-genome random sequencing and assembly of *Haemophilus influenzae* Rd. *Science*, 269:496–498, 1995.

[124] O. Folger *et al.* Predicting selective drug targets in cancer through metabolic networks. *Molecular Systems Biology*, 7:501, 2011.

[125] A. Folkesson *et al.* Adaptation of *Pseudomonas aeruginosa* to the cystic fibrosis airway: An evolutionary perspective. *Nature Reviews Microbiology*, 10:841–851, 2012.

[126] S. S. Fong *et al. In silico* design and adaptive evolution of *Escherichia coli* for production of lactic acid. *Biotechnology and Bioengineering*, 91:643–648, 2005.

[127] S. S. Fong and B. O. Palsson. Metabolic gene-deletion strains of *Escherichia coli* evolve to computationally predicted growth phenotypes. *Nature Genetics*, 36:1056–1058, 2004.

[128] American Society for Microbiology. Cover. *Journal of Bacteriology*, 190:1, 2008.

[129] J. Forster, I. Famili, P. C. Fu, B. O. Palsson, and J. Nielsen. Genome-scale reconstruction of the *Saccharomyces cerevisiae* metabolic network. *Genome Research*, 13:244–253, 2003.

[130] J. Forster, I. Famili, B. O. Palsson, and J. Nielsen. Large-scale evaluation of *in silico* gene deletions in *Saccharomyces cerevisiae*. *OMICS*, 7(2):193–202, 2003.

[131] Z. L. Fowler, W. W. Gikandi, and M. A. G. Koffas. Increased malonyl coenzyme A biosynthesis by tuning the *Escherichia coli* metabolic network and its application to flavanone production. *Applied Environmental Microbiology*, 75:5831–5839, 2009.

[132] P. L. Freddolino and S. Tavazoie. Beyond homeostasis: A predictive-dynamic-framework for understanding cellular behavior. *Annual Review of Cell and Developmental Biology*, 28:363–384, 2012.

[133] C. Frezza *et al.* Haem oxygenase is synthetically lethal with the tumour suppressor fumarate hydratase. *Nature*, 477:225–228, 2011.

[134] P. Fu. Genome-scale modeling of *Synechocystis* sp. PCC 6803 and prediction of pathway insertion. *Journal of Chemical Technology and Biotechnology*, 84:473–483, 2009.

[135] T. Fuhrer, L. Chen, U. Sauer, and D. Vitkup. Computational prediction and experimental verification of the gene encoding the NAD+/NADP+-dependent succinate semialdehyde dehydrogenase in *Escherichia coli*. *Journal of Bacteriology*, 189:8073–8088, 2007.

[136] R. A. Gatenby and R. J. Gillies. A microenvironmental model of carcinogenesis. *Nature Reviews Cancer*, 8:56–61, 2008.

[137] D. R. Georgianna and S. P. Mayfield. Exploiting diversity and synthetic biology for the production of algal biofuels. *Nature*, 488:329–335, 2012.

[138] S. Y. Gerdes *et al.* Experimental determination and system level analysis of essential genes in *Escherichia coli* MG1655. *Journal of Bacteriology*, 185:5673–5684, 2003.

[139] J. Gerhart and M. Kirschner, editors. *Cells, Embryos, and Evolution: Toward a Cellular and Developmental Understanding of Phenotypic Variation and Evolutionary Adaptability*. Blackwell Science, Malden, MA, 1997.

[140] E. P. Gianchandani *et al.* Predicting biological system objectives *de novo* from internal state measurements. *BMC Bioinformatics*, 9:43, 2008.

[141] E. P. Gianchandi *et al.* Functional states of the genome-scale *Escherichia coli* transcriptional regulatory system. *PLoS Computational Biology*, 5:e100403, 2009.

[142] A. Giraud *et al.* Dissecting the genetic components of adaptation of *Escherichia coli* to the mouse gut. *PLoS Genetics*, 4:e2, 2008.

[143] A. Goffeau *et al.* The yeast genome directory. *Nature*, 387:5–6, 1997.

[144] Y. Gong *et al.* Sulfide-driven microbial electrosynthesis. *Environmental Science and Technology*, 47:568–573, 2013.

[145] O. Gonzalez *et al.* Characterization of growth and metabolism of the haloalkaliphile *Natronomonas pharaonis*. *PLoS Computational Biology*, 6:e1000799, 2010.

[146] D. S. Goodsell. Inside a living cell. *TIBS*, 16:203–6, 1991.

[147] D. S. Goodsell. *Our Molecular Nature: The Body's Motors, Machines and Messages*. Copernicus, New York, 1997.

[148] D. S. Goodsell. *The Machinery of Life*. Springer-Verlag, New York, 2009.

[149] R. L. Gorsuch, editor. *Factor Analysis*. Erlbaum Associates, Hillsdale, NJ, 1983.

[150] M. L. Green and P. D. Karp. A Bayesian method for identifying missing enzymes in predicted metabolic pathway databases. *BMC Bioinformatics*, 5:76, 2004.

[151] A. R. Grossman *et al.* Novel metabolism in *Chlamydomonas* through the lens of genomics. *Current Opinion in Plant Biology*, 10:190–198, 2007.

[152] S. Gupta and D. P. Clark. *Escherichia coli* derivatives lacking both alcohol dehydrogenase and phosphotransacetylase grow anaerobically by lactate fermentation. *Journal of Bacteriology*, 171:3650–3655, 1989.

[153] H. Gutfreund. *Enzymes: Physical Principles*. John Wiley & Sons, London, 1972.

[154] R. L. Hanson and C. Rose. Effects of an insertion mutation in a locus affecting pyridine nucleotide transhydrogenase (PNT::TN5) on the growth of *Escherichia coli*. *Journal of Bacteriology*, 141:401–404, 1980.

[155] E. Harris, D. Stern, and G. Witman, editors. *The Chlamydomonas Sourcebook: Introduction to Chlamydomonas and its Laboratory Use*. Academic Press, Oxford, 2008.

[156] E. H. Harris. *Chlamydomonas* as a model organism. *Annual Review of Plant Physiology and Plant Molecular Biology*, 52:363–406, 2001.

[157] R. Harrison *et al.* Plasticity of genetic interactions in metabolic networks of yeast. *Proceedings of the National Academy of Sciences, USA*, 104:2307–2312, 2007.

[158] V. Hatzimanikatis *et al.* Exploring the diversity of complex metabolic networks. *Bioinformatics*, 21:1603–1609, 2005.

[159] X. He *et al.* Prevalent positive epistasis in *Escherichia coli* and *Saccharomyces cerevisiae* metabolic networks. *Nature Genetics*, 42:272–276, 2010.

[160] B. D. Heavner *et al.* Yeast 5 – an expanded reconstruction of the *Saccharomyces cerevisiae* metabolic network. *BMC Systems Biology*, 6:55, 2012.

[161] M. Heinemann *et al. In silico* genome-scale reconstruction and validation of the *Staphylococcus aureus* metabolic network. *Biotechnology and Bioengineering*, 92:850–864, 2005.

[162] C. S. Henry *et al.* ibsu1103: A new genome-scale metabolic model of *Bacillus subtilis* based on SEED annotations. *Genome Biology*, 10:R69, 2009.

[163] C. S. Henry *et al.* High-throughput generation, optimization and analysis of genome-scale metabolic models. *Nature Biotechnology*, 28:977–982, 2010.

[164] M. J. Herrgard, M. W. Covert, and B. O. Palsson. Reconstruction of microbial transcriptional regulatory networks. *Current Opinion in Biotechnology*, 15:70–77, 2004.

[165] M. J. Herrgard *et al.* A consensus yeast metabolic network obtained from a community approach to systems biology. *Nature Biotechnology*, 26:1155–1160, 2008.

[166] M. J. Herrgard, S. S. Fong, and B. O. Palsson. Identification of genome-scale metabolic network models using experimentally measured flux profiles. *PLoS Computational Biology*, 2:e72, 2006.

[167] C. D. Herring *et al.* Comparative genome sequencing of *Escherichia coli* allows observation of bacterial evolution on a laboratory timescale. *Nature Genetics*, 38:1406–1412, 2006.

[168] T. Hindre *et al.* New insights into bacterial adaptation through *in vivo* and *in silico* experimental evolution. *Nature Reviews Microbiology*, 10:352–365, 2012.

[169] N. S. Holter *et al.* Fundamental patterns underlying gene expression profiles: Simplicity from complexity. *Proceedings of the National Academy of Sciences of the United States of America*, 97:8409–8414, 2000.

[170] H. G. Holzhutter. The principle of flux minimization and its application to estimate stationary fluxes in metabolic networks. *European Journal of Biochemistry*, 271:2905–2922, 2004.

[171] S. J. Hong and C. G. Lee. Evaluation of central metabolism based on a genomic database of *Synechocystis* PCC6803. *Biotechnology and Bioprocess Engineering*, 12:165–173, 2007.

[172] N. H. Horowitz. On the evolution of biochemical syntheses. *Proceedings of the National Academy of Sciences, USA*, 31:153–157, 1945.

[173] D. Houle, D. R. Govindaraju, and S. Omholt. Phenomics: The next challenge. *Nature Reviews Genetics*, 11:855–866, 2010.

[174] C. Huthmacher *et al.* Antimalarial drug targets in *Plasmodium falciparum* predicted by stage-specific metabolic network analysis. *BMC Systems Biology*, 4:120, 2010.

[175] D. R. Hyduke, N. E. Lewis, and B. O. Palsson. Analysis of omics data with genome-scale models of metabolism. *Molecular Biosystems*, 9:167–174, 2013.

[176] D. R. Hyduke and B. O. Palsson. Towards genome-scale signalling – network reconstructions. *Nature Reviews Genetics*, 11:297–307, 2010.

[177] R. U. Ibarra, J. S. Edwards, and B. O. Palsson. *Escherichia coli* K-12 undergoes adaptive evolution to achieve *in silico* predicted optimal growth. *Nature*, 420:186–189, 2002.

[178] T. Ideker, T. Galitski, and L. Hood. A new approach to decoding life: Systems biology. *Annual Review of Genomics and Human Genetics*, 2:343–372, 2001.

[179] F. Jacob. Evolution and tinkering. *Science*, 196:1161–1166, 1977.

[180] M. Jain *et al.* Metabolite profiling identifies a key role for glycine in rapid cancer cell proliferation. *Science*, 336:1040–1044, 2012.

[181] N. Jamshidi and B. O. Palsson. Systems biology of SNPs. *Molecular Systems Biology*, 2:38, 2006.

[182] N. Jamshidi and B. O. Palsson. Investigating the metabolic capabilities of *Mycobacterium tuberculosis* H37Rv using the *in silico* strain *i*NJ661 and proposing alternative drug targets. *BMC Systems Biology*, 1:26, 2011.

[183] N. Jamshidi, S. J. Wiback, and B. O. Palsson. *In silico* model-driven assessment of the effects of single nucleotide polymorphisms (SNPs) on human red blood cell metabolism. *Genome Research*, 12:1687–1692, 2002.

[184] P. Janssen, L. Goldovsky, V. Kunin, N. Darzentas, and C. A. Ouzounis. Genome coverage, literally speaking. The challenge of annotating 200 genomes with 4 million publications. *EMBO Reports*, 6(5):397–399, 2005.

[185] K. Jantama *et al.* Combining metabolic engineering and metabolic evolution to develop nonrecombinant strains of *Escherichia coli* C that produce succinate and malate. *Biotechnology and Bioengineering*, 99:1140–1153, 2008.

[186] J. Jantsch, D. Chikkaballi, and M. Hensel. Cellular aspects of immunity to intracellular *Salmonella enterica. Immunological Reviews*, 240:185–195, 2011.

[187] P. A. Jensen and J. A. Papin. Functional integration of a metabolic network model and expression data without arbitrary thresholding. *Bioinformatics*, 27:541–547, 2011.

[188] R. A. Jensen. Enzyme recruitment in evolution of new function. *Annual Reviews in Microbiology*, 30:409–425, 1976.

[189] H. Jeong, S. P. Mason, A. L. Barabasi, and Z. N. Oltvai. Lethality and centrality in protein networks. *Nature*, 411:41–42, 2001.

[190] H. Jeong, B. Tombor, R. Albert, Z. N. Oltvai, and A. L. Barabasi. The large-scale organization of metabolic networks. *Nature*, 407:651–654, 2000.

[191] L. Jerby, T. Shlomi, and E. Ruppin. Computational reconstruction of tissue-specific metabolic models: Application to human liver metabolism. *Molecular Systems Biology*, 6:401, 2010.

[192] A. Joshi and B. O. Palsson. Metabolic dynamics in the human red cell, Part I – a comprehensive kinetic model. *Journal of Theoretical Biology*, 141:515–528, 1989.

[193] A. Joshi and B. O. Palsson. Metabolic dynamics in the human red cell. Part II – Interactions with the environment. *Journal of Theoretical Biology*, 141:529–545, 1989.

[194] A. R. Joyce *et al*. Experimental and computational assessment of conditionally essential genes in *Escherichia coli. Journal of Bacteriology*, 188(23):8259–8271, 2006.

[195] A. R. Joyce and B. O. Palsson. The model organism as a system: Integrating 'omics' data sets. *Nature Reviews Molecular Cell Biology*, 7:198–210, 2006.

[196] P. D. Karp *et al*. The pathway tools software. *Bioinformatics*, 18:S225–S232, 2002.

[197] P. D. Karp and S. Paley. Call for an enzyme genomics initiative. *Genome Biology*, 5:401, 2004.

[198] K. J. Kauffman, J. D. Pajerowski, N. Jamshidi, B. O. Palsson, and J. S. Edwards. Description and analysis of metabolic connectivity and dynamics in the human red blood cell. *Biophysical Journal*, 83:646–662, 2002.

[199] D. E. Kaufman and R. L. Smith. Direction choice for accelerated convergence in hit-and-run sampling. *Operations Research*, 46:84–95, 1998.

[200] I. M. Keseler *et al*. EcoCyc: A comprehensive view of *Escherichia coli* biology. *Nucleic Acids Research*, 37:D464–D470, 2009.

[201] P. Kharchenko, L. Chen, Y. Freund, D. Vitkup, and G. M. Church. Identifying metabolic enzymes with multiple types of association evidence. *BMC Bioinformatics*, 7(177), 2006.

[202] P. Kharchenko, G.M. Church, and D. Vitkup. Expression dynamics of a cellular metabolic network. *Molecular Systems Biology*, 1:10.1038/msb4100023, 2005.

[203] P. Kharchenko, D. Vitkup, and G. M. Church. Filling gaps in a metabolic network using expression information. *Bioinformatics*, 20 Suppl 1:I178–I185, 2004.

[204] S. Kikuchi, I. Shibuya, and K. Matsumoto. Viability of an *Escherichia coli pgsA* null mutant lacking detectable phosphatidylglycerol and cardiolipin. *Journal of Bacteriology*, 182:371–376, 2000.

[205] H. U. Kim *et al*. Genome-scale metabolic network analysis and drug targeting of multi-drug resistant pathogen *Acinetobacter baumannii* AYE. *Biosystems*, 6:339–348, 2010.

[206] H. U. Kim *et al*. Integrative genome-scale metabolic analysis of *Vibrio vulnificus* for drug targeting and discovery. *Molecular Systems Biology*, 7:460, 2011.

[207] H. U. Kim, T. Y. Kim, and S. Y. Lee. Metabolic flux analysis and metabolic engineering of microorganisms. *Molecular Biosystems*, 4:113–120, 2008.

[208] T. Y. Kim *et al*. Recent advances in reconstruction and applications of genome-scale metabolic models. *Current Opinion in Biotechnology*, 23:617–623, 2012.

[209] T. Y. Kim, H. U. Kim, and S. Y. Lee. Data integration and analysis of biological networks. *Current Opinion in Biotechnology*, 21:78–84, 2010.

[210] Y. M. Kim *et al*. *Salmonella* modulates metabolism during growth under conditions that induce expression of virulence genes. *Molecular Biosystems*, 9:1522–1534, 2013.

[211] N. Klitgord and D. Segre. Environments that induce synthetic microbial ecosystems. *PLoS Computational Biology*, 6:e1001002, 2010.

[212] H. Knoop *et al*. The metabolic network of *Synechocystis* sp. PCC 6803: Systemic properties of autotrophic growth. *Plant Physiology*, 154:410–422, 2010.

[213] A. L. Knorr, R. Jain, and R. Srivastava. Bayesian-based selection of metabolic objective functions. *Bioinformatics*, 23:351–357, 2007.

[214] S. D. Kobayashi *et al*. Essential *Bacillus subtilis* genes. *Proceedings of the National Academy of Sciences, USA*, 100:4678–4683, 2003.

[215] M. A. Kohansky, D. J. Dwyer, and J. J. Collins. How antibiotics kill bacteria: From targets to networks. *Nature Reviews Microbiology*, 8:423–435, 2010.

[216] J. Koolman and K. H. Roehm. *Color Atlas of Biochemistry*. Thieme, New York, 2005.

[217] V. S. Kumar and C. D. Maranas. GrowMatch: An automated method for reconciling *in silico/in vivo* growth predictions. *PLoS Computational Biology*, 5:e1000308, 2009.

[218] S. Kumari *et al*. Cloning, characterization, and functional expression of acs, the gene which encodes acetyl coenzyme A synthetase in *Escherichia coli*. *Journal of Bacteriology*, 177:2878–2886, 1995.

[219] B. I. Kurganov. Specific ligand-induced association of an enzyme. A new model of dissociating allosteric enzyme. *Journal of Theoretical Biology*, 103:227–245, 1983.

[220] H. Latif *et al*. The genome organization of *Thermotoga maritima* reflects its lifestyle. *PLoS Genetics*, 9(4):e1003485, 2013.

[221] D-H Lee, A. M. Feist, C. L. Barrett, and B. O. Palsson. Cumulative number of cell divisions as a meaningful timescale for adaptive laboratory evolution of *Escherichia coli*. *PLoS One*, 6:e26172, 2011.

[222] D. S. Lee *et al*. Comparative genome-scale metabolic reconstruction and flux balance analysis of multiple *Staphylococcus aureus* genomes identify novel antimicrobial drug targets. *Journal of Bacteriology*, 191:4015–4024, 2009.

[223] J. W. Lee *et al*. Systems metabolic engineering of microorganisms for natural and non-natural chemicals. *Nature Chemical Biology*, 8:536–546, 2012.

[224] J. W. Lee, H. U. Kim, S. Choi, J. Yi, and S. Y. Lee. Microbial production of building block chemicals and polymers. *Current Opinion in Biotechnology*, 22:758–767, 2011.

[225] K. H. Lee *et al*. Systems metabolic engineering of *Escherichia coli* for L-threonine production. *Molecular Systems Biology*, 3, 2007.

[226] S. Lee, C. Phalakornkule, M. M. Domach, and I. E. Grossmann. Recursive MILP model for finding all the alternate optima in LP models for metabolic networks. *Computers and Chemical Engineering*, 24:711–716, 2000.

[227] S. Y. Lee *et al*. Systems-level analysis of genome-scale *in silico* metabolic models using MetaFluxNet. *Biotechnology and Bioprocess Engineering*, 10:425–431, 2005.

[228] R. E. Lenski and M. Travisano. Dynamics of adaptation and diversification: A 10,000-generation experiment with bacterial populations. *Proceedings of the National Academy of Sciences, USA*, 91:6808–6814, 1994.

[229] J. A. Lerman *et al*. *In silico* method for modeling metabolism and gene product expression at genome scale. *Nature Communications*, 3:929, 2012.

[230] O. Lespinet and B. Labedan. ORENZA: A web resource for studying ORphan ENZyme activities. *BMC Bioinformatics*, 7:436, 2006.

[231] O. Lespinet and B. Labedan. Orphan enzymes could be an unexplored reservoir of new drug targets. *Drug Discovery Today*, 11:300–305, 2006.

[232] N. E. Lewis and A. M. Abdel-Haleem. The evolution of genome-scale models of cancer metabolism. *Frontiers in Physiology*, 4:doi10.3389/fphys.2013.00237, 2013.

[233] N. E. Lewis *et al*. Large-scale *in silico* modeling of metabolic interactions between cell types in the human brain. *Nature Biotechnology*, 28:1279–1285, 2010.

[234] N. E. Lewis *et al*. Omic data from evolved *E. coli* are consistent with computed optimal growth from genome-scale models. *Molecular Systems Biology*, 6:390, 2010.

[235] N. E. Lewis, N. Jamshidi, I. Thiele, and B. O. Palsson. Metabolic systems biology: A constraint-based approach. In *Encyclopedia of Complexity and Systems Science*, pp. 5535–5552. Springer, New York, 2009.

[236] N. E. Lewis, H. Nagarajan, and B. O. Palsson. Constraining the metabolic genotype–phenotype relationship using a phylogeny of *in silico* methods. *Nature Reviews Microbiology*, 10:291–305, 2012.

[237] Y. C. Liao *et al.* An experimentally validated genome-scale metabolic reconstruction of *Klebsiella pneumoniae* MGH 78578, *i*YL1228. *Journal of Bacteriology*, 193:1710–1717, 2011.

[238] D. Liu and P. R. Reeves. *Escherichia coli* K12 regains its O antigen. *Microbiology*, 140:49–57, 1994.

[239] L. Lobel *et al.* Integrative genomic analysis identifies isoleucine and codY as regulators of *Listeria monocytogenes* virulence. *PLoS Genetics*, 8:e1002887, 2012.

[240] W. R. Loewenstein. *The Touchstone of Life*. Oxford University Press, Oxford, 1999.

[241] K. D. Loh *et al.* A previously undescribed pathway for pyrimidine catabolism. *Proceedings of the National Academy of Sciences, USA*, 103:5114–5119, 2006.

[242] P. Lorenz and J. Eck. Metagenomics and industrial applications. *Nature Reviews Microbiology*, 3:510–516, 2005.

[243] M. Lotierzo *et al.* Biotin synthase mechanism: An overview. *Biochemical Society Transactions*, 33:820–823, 2005.

[244] L. Lovasz. Hit-and-run mixes fast. *Mathematical Programming*, 86:443–461, 1999.

[245] D. R. Lovley. Bug juice: Harvesting electricity with microorganisms. *Nature Reviews Microbiology*, 4:497–508, 2006.

[246] D. R. Lovley *et al.* Anaerobic production of magnetite by a dissimilatory iron-reducing microorganism. *Nature*, 330:252–254, 1987.

[247] D. R. Lovley *et al.* A detailed introduction to the physiology and ecology of *Geobacter* spp. as well as other metal reducers. *Microbial Physiology*, 49:219–286, 2004.

[248] D. R. Lovley and D. Lonergan. Anaerobic oxidation of toluene, phenol, and *para*-cresol by the dissimilatory iron-reducing organism, GS-15. *Applied Environmental Microbiology*, 56:1858–1864, 1990.

[249] D. R. Lovley and E. J. P. Phillips. Novel mode of microbial energy-metabolism – organic-carbon oxidation coupled to dissimilatory reduction of iron or manganese. *Applied Environmental Microbiology*, 54:1472–1480, 1988.

[250] C. Lu *et al.* IDH mutation impairs histone demethylation and results in a block to cell differentiation. *Nature*, 483:474–478, 2012.

[251] O. Lukjancenko, T. M. Wassenaar, and D. W. Ussery. Comparison of 61 sequenced *Escherichia coli* genomes. *Microbial Ecology*, 60:708–720, 2010.

[252] D. S. Lun *et al.* Large-scale identification of genetic design strategies using local search. *Molecular Systems Biology*, 5:296, 2009.

[253] D. Machado and M. Herrgard. Systematic evaluation of methods for integration of transcriptomic data into constraint-based models of metabolism. *PLoS Computational Biology*, 10:e1003580, 2014.

[254] K. Mahadevan, B. O. Palsson, and D. R. Lovley. *In situ* to *in silico* and back: Elucidating the physiology and ecology of *Geobacter* spp. using genome-scale modeling. *Nature Reviews Microbiology*, 9:39–50, 2011.

[255] R. Mahadevan and B. O. Palsson. Properties of metabolic networks: Structure versus function. *Biophysical Journal*, 88:L7–L9, 2005.

[256] R. Mahadevan and C. H. Schilling. The effects of alternate optimal solutions in constraint-based genome-scale metabolic models. *Metabolic Engineering*, 5:264–276, 2003.

[257] T. Maier *et al.* Quantification of mRNA and protein and integration with protein turnover in a bacterium. *Molecular Systems Biology*, 7:511, 2011.

[258] R. A. Majewski and M. M. Domach. Simple constrained optimization view of acetate overflow in *E. coli*. *Biotechnology and Bioengineering*, 35:732–738, 1990.

[259] A. Manichaikul *et al.* Metabolic network analysis integrated with transcript verification for sequenced genomes. *Nature Methods*, 6:589–592, 2009.

[260] S. L. Marcus *et al.* *Salmonella* pathogenicity islands: Big virulence in small packages. *Microbes and Infection*, 2:145–156, 2000.

[261] E. Mayr. *This is Biology: The Science of the Living World*. Belknap Press of Harvard University Press, Cambridge, MA, 1997.

[262] V. Mazumdar *et al.* Metabolic network model of a human oral pathogen. *Journal of Bacteriology*, 191:74–90, 2009.

[263] D. McCloskey, B. O. Palsson, and A. M. Feist. Basic and applied uses of genome-scale metabolic network reconstructions of *Escherichia coli*: The good, the bad, and the likely. *Molecular Systems Biology*, 9:661, 2013.

[264] M. H. Medema, R. van Raaphorst, E. Takano, and R. Breitling. Computational tools for the synthetic design of biochemical pathways. *Nature Reviews Microbiology*, 10:191–202, 2012.

[265] S. S. Merchant *et al.* The *Chlamydomonas* genome reveals the evolution of key animal and plant functions. *Science*, 318:245–250, 2007.

[266] M. L. Mo, B. O. Palsson, and M. J. Herrgard. Connecting extracellular metabolomic measurements to intracellular flux states in yeast. *BMC Systems Biology*, 3:37–54, 2009.

[267] M. Mollney, W. Wiechert, D. Kownatzki, and A. A. de Graaf. Bidirectional reaction steps in metabolic networks: IV. Optimal design of isotopomer labeling experiments. *Biotechnology and Bioengineering*, 663:86–103, 2007.

[268] J. M. Monk *et al.* Genome-scale metabolic reconstructions of multiple *Escherichia coli* strains highlight strain-specific adaptations to nutritional environments. *Proceedings of the National Academy of Sciences, USA*, 110:20338–20343, 2013.

[269] C. Monnet *et al.* The *Arthrobacter arilaitensis* Re117 genome sequence reveals its genetic adaptation to the surface of cheese. *PloS One*, 5:e15489, 2010.

[270] A. Montagud *et al.* Reconstruction and analysis of genome-scale metabolic model of a photosynthetic bacterium. *BMC Systems BIology*, 4:156, 2010.

[271] A. Montagud *et al.* Flux coupling and transcriptional regulation within the metabolic network of the photosynthetic bacterium *Synechocystis* sp. PCC 6803. *Biotechnology Journal*, 6:330–342, 2011.

[272] T. Moore and D. Haig. Genomic imprinting in mammalian development: A parental tug-of-war. *Trends in Genetics*, 7:45–49, 1991.

[273] V. Mozhayskiy and I. Tagkopoulos. Microbial evolution *in vivo* and *in silico*: Methods and applications. *Integrative Biology*, 5:262–277, 2013.

[274] H. Nagarajan *et al.* Characterization and modelling of interspecies electron transfer mechanisms and microbial community dynamics of a syntrophic association. *Nature Communications*, 4:2809, 2013.

[275] K. Nakahigashi *et al.* Systematic phenome analysis of *Escherichia coli* multiple-knockout mutants reveals hidden reactions in central carbon metabolism. *Molecular Systems Biology*, 5:306, 2009.

[276] H. Nam, T. M. Conrad, and N. E. Lewis. The role of cellular objectives and selective pressures in metabolic pathway evolution. *Current Opinion in Biotechnology*, 22:595–600, 2011.

[277] H. Nam *et al.* Network context and selection in the evolution to enzyme specificity. *Science*, 337:1101–1104, 2012.

[278] E. Navarro *et al.* Metabolic flux analysis of the hydrogen production potential in *Synechocystis* sp. PCC 6803. *International Journal of Hydrogen Energy*, 34:8828–8838, 2009.

[279] A. Navid and E. Almaas. Genome-scale reconstruction of the metabolic network in *Yersinia pestis*, strain 91001. *Molecular Biosystems*, 5:368–375, 2009.

[280] F. C. Neidhardt, editor. *Escherichia coli and Salmonella: Cellular and Molecular Biology*, pp. 189–198. American Society of Microbiology, Washington, DC, 1996.

[281] F. C. Neidhardt, editor. *Escherichia coli and Salmonella: Cellular and Molecular Biology*, pp. 283–306. American Society of Microbiology, Washington, DC, 1996.

[282] F. C. Neidhardt, J. L. Ingraham, and M. Schaechterm. *Physiology of the Bacterial Cell: A Molecular Approach*. Sinauer Associates, Sunderland, MA, 1990.

[283] A. L. N'Guessan *et al.* Sustained removal of uranium from contaminated groundwater following stimulation of dissimilatory metal reduction. *Environmental Science and Technology*, 42:2999–3004, 2008.

[284] R. J. Nichols *et al.* Phenotypic landscape of a bacterial cell. *Cell*, 144:143–156, 2011.

[285] J. Nielsen and J. Villadsen. *Bioreaction Engineering Principles*. Plenum Press, New York, 1994.

[286] J. Nogales *et al.* Detailing the optimality of photosynthesis in cyanobacteria through systems biology analysis. *Proceedings of the National Academy of Sciences, USA*, 109:2678–2683, 2012.

[287] J. Nogales, B. O. Palsson, and I. Thiele. A genome-scale metabolic reconstruction of *Pseudomonas putida* KT2440: IJN746 as a cell factory. *BMC Systems Biology*, 2:79, 2008.

[288] Y. Noguchi *et al.* The energetic conversion competence of *Escherichia coli* during aerobic respiration studied by 31p NMR using a circulating fermentation system. *Journal of Biochemistry (Tokyo)*, 136:509–515, 2004.

[289] M. A. Oberhardt *et al.* Genome-scale metabolic network analysis of the opportunistic pathogen *Pseudomonas aeruginosa* PAO1. *Journal of Bacteriology*, 190:2790–2803, 2008.

[290] M. A. Oberhardt, B. O. Palsson, and J. A. Papin. Applications of genome-scale metabolic reconstructions. *Molecular Systems Biology*, 5:320, 2009.

[291] M. K. Oh and J. C. Liao. Gene expression profiling by DNA microarrays and metabolic fluxes in *Escherichia coli*. *Biotechnology Progress*, 16:278–286, 2000.

[292] Y.-K. Oh *et al.* Genome-scale reconstruction of metabolic network in *Bacillus subtilis* based on high-throughput phenotyping and gene essentiality data. *Journal of Biological Chemistry*, 282:28791–28799, 2007.

[293] D. A. Okar *et al.* PFK-2/FBPase-2: Maker and breaker of the essential biofactor fructose-2,6-bisphosphate. *Trends in Biochemical Sciences*, 26(1):30–35, 2001.

[294] S. Ong *et al.* Analysis and construction of pathogenicity island regulatory pathways in *Salmonella enterica* serovar *typhi*. *Journal of Integrative Bioinformatics*, 7:1–34, 2010.

[295] J. D. Orth *et al.* A comprehensive genome-scale reconstruction of *Escherichia coli* metabolism – 2011. *Molecular Systems Biology*, 7:535, 2011.

[296] J. D. Orth, R. M. T. Fleming, and B. O. Palsson. Reconstruction and use of microbial metabolic networks: The core *Escherichia coli* metabolic model as an educational guide. In *EcoSal – Escherichia coli and Salmonella: Cellular and Molecular Biology*, Chapter 10.2.1 ASM Press, Washington, DC, 2010.

[297] J. D. Orth and B. O. Palsson. Systematizing the generation of missing metabolic knowledge. *Biotechnology and Bioengineering*, 107(3):403–412, 2010.

[298] J. D. Orth and B. O. Palsson. Gap-filling analysis of the *i*JO1366 *Escherichia coli* metabolic network reconstruction for discovery of metabolic functions. *BMC Systems Biology*, 6:30, 2012.

[299] J. D. Orth, I. Thiele, and B. O. Palsson. What is flux balance analysis? *Nature Biotechnology*, 28:245–248, 2010.

[300] T. Osterlund, I. Nookaew, S. Bordel, and J. Nielsen. Mapping condition-dependent regulation of metabolism in yeast through genome-scale modeling. *BMC Systems Biology*, 7:36, 2013.

[301] T. Osterlund, I. Nookaew, and J. Nielsen. Fifteen years of large scale metabolic modeling of yeast: Developments and impacts. *Biotechnology Advances*, 30:979–988, 2012.

[302] A. Osterman. A hidden metabolic pathway exposed. *Proceedings of the National Academy of Sciences, USA*, 103:5637–5638, 2006.

[303] A. Osterman and R. Overbeek. Missing genes in metabolic pathways: A comparative genomics approach. *Current Opinion in Chemical Biology*, 7:238–251, 2003.

[304] R. Overbeek, T. Disz, and R. Stevens. The SEED: A peer-to-peer environment for genome annotation. *Communications of the ACM*, 47:46–51, 2004.

[305] R. Overbeek *et al.* The subsystems approach to genome annotation and its use in the project to annotate 1000 genomes. *Nucleic Acids Research*, 33:5691–5702, 2005.

[306] D. S. Ow *et al.* Identification of cellular objective for elucidating the physiological state of plasmid-bearing *Escherichia coli* using genome-scale *in silico* analysis. *Biotechnology Progress*, 25:61–67, 2009.

[307] C. Pal *et al.* Chance and necessity in the evolution of minimal metabolic networks. *Nature*, 440(7084):667–70, 2006.

[308] S. M. Paley and P. D. Karp. Evaluation of computational metabolic pathway predictions for *Helicobacter pylori*. *Bioinformatics*, 18:715–724, 2002.

[309] B. O. Palsson and A. Joshi. On the dynamic orders of structured *E. coli* growth models. *Biotechnology and Bioengineering*, 29:789, 1987.

[310] B. O. Palsson, A. Joshi, and S. S. Ozturk. Reducing complexity in metabolic networks: Making metabolic meshes manageable. *Federation Proceedings*, 46:2485, 1987.

[311] B. O. Palsson. The challenges of *in silico* biology. *Nature Biotechnology*, 18:1147–1150, 2000.

[312] B. O. Palsson. *In silico* biology through 'omics'. *Nature Biotechnology*, 20:649–650, 2002.

[313] B. O. Palsson. Two-dimensional annotation of genomes. *Nature Biotechnology*, 22:1218–1219, 2004.

[314] B. O. Palsson. *Systems Biology: Properties of Reconstructed Networks*. Cambridge University Press, New York, 2006.

[315] B. O. Palsson. Metabolic systems biology. *FEBS Letters*, 583:3900–3904, 2009.

[316] B. O. Palsson. Adaptive laboratory evolution. *Microbe*, 6:69–74, 2011.

[317] B. O. Palsson. *Systems Biology: Simulation of Dynamic Network States*. Cambridge University Press, New York, 2011.

[318] B. O. Palsson and S. N. Bhatia. *Tissue Engineering*. Prentice Hall, New Jersey, 2003.

[319] B. O. Palsson and K. Zengler. The challenges of integrating multi-omic data sets. *Nature Chemical Biology*, 6:787–789, 2010.

[320] J. A. Papin *et al.* Comparison of network-based pathway analysis methods. *Trends in Biotechnology*, 22:400–405, 2004.

[321] J. A. Papin *et al.* Reconstruction of cellular signalling networks and analysis of their properties. *Nature Reviews Molecular Cell Biology*, 6:99–111, 2005.

[322] J. A. Papin and B. O. Palsson. Topological analysis of mass-balanced signaling networks: A framework to obtain emergent properties including crosstalk. *Journal of Theoretical Biology*, 227:283–297, 2004.

[323] J. A. Papin, N. D. Price, and B. O. Palsson. Extreme pathway lengths and reaction participation in genome-scale metabolic networks. *Genome Research*, 12:1889–1900, 2002.

[324] J.A. Papin *et al*. Metabolic pathways in the post-genome era. *Trends in Biochemical Sciences*, 28:250–258, 2003.

[325] J. A. Papin, J. L. Reed, and B. O. Palsson. Hierarchical thinking in network biology: The unbiased modularization of biochemical networks. *Trends in Biochemical Sciences*, 29:641–647, 2004.

[326] B. Papp, R. A. Notebaart, and C. Pal. Systems-biology approaches for predicting genomic evolution. *Nature Reviews Genetics*, 12:591–602, 2011.

[327] J. H. Park and S. Y. Lee. Towards systems metabolic engineering of microorganisms for amino acid production. *Current Opinion in Biotechnology*, 19:454–460, 2008.

[328] J. M. Park, T. Y. Kim, and S. Y. Lee. Constraints-based genome-scale metabolic simulation for systems metabolic engineering. *Biotechnology Advances*, 27:979–988, 2009.

[329] K. R. Patil and J. Nielsen. Uncovering transcriptional regulation of metabolism by using metabolic network topology. *Proceedings of the National Academy of Sciences, USA*, 102:2685–2689, 2005.

[330] K. R. Patil, I. Rocha, J. Forster, and J. Nielsen. Evolutionary programming as a platform for *in silico* metabolic engineering. *BMC Bioinformatics*, 6:308, 2005.

[331] A. Perrenoud and U. Sauer. Impact of global transcriptional regulation by ArcA, ArcB, Cra, Crp, Cya, Fnr, and Mlc on glucose catabolism in *Escherichia coli*. *Journal of Bacteriology*, 187:3171–3179, 2005.

[332] T. Pfeiffer, I. Sanchez-Valdenebro, J. C. Nuno, F. Montero, and S. Schuster. Metatool: For studying metabolic networks. *Bioinformatics*, 15:251–257, 1999.

[333] C. Phalakornkule *et al*. A MILP-based flux alternative generation and NMR experimental design strategy for metabolic engineering. *Metabolic Engineering*, 3:124–137, 2001.

[334] P. Pharkya, A. P. Burgard, and C. D. Maranas. Exploring the overproduction of amino acids using the bilevel optimization framework OptKnock. *Biotechnology and Bioengineering*, 84:887–899, 2003.

[335] P. Pharkya, A. P. Burgard, and C. D. Maranas. OptStrain: A computational framework for redesign of microbial production systems. *Genome Research*, 14:2367–2376, 2004.

[336] G. Plata *et al*. Reconstruction and flux-balance analysis of the *Plasmodium falciparum* metabolic network. *Molecular Systems Biology*, 6:408, 2010.

[337] V. Potapov, M. Cohen, and G. Schreiber. Assessing computational methods for predicting protein stability upon mutation: Good on average but not in the details. *Protein Engineering Design and Selection*, 22:553–560, 2009.

[338] Y. Pouliot and P. D. Karp. A survey of orphan enzyme activities. *BMC Bioinformatics*, 8:244, 2007.

[339] J. Pramanik and J. D. Keasling. Stoichiometric model of *Escherichia coli* metabolism: Incorporation of growth-rate dependent biomass composition and mechanistic energy requirements. *Biotechnology and Bioengineering*, 56:398–421, 1997.

[340] J. Pramanik and J. D. Keasling. Effect of *Escherichia coli* biomass composition on central metabolic fluxes predicted by a stoichiometric model. *Biotechnology and Bioengineering*, 60:230–238, 1998.

[341] W. H. Press. *Numerical Recipes in C*. Cambridge University Press, New York, 1994.

[342] N. D. Price, I. F. Famili, D. A. Beard, and B. O. Palsson. Extreme pathways and Kirchhoff's second law. *Biophysical Journal*, 83:2879–2882, 2002.

[343] N. D. Price, J. A. Papin, and B. O. Palsson. Determination of redundancy and systems properties of the metabolic network of *Helicobacter pylori* using genome-scale extreme pathway analysis. *Genome Research*, 12:760–769, 2002.

[344] N. D. Price, J. L. Reed, and B. O. Palsson. Genome-scale models of microbial cells: Evaluating the consequences of constraints. *Nature Reviews Microbiology*, 2:886–897, 2004.

[345] N. D. Price, J. L. Reed, J. A. Papin, I. Famili, and B. O. Palsson. Analysis of metabolic capabilities using singular value decomposition of extreme pathway matrices. *Biophysical Journal*, 84:794–804, 2003.

[346] N. D. Price, J. L. Reed, J. A. Papin, S. J. Wiback, and B. O. Palsson. Network-based analysis of regulation in the human red blood cell. *Journal of Theoretical Biology*, 225:1985–1994, 2003.

[347] N. D. Price, J. Schellenberger, and B. O. Palsson. Uniform sampling of steady state flux spaces: Means to design experiments and to interpret enzymopathies. *Biophysical Journal*, 87:2172–2186, 2004.

[348] N. D. Price and I. Shmulevich. Biochemical and statistical network models for systems biology. *Current Opinion in Biotechnology*, 18:365–370, 2007.

[349] Y. Qiu *et al.* Structural and operational complexity of the *Geobacter sulfurreducens* genome. *Genome Research*, 20:1304–1311, 2010.

[350] J.D. Rabinowitz. Cellular metabolomics of *Escherichia coli*. *Expert Reviews in Proteomics*, 4:187 198, 2007.

[351] A. Raghunathan *et al.* Constraint-based analysis of metabolic capacity of *Salmonella typhimurium* during host–pathogen interaction. *BMC Systems Biology*, 3:38, 2009.

[352] A. Raghunathan *et al.* Systems approach to investigating host–pathogen interactions in infections with the biothreat agent *Francisella*. Constraints-based model of *Francisella tularensis*. *BMC Systems Biology*, 4:118, 2010.

[353] R. Ramakrishna *et al.* Flux balance analysis of mitochondrial energy metabolism: Consequences of systemic stoichiometric constraints. *The American Journal of Physiology: Regulatory, Integrative and Comparative Physiology*, 280:695–704, 2001.

[354] B. A. Rasala and S. P. Mayfield. The microalga *Chlamydomonas reinhardtii* as a platform for the production of human protein therapeutics. *Bioengineered Bugs*, 2:50–54, 2011.

[355] J. L. Reed, I. Famili, I. Thiele, and B. O. Palsson. Towards multidimensional genome annotation. *Nature Reviews Genetics*, 7(2):130–41, 2006.

[356] J. L. Reed. Shrinking the metabolic solution space using experimental datasets. *PLoS Computational Biology*, 8:e1002662, 2012.

[357] J. L. Reed *et al.* Systems approach to refining genome annotation. *Proceedings of the National Academy of Sciences, USA*, 103:17480–17484, 2006.

[358] J. L. Reed and B. O. Palsson. Thirteen years of building constraints-based *in silico* models of *Escherichia coli*. *Journal of Bacteriology*, 185:2692–2699, 2003.

[359] J. L. Reed and B. O. Palsson. Genome-scale *in silico* models of *E. coli* have multiple equivalent phenotypic states: Assessment of correlated reaction subsets that comprise network states. *Genome Research*, 14:1797–1805, 2004.

[360] J. L. Reed, T. D. Vo, C. H. Schilling, and B. O. Palsson. An expanded genome-scale model of *Escherichia coli* K-12 (*i*JR904 GSM/GPR). *Genome Biology*, 4:R54.1–R54.12, 2003.

[361] J. G. Reich and E. E. Sel'kov. *Energy Metabolism of the Cell*. Academic Press, New York, 1981.

[362] M. Riley *et al.* *Escherichia coli* K-12: A cooperatively developed annotation snapshot–2005. *Nucleic Acids Research*, 34(1):1–9, 2006.

[363] S. B. Roberts *et al.* Proteomic and network analysis characterize stage-specific metabolism in *Trypanosoma cruzi*. *BMC Systems Biology*, 3:52, 2009.

[364] R. T. Rockafellar. *Convex Analysis*. Princeton University Press, New Haven, CT, 1996.

[365] D. A. Rodionov *et al.* Genomic identification and *in vitro* reconstitution of a complete biosynthetic pathway for the osmolyte di-*myo*-inositol-phosphate. *Proceedings of the National Academy of Sciences, USA*, 104:4279–4284, 2007.

[366] J. N. Rooney-Varga *et al.* Microbial communities associated with anaerobic benzene degradation in a petroleum-contaminated aquifer. *Applied Environmental Microbiology*, 65:3056–3063, 1999.

[367] X. Rubires *et al.* A gene (*wbbL*) from *Serratia marcescens* N28b (O4) complements the rfb-50 mutation of *Escherichia coli* K-12 derivatives. *Journal of Bacteriology*, 179:7581–7586, 1997.

[368] S. Sahoo *et al.* A compendium of inborn errors of metabolism mapped onto the human metabolic network. *Molecular Biosystems*, 8:2545–2558, 2012.

[369] R. Samson and J. M. Deutch. Diffusion-controlled reaction rate to a buried active site. *The Journal of Chemical Physics*, 68:285–290, 1978.

[370] V. Satish Kumar, M. S. Dasika, and C. D. Maranas. Optimization based automated curation of metabolic reconstructions. *BMC Bioinformatics*, 8:212, 2007.

[371] U. Sauer. Metabolic networks in motion: ^{13}C-based flux analysis. *Molecular Systems Biology*, 2:62, 2006.

[372] J. M. Savinell and B. O. Palsson. Network analysis of intermediary metabolism using linear optimization: I. Development of mathematical formalism. *Journal of Theoretical Biology*, 154:455–473, 1992.

[373] J. M. Savinell and B. O. Palsson. Network analysis of intermediary metabolism using linear optimization: II. Interpretation of hybridoma cell metabolism. *Journal of Theoretical Biology*, 154:455–473, 1992.

[374] A. Sboner. The real cost of sequencing: Higher than you think! *Genome Biology*, 12:125, 2011.

[375] J. Schellenberger *et al*. Quantitative prediction of cellular metabolism with constraint-based models: The COBRA toolbox v2.0. *Nature Protocols*, 6:1290–1307, 2011.

[376] J. Schellenberger *et al*. Predicting outcomes of steady-state ^{13}C isotope tracing experiments with Monte Carlo sampling. *BMC Systems Biology*, 6:9, 2012.

[377] J. Schellenberger, N. E. Lewis, and B. O. Palsson. Elimination of thermodynamically infeasible loops in steady-state metabolic models. *Biophysical Journal*, 100:544–553, 2011.

[378] J. Schellenberger and B. O. Palsson. Use of randomized sampling for analysis of metabolic networks. *Journal of Biological Chemistry*, 284:5457–5461, 2009.

[379] C. H. Schilling, D. Letscher, and B. O. Palsson. Theory for the systemic definition of metabolic pathways and their use in interpreting metabolic function from a pathway-oriented perspective. *Journal of Theoretical Biology*, 203:229–248, 2000.

[380] C. H. Schilling *et al*. Genome-scale metabolic model of *Helicobacter pylori* 26695i. *Journal of Bacteriology*, 184:4582–4593, 2002.

[381] C. H. Schilling and B. O. Palsson. Assessment of the metabolic capabilities of *Haemophilus influenzae* Rd through a genome-scale pathway analysis. *Journal of Theoretical Biology*, 203:249–283, 2000.

[382] C. H. Schilling, S. Schuster, B. O. Palsson, and R. Heinrich. Metabolic pathway analysis: Basic concepts and scientific applications in the post-genomic era. *Biotechnology Progress*, 15:296–303, 1999.

[383] B. J. Schmidt *et al*. Metabolic systems analysis to advance algal biotechnology. *Biotechnology Journal*, 5:660–670, 2010.

[384] K. Schmidt, M. Carlsen, J. Nielsen, and J. Villadsen. Modeling isotopomer distributions in biochemical networks using isotopomer mapping matrices. *Biotechnology and Bioengineering*, 55:831–840, 1997.

[385] K. Schmidt, J. Nielsen, and J. Villadsen. Quantitative analysis of metabolic fluxes in *Escherichia coli*, using two-dimensional NMR spectroscopy and complete isotopomer models. *Journal of Biotechnology*, 71:175–189, 1999.

[386] R. Schuetz, L. Kuepfer, and U. Sauer. Systematic evaluation of objective functions for predicting intracellular fluxes in *Escherichia coli*. *Molecular Systems Biology*, 3:119–134, 2007.

[387] J. M. Schurr and K. S. Schmitz. Orientation constraints and rotational diffusion in bimolecular solution kinetics. a simplification. *Journal of Physical Chemistry*, 80:1934–1936, 1976.

[388] D. Segre, D. Vitkup, and G. M. Church. Analysis of optimality in natural and perturbed metabolic networks. *Proceedings of the National Academy of Sciences of the United States of America*, 99:15112–15117, 2002.

[389] M. H. Serres *et al.* A functional update of the *Escherichia coli* K-12 genome. *Genome Biology*, 2:35.1–35.7, 2001.

[390] A. S. Seshasayee *et al.* Transcriptional regulatory networks in bacteria: From input signals to output responses. *Current Opinion in Microbiology*, 9:511–519, 2006.

[391] C. M. Sharma *et al.* The primary transcriptome of the major human pathogen *Helicobacter pylori*. *Nature*, 464:250–255, 2010.

[392] A. A. Shastri and J. A. Morgan. Flux balance analysis of photoautotrophic metabolism. *Biotechnology Progress*, 21:1617–1626, 2005.

[393] Y. Shen *et al.* Blueprint for antimicrobial hit discovery targeting metabolic networks. *Proceedings of the National Academy of Sciences, USA*, 107:1082–1087, 2010.

[394] T. Shlomi, O. Berkman, and E. Ruppin. Regulatory on/off minimization of metabolic flux changes after genetic perturbations. *Proceedings of the National Academy of Sciences, USA*, 102:7695–7700, 2005.

[395] T. Shlomi, M. N. Cabili, and E. Ruppin. Predicting metabolic biomarkers of human inborn errors of metabolism. *Molecular Systems Biology*, 5:263, 2009.

[396] T. Shlomi *et al.* Network-based prediction of human tissue-specific metabolism. *Nature Biotechnology*, 26:1003–1010, 2008.

[397] M. Sigurdsson *et al.* A detailed genome-wide reconstruction of mouse metabolism based on human recon 1. *BMC Systems Biology*, 4:140, 2010.

[398] M. Sigurdsson, N. Jamshidi, J. J. Jonsson, and B. O. Palsson. Genome-scale network analysis of imprinted human metabolic genes. *Epigenetics*, 4:43–46, 2009.

[399] J. M. Skerker *et al.* Two-component signal transduction pathways regulating growth and cell cycle progression in a bacterium: A system-level analysis. *PLoS Computational Biology*, 3:e334, 2005.

[400] P. D Sniegowski, P. J. Gerrish, and R. E. Lenski. Evolution of high mutation rates in experimental populations of *E. coli*. *Nature*, 387:703–705, 1997.

[401] E. S. Snitkin *et al.* Model-driven analysis of experimentally determined growth phenotypes for 465 yeast gene deletion mutants under 16 different conditions. *Genome Biology*, 9:R140, 2008.

[402] S. R. Neves and R. Iyengar. Modeling of signaling networks. *Bioessays*, 12:1110–1117, 2002.

[403] K. Srinivasan and K. Mahadevan. Characterization of proton production and consumption associated with microbial metabolism. *BMC Biotechnology*, 10:2, 2010.

[404] B. Steeb *et al.* Parallel exploitation of diverse host nutrients enhances *Salmonella* virulence. *PLoS Pathogens*, 9:e1003301, 2013.

[405] D. Stern, E. Harris, and G. Witman, editors. *The Chlamydomonas Sourcebook: Organellar and Metabolic Processes*. Academic Press, Oxford, 2008.

[406] S. Stolyar *et al*. Metabolic modeling of a mutualistic microbial community. *Molecular Systems Biology*, 3:92, 2007.

[407] G. Strang. *Linear Algebra and its Applications*, 3rd Edition. Harcourt Brace, San Diego, 1988.

[408] P. F. Suthers *et al*. Metabolic flux elucidation for large-scale models using ^{13}C labeled isotopes. *Metabolic Engineering*, 9(5-6):387–405, 2007.

[409] P. F. Suthers *et al*. A genome-scale metabolic reconstruction of *Mycoplasma genitalium*, ips189. *PLoS Computational Biology*, 5:e1000285, 2009.

[410] P. F. Suthers, A. Zomorrodi, and C. D. Maranas. Genome-scale gene/reaction essentiality and synthetic lethality analysis. *Molecular Systems Biology*, 5:301, 2009.

[411] N. Swainston *et al*. The SuBliMinaL toolbox: Automating steps in the reconstruction of metabolic networks. *Journal of Integrative Bioinformatics*, 8:186, 2011.

[412] B. Szappanos *et al*. An integrated approach to characterize genetic interaction networks in yeast metabolism. *Nature Genetics*, 43:656–662, 2011.

[413] R. Taffs *et al*. *In silico* approaches to study mass and energy flows in microbial consortia: A syntrophic case study. *BMC Systems Biology*, 3:114, 2009.

[414] R. Tanaka. Scale-rich metabolic networks. *Physical Review Letters*, 94:168101–168104, 2005.

[415] K. Temme, D. Zhao, and C. A. Voigt. Refactoring the nitrogen fixation gene cluster from *Klebsiella oxytoca*. *Proceedings of the National Academy of Sciences, USA*, 109:7085–7090, 2012.

[416] N. Tepper and T. Shlomi. Predicting metabolic engineering knockout strategies for chemical production: Accounting for competing pathways. *Bioinformatics*, 26:536–543, 2010.

[417] B. Teusink *et al*. Analysis of growth of *Lactobacillus plantarum* WCFS1 on a complex medium using a genome-scale metabolic model. *Journal of Biological Chemistry*, 281:40041–40048, 2006.

[418] B. Teusink *et al*. Understanding the adaptive growth strategy of *Lactobacillus plantarum* by *in silico* optimisation. *PLoS Computational Biology*, 5:e1000410, 2009.

[419] I. Thiele *et al*. An expanded metabolic reconstruction of *Helicobacter pylori* (*i*IT341 GSM/GPR): An *in silico* genome-scale characterization of single and double deletion mutants. *Journal of Bacteriology*, 187:5818–5830, 2005.

[420] I. Thiele *et al*. Genome-scale reconstruction of *Escherichia coli*'s transcriptional and translational machinery: a knowledgebase, its mathematical formulation, and its functional characterization. *PLoS Computational Biology*, 5:e1000312, 2009.

[421] I. Thiele *et al*. Functional characterization of alternate optimal solutions of *Escherichia coli*'s transcriptional and translational machinery. *Biophysical Journal*, 98:2072–2081, 2010.

[422] I. Thiele *et al*. A community effort towards a knowledge-base and mathematical model of the human pathogen *Salmonella typhimurium* LT2. *BMC Systems Biology*, 5:8, 2011.

[423] I. Thiele *et al*. Multiscale modeling of metabolism and macromolecular synthesis in *E. coli* and its application to the evolution of codon usage. *PLoS ONE*, 7:e45635, 2012.

[424] I. Thiele *et al*. A community-driven global reconstruction of human metabolism. *Nature Biotechnology*, 31:419–425, 2013.

[425] I. Thiele, A. Heinken, and M. T. Fleming. A systems biology approach to studying the role of microbes in human health. *Current Opinion in Biotechnology*, 24:4–12, 2012.

[426] I. Thiele and B. O. Palsson. Bringing genomes to life: The use of genome-scale *in silico* models. In *Introduction to Systems Biology*, pp. 14–36. Humana Press, New Jersey, 2007.

[427] I. Thiele and B. O. Palsson. A protocol for generating a high-quality genome-scale metabolic reconstruction. *Nature Protocols*, 5:93–121, 2010.

[428] I. Thiele and B. O. Palsson. Reconstruction annotation jamborees: A community approach to systems biology. *Molecular Systems Biology*, 6:361, 2010.

[429] I. Thiele, N. D. Price, T. D. Vo, and B. O. Palsson. Candidate metabolic network states in human mitochondria: Impact of diabetes, ischemia, and diet. *Journal of Biological Chemistry*, 280:11683–11695, 2005.

[430] K. Tornheim. Co-ordinate control of phosphofructokinase and pyruvate kinase by fructose diphosphate: A mechanism for amplification and step changes in the regulation of glycolysis in liver. *Journal of Theoretical Biology*, 85:199–222, 1980.

[431] Q. H. Tran *et al*. Requirement for the proton-pumping NADH dehydrogenase I of *Escherichia coli* in respiration of NADH to fumarate and its bioenergetic implications. *The FEBS Journal*, 244:155–160, 1997.

[432] W. M. van Gulik. Fast sampling for quantitative microbial metabolomics. *Current Opinion in Biotechnology*, 21:27–34, 2010.

[433] W. A. van Winden *et al*. Correcting mass isotopomer distributions for naturally occurring isotopes. *Biotechnology and Bioengineering*, 80:477–479, 2002.

[434] N. Vanee *et al*. A genome-scale metabolic model of *Cryptosporidium hominis*. *Chemistry and Biodiversity*, 7:1026–1039, 2010.

[435] A. Varma. *Flux balance analysis of* Escherichia coli *Metabolism*. PhD thesis, The University of Michigan, 1994.

[436] A. Varma, B. W. Boesch, and B. O. Palsson. Biochemical production capabilities of *Escherichia coli*. *Biotechnology and Bioengineering*, 42:59–73, 1993.

[437] A. Varma, B. W. Boesch, and B. O. Palsson. Stoichiometric interpretation of *Escherichia coli* glucose catabolism under various oxygenation rates. *Applied and Environmental Microbiology*, 59:2465–2473, 1993.

[438] A. Varma and B. O. Palsson. Stoichiometric flux balance models quantitatively predict growth and metabolic by-product secretion in wild-type *Escherichia coli* W3110. *Applied Environmental Microbiology*, 60(10):3724–3731, 1994.

[439] A. Varma and B. O. Palsson. Metabolic capabilities of *Escherichia coli*. I. Synthesis of biosynthetic precursors and cofactors. *Journal of Theoretical Biology*, 165:477–502, 1993.

[440] A. Varma and B. O. Palsson. Metabolic capabilities of *Escherichia coli*. II. Optimal growth patterns. *Journal of Theoretical Biology*, 165:503–522, 1993.

[441] A. Varma and B. O. Palsson. Metabolic flux balancing – Basic concepts, scientific and practical use. *Bio-Technology*, 12:994–998, 1994.

[442] A. Varma and B. O. Palsson. Parametric sensitivity of stoichiometric flux balance models applied to wild type *Escherichia coli* metabolism. *Biotechnology and Bioengineering*, 45:69–79, 1995.

[443] G. Vemuri, M. Eiteman, and E. Altman. Effects of growth mode and pyruvate carboxylase on succinic acid production by metabolically engineered strains of *Escherichia coli*. *Applied and Environmental Microbiology*, 68:1715–1727, 2002.

[444] R. T. Vinopal and D. G. Fraenkel. Phenotypic suppression of phosphofructokinase mutations in *Escherichia coli* by constitutive expression of the glyoxylate shunt. *Journal of Bacteriology*, 118(1):1090–1100, 1974.

[445] A. Nanchen, T. Fuhrer, and U. Sauer. Determination of metabolic flux ratios from 13C-experiments and gas chromatography–mass spectrometry data: protocol and principles. *Methods in Molecular Biology*, 358:177–197, 2007.

[446] T. D. Vo, H. J. Greenberg, and B. O. Palsson. Reconstruction and functional characterization of the human mitochondrial metabolic network based on proteomic and biochemical data. *Journal of Biological Chemistry*, 279:39532–39540, 2004.

[447] K. von Meyenburg *et al.* Promoters of the ATP operon coding for the membrane-bound ATP synthase of *Escherichia coli* mapped by tn10 insertion mutations. *Molecular and General Genetics*, 188:240–248, 1982.

[448] A. Wagner and D. A. Fell. The small world inside large metabolic networks. *Proceedings of the Royal Society B: Biological Sciences*, 268:1803–1810, 2001.

[449] K. Walsh and D. E. Koshland Jr. Branch point control by the phosphorylation state of isocitrate dehydrogenase. A quantitative examination of fluxes during a regulatory transition. *Journal of Biological Chemistry*, 260:8430–8437, 1985.

[450] H. H. Wang *et al.* Programming cells by multiplex genome engineering and accelerated evolution. *Nature*, 460:894–898, 2009.

[451] K. Watson and D. Holden. Dynamics of growth and dissemination of *Salmonella in vivo*. *Cellular Microbiology*, 12:1389–1397, 2010.

[452] S. J. Wiback, I. Famili, H. J. Greenberg, and B. O. Palsson. Monte Carlo sampling can be used to determine the size and shape of the steady-state flux space. *Journal of Theoretical Biology*, 228:437–447, 2004.

[453] S. J. Wiback and B. O. Palsson. Extreme pathway analysis of human red blood cell metabolism. *Biophysical Journal*, 83:808–818, 2002.

[454] W. Wiechert. ^{13}C metabolic flux analysis. *Metabolic Engineering*, 3:195–206, 2001.

[455] W. Wiechert *et al*. Bidirectional reaction steps in metabolic networks: III. Explicit solution and analysis of isotopomer labeling systems. *Biotechnology and Bioengineering*, 66:69–85, 1999.

[456] E. H. Wintermute and P. A. Silver. Emergent cooperation in microbial metabolism. *Molecular Systems Biology*, 6:407, 2010.

[457] D. S. Wishart *et al*. DrugBank: A knowledgebase for drugs, drug actions and drug targets. *Nucleic Acids Research*, 36:D901–906, 2008.

[458] D. S. Wishart *et al*. HMDB: A knowledgebase for the human metabolome. *Nucleic Acids Research*, 37:D603–610, 2009.

[459] C. H. Wu *et al*. The protein information resource. *Nucleic Acids Research*, 31:345–347, 2003.

[460] C. Xu *et al*. Genome-scale metabolic model in guiding metabolic engineering of microbial improvement. *Applied Microbiology and Biotechnology*, 97:519–539, 2013.

[461] W. Xu *et al*. Oncometabolite 2-hydroxyglutarate is a competitive inhibitor of alpha-ketoglutarate-dependent dioxygenases. *Cancer Cell*, 19:17–30, 2011.

[462] K. Yamamoto *et al*. Functional characterization *in vitro* of all two-component signal transduction systems from *Escherichia coli*. *Journal of Biological Chemistry*, 280:1448–1456, 2005.

[463] E. Yang *et al*. Decay rates of human mRNAs: Correlation with functional characteristics and sequence attributes. *Genome Research*, 13:1863–1872, 2003.

[464] P. J. Yeh *et al*. Drug interactions and the evolution of antibiotic resistance. *Nature Reviews Microbiology*, 7:460–466, 2009.

[465] M. Yeung, I. Thiele, and B. O. Palsson. Estimation of the number of extreme pathways for metabolic networks. *BMC Bioinformatics*, 8:363, 2007.

[466] M. A. Yildirim *et al*. Drug-target network. *Nature Biotechnology*, 25:1119–1126, 2007.

[467] H. Yim *et al*. Metabolic engineering of *Escherichia coli* for direct production of 1,4-butanediol. *Nature Chemical Biology*, 7:445–452, 2011.

[468] K. Yizhak *et al*. Metabolic modeling of endosymbiont genome reduction on a temporal scale. *Molecular Systems Biology*, 7:479, 2011.

[469] K. Yoshikawa *et al*. Reconstruction and verification of a genome-scale metabolic model for *Synechocystis* sp. PCC6803. *Applied Microbiology and Biotechnology*, 92:347–358, 2011.

[470] N. Zamboni *et al*. (13)c-based metabolic flux analysis. *Nature Protocols*, 4:878–892, 2009.

[471] K. Zengler and B. O. Palsson. A road map for the development of community systems (CoSy) biology. *Nature Reviews Microbiology*, 10:366–372, 2012.

[472] X. Zhang *et al.* Production of L-alanine by metabolically engineered *Escherichia coli*. *Applied and Environmental Microbiology*, 77:355–366, 2007.

[473] Y. Zhang *et al.* Three-dimensional structural view of the central metabolic network of *Thermotoga maritima*. *Science*, 325:1544–1549, 2009.

[474] D. Zhou *et al.* Experimental selection of hypoxia-tolerant *Drosophila melanogaster*. *Proceedings of the National Academy of Sciences, USA*, 108:2349–2354, 2011.

[475] S. Zhou *et al.* Production of optically pure D-lactic acid in mineral salts medium by metabolically engineered *Escherichia coli* W3110. *Applied and Environmental Microbiology*, 39:399, 2003.

[476] K. Zhuang *et al.* Genome-scale dynamic modeling of the competition between *Rhodoferax* and *Geobacter* in anoxic subsurface environments. *ISME Journal*, 5:305–316, 2011.

[477] K. Zhuang, G. N. Vemuri, and K. Mahadevan. Economics of membrane occupancy and respiro-fermentation. *Molecular Systems Biology*, 7:500, 2011.

[478] A. R. Zomorrodi *et al.* Mathematical optimization applications in metabolic networks. *Metabolic Engineering*, 14:672–686, 2012.

Index